*Nonlinear Oscillations
in Physical Systems*

NONLINEAR OSCILLATIONS IN PHYSICAL SYSTEMS

Chihiro Hayashi,   Professor of Electrical Engineering,   Kyoto University

PRINCETON UNIVERSITY PRESS
PRINCETON, NEW JERSEY

Published by Princeton University Press, 41 William Street,
Princeton, New Jersey 08540
In the United Kingdom: Princeton University Press, Guildford, Surrey
Copyright © 1964 by McGraw-Hill, Inc.
All rights reserved
First Princeton printing, 1985
LCC 85-42664
ISBN 0-691-08383-5 (pbk.)
Reprinted by arrangement with McGraw-Hill, Inc.

Clothbound editions of Princeton University Press books are printed
on acid-free paper, and binding materials are chosen for strength
and durability. Paperbacks, while satisfactory for personal collections,
are not usually suitable for library rebinding.

Printed in the United States of America by Princeton University Press,
Princeton, New Jersey

Preface

The purpose of this book is to provide engineers and scientists with fundamental knowledge concerning the important subject of nonlinear oscillations in physical systems. Since differential equations which describe the systems are nonlinear and generally cannot be solved exactly, we must be satisfied with approximate solutions that furnish adequate information about the nonlinear oscillations. Accordingly, this book is also intended to give an account of the basic techniques for finding approximate solutions for nonlinear differential equations. It should first be mentioned that the problems discussed in this book are mostly concerned with forced oscillations; furthermore, in most of the text, only systems with one degree of freedom are treated. However, a wide variety of oscillations may occur in those systems under the influence of external forces.

The present book is essentially a revised and enlarged edition of the original text "Forced Oscillations in Nonlinear Systems," published in 1953. The material treated in this book includes that of the previous edition plus reports by the author which have appeared in various technical journals since 1953. As mentioned above, the text is devoted largely to an exposition of forced oscillations and encompasses accounts of corresponding experimental investigation and substantiation of various phases of the theory. Some of the material presented in this book has been given to graduate students at Kyoto University and also at the Massachusetts Institute of Technology.

In writing this book, the author has tried to present the underlying principle and theory in such a way that they will be of use to readers whose primary interest lies in applying these ideas and methods to concrete physical problems. He has therefore stressed the relationship between analysis and experiment. Practically, in so far as is discussed in this book, the analytical results have shown good agreement with experimental facts. The author has not aimed at a rigorous mathematical treatment of nonlinear differential equations. Therefore, from the mathematical point of view, there will be some points open to discussion in proving the validity, particularly in the case where the departure from linearity is not small. Although this book is not addressed in any sense to the pure mathematicians, it will be the author's pleasure if the results in this book attract the attention of mathematicians so that he could expect more rigorous justification of the analysis.

The author wishes to acknowledge his indebtedness to Dr. R. Torikai, former President of Kyoto University, whose influence upon the author's scientific development is great. The author is grateful to the late Dr. R. Rüdenberg of Harvard University and Prof. J. P. Den Hartog and Prof. H. M. Paynter of the Massachusetts Institute of Technology for their encouragement and invaluable suggestions. He is indebted to many who have facilitated the preparation of this text. In particular, Dr. Y. Nishikawa has joined in the author's work on nonlinear mechanics, supervised a considerable number of calculations, and given much good advice of all kinds. Acknowledgment must also be made to Dr. H. Shibayama, Mr. M. Abe, and Mr. Y. Ueda for their excellent cooperation. The author's thanks are due to Prof. T. J. Higgins of the University of Wisconsin, who read and polished the manuscript. His careful examination of the text is deeply appreciated. Finally, the author appreciates the assistance he received from Miss S. Ikari, who typed the manuscript.

CHIHIRO HAYASHI

Contents

Nonlinear Oscillations
in Physical Systems

Introduction

Basically, all the problems in mechanics are nonlinear from the outset. The linearizations commonly practiced are approximating devices that are good enough or quite satisfactory for most purposes. There are, however, also certain cases in which linear treatments may not be applicable at all. Frequently, essentially new phenomena occur in nonlinear systems which cannot, in principle, occur in linear systems [15, 66, 111].[1] The principal aim of this book is not to introduce methods of improving the accuracy obtainable by linearization, but rather to focus attention on those features of the problems in which the nonlinearity results in distinctive new phenomena. The basic mathematical idea in this book is due to the works of Poincaré [81–84], Bendixson [6], and Liapunov [28, 29, 57, 61]. The technique applied is also due to Rayleigh [91], van der Pol [85–88], Duffing [21], Mandelstam and Papalexi [72], and Andronow, Chaikin, and Witt [3, 4].

Before the actual contents of the book are outlined, some of the literature in the field of nonlinear mechanics will be mentioned. Although the phenomena of nonlinear oscillations have long been recognized by many scientists, the recent developments in the theory and methods of nonlinear analysis have been stimulated by the works of Duffing [21] and van der Pol [85–88]. A great volume of subsequent investigation was carried out in the Soviet Union. In the main, two groups of workers are to be distinguished. One comprises Mandelstam, Papalexi, Andronow, Chaikin, and Witt, and the other comprises Kryloff, Bogoliuboff, and Mitropolsky. Their contributions are to be found in Refs. 3, 10, 55, and 72. Minorsky's book [75] includes those works of Russian scholars as well as his own contribution. Among other books in this field, those of Stoker [100], McLachlan [70], Cunningham [18], Ku [56], and Kauderer [53] are to be mentioned. Minorsky's recent book [76] is another excellent addition to the bibliography of nonlinear mechanics.

[1] Numbers in brackets indicate references on pages 378 to 383.

The text is divided into four parts. Part I, consisting of three chapters, describes the principal methods of nonlinear analysis. There are two major trends in modern developments of nonlinear analysis: the analytical methods of successive approximations and the topological methods of graphical integration. The first two chapters describe these methods. The analytical methods described in Chap. 1 include the perturbation method, the iteration method, the averaging method, and the method of harmonic balance. These methods are usually used for finding periodic solutions of nonlinear differential equations. However, they are also applicable to the study of transient solutions under certain conditions.

The perturbation method is applicable to equations in which a small parameter is associated with the nonlinear term. An approximate solution is found as a power series, with terms of the series involving the small parameter raised to successively higher powers. A method of solving differential equations based on the process of successive iteration is called the iteration method. Although iteration may be performed in a number of ways, the procedure described in the text may be considered to be legitimate, and it agrees with the result obtained by the perturbation method. The averaging method is applicable when the amplitude and phase of an oscillation vary slowly with time. The method of harmonic balance is a direct consequence of the averaging method and is frequently used to obtain periodic solutions in forced oscillatory systems. In connection with the principle of harmonic balance, a method which improves the accuracy of approximation of the periodic solutions is described.

The analytical methods mentioned above are mathematically legitimate for equations of small nonlinearity. However, they may still be applicable even to equations with nonlinearity not too large. Applicability of the methods is examined by solving numerical examples with an associated large nonlinearity. Upon comparison with solutions calculated by using a digital computer, a satisfactory accuracy of the numerical solutions is obtained.

Chapter 2 is concerned with the topological analysis of nonlinear systems. The method is based on the study of the representation of solutions of differential equations in the phase plane or phase space. Periodic solutions in nonlinear systems can be correlated with limit cycles or singular points, transient oscillations with integral curves which tend to them with increasing time. Following Poincaré [81–84] and Bendixson [6], the types of singular points and their associated properties are investigated; some criteria for the existence of limit cycles are also discussed.

The graphical methods of analysis discussed in Chap. 2 are useful for the construction of integral curves. The isocline method is of widest utility but is relatively time-consuming. Liénard's method is useful for dealing with what are called self-excited oscillations. In starting a solution by either the isocline method or the Liénard method, the entire plane must be filled with line segments fixing the slope of a solution curve. If only a single solution curve is needed, the delta method and the slope-line method lead more directly to the desired solution, since they use only information related directly to the desired solution curve. It is also mentioned that solution curves obtained by these methods are easily conversible to the time-response curves, since the lapse of time is indicated on the solution curve because of the way in which the curve is constructed. The delta method may be considered as a generalization of Liénard's principle. The double-delta method, an extension of the delta method, is also investigated. The slope-line method, with modification of the basic principle, can be applied to the graphical solution of non-autonomous as well as autonomous equations.

Chapter 3 is concerned with the investigation of stability of nonlinear systems; in it are described some of the criteria for stability which will be used in later chapters. An application of the Routh-Hurwitz criterion for nonlinear systems is also described. A fundamental theorem in this connection is due to Liapunov [61], whose criterion for stability is introduced. An essential part of this method lies in the construction of the Liapunov function; however, no straightforward technique is available on this point.

The question of stability is particularly important in the study of periodic oscillations in nonlinear systems. It may be discussed by solving a variational equation which characterizes small deviations from the periodic states of equilibrium. The variational equation leads to a linear equation with periodic coefficients in time. Typical examples of such equations are Mathieu's equation and Hill's equation. The theory of these equations is discussed at some length. The characteristic exponent of the solution is closely related to the stability of periodic oscillations and is calculated by using Whittaker's method [112–114].

Part II, consisting of four chapters, is devoted to the study of forced oscillations in steady states. Particular attention is directed toward the investigation of the stability of equilibrium states by applying Mathieu's or Hill's equation as a stability criterion. In Chap. 4, stability conditions are derived by comparison of the characteristic exponent (obtained in the preceding chapter) with the damping of the system. These conditions secure the stability not only of the oscillation having the fundamental frequency under consideration, but also of the oscillation with higher-harmonic or subharmonic frequency. Hence, the generalized stability conditions obtained in this way are particularly useful for the study of periodic oscillations in nonlinear systems, since these states of equilibrium might become unstable owing to the buildup of self-excited oscillations having a higher-harmonic or subharmonic frequency. It is noted that the stability condition obtained by the Routh-Hurwitz criterion is included in the above conditions, and it is shown that the vertical tangency of characteristic curves results at the stability limit.

The three chapters that follow are devoted to the investigation of different types of periodic oscillation. Chapter 5 treats the harmonic oscillations in which the fundamental component having the same frequency as that of the external force predominates over higher harmonics. The nonlinearity is provided with symmetrical and unsymmetrical characteristics. In Chap. 6 the higher-harmonic oscillations in series-resonance circuits are studied. These oscillations occur when the amplitude of the external force is very large. The self-excitation of higher-harmonic oscillations in parallel-resonance circuits is also discussed. Chapter 7 deals with the subharmonic oscillations in which the smallest period of the oscillations is an integral multiple of the period of the external force. The relationship between nonlinear characteristics and the order of subharmonics is discussed. Then the subharmonic oscillations of order $\frac{1}{3}$ and of order $\frac{1}{2}$ are investigated at full length.

For each type of oscillation mentioned above, the stability condition derived in Chap. 4 is applied to examine the stability of equilibrium states. This condition is particularly effective in studying the stability problems of the higher-harmonic and subharmonic oscillations. Experimental investigations are also carried out by making use of electric circuits which contain a saturable-core inductor as a nonlinear element, and satisfactory agreement between the theoretical analysis and experimental results is found in the respective cases.

Part III, consisting of four chapters, is devoted to the study of nonlinear oscillations in transient states. The topological method of analysis is used for this purpose. In the first two chapters the phase-plane analysis is used: the coordinates of the phase plane are the time-varying components of the oscillation which are in phase and in quadrature with the external force. In order to derive autonomous equations that permit the use of the phase-plane analysis, an assumption is made that these components vary slowly with time. The integral curves of the autonomous equations are studied with the basic idea that singular points are correlated with steady states and integral curves with transient states.

The transient states of harmonic oscillations are treated in Chap. 8. The singularities correlated with periodic states of equilibrium are discussed. Typical examples of the phase-plane diagram are then illustrated; the illustrations provide a general view of harmonic oscillations in both transient and steady-state conditions. In this way the relationship between initial conditions and the resulting periodic oscillations is discussed. Furthermore, following Poincaré and Bendixson, a detailed investigation of the singularities is carried out, particularly in the case when two singular points coalesce to form a singularity of higher order. Experimental investigation using an electric circuit with saturable core conforms with the result of the foregoing analysis.

Proceeding analogously, the transient states of subharmonic oscillations of orders $\frac{1}{2}$, $\frac{1}{3}$, and $\frac{1}{5}$ are investigated in Chap. 9. The situation is rather complicated for the $\frac{1}{2}$-harmonic oscillation, since a number of responses with different amplitudes exist in this case. The theoretical results are also confirmed by experimental investigation and analog-computer analysis.

In Chap. 10 the relationship between initial conditions and the resulting periodic solutions for Duffing's equation is investigated. This problem has already been studied in the preceding chapters by making use of the phase-plane analysis. It was assumed there that the amplitude and phase of the oscillation vary slowly with time. Although this method of analysis has been used with success for the study of harmonic and subharmonic oscillations, it may well not be applicable to the following cases. First, if the initial conditions are prescribed at values which are far different from those of the steady state, the assumption that the amplitude and phase of the oscillation vary slowly does not hold; therefore, the result obtained by this method is not very accurate. The second and more serious drawback is that, if a number of steady-state responses are to be expected, this method is practically inapplicable, since the analysis is obliged to rely on the graphical solution in a higher-dimensional space—a practical impossibility.

In this chapter is described the method of analysis which is applicable in such situations. Again the behavior of the representative point is considered in the phase plane; however, the coordinates of the phase plane are the dependent variable and its time derivative of the differential equation. The initial condition is prescribed by a point in the phase plane. Special attention is directed toward the location of the representative points at the beginning of every cycle of the external force. Mathematically, these points will be obtained as successive images of the initial point under iterations of the mapping. Then fixed points are correlated with the periodic solutions of the above equation. The fixed point that is directly unstable is of particular interest. Through this point there are two curves which are invariant under the mapping. Successive images approach the unstable fixed point along one of these curves, while the images depart from the fixed point along the other curve. The former is the boundary curve between the domains of attraction, in each of which are initial conditions leading to a particular type of periodic solution.

Two examples of the domains of attraction are illustrated. The first deals with the domains of attraction leading to the harmonic oscillation and the subharmonic oscillation of order $\frac{1}{3}$ in a symmetrical system. The second example is concerned with the domains of attraction for the harmonic oscillation, the subharmonic oscillations of order $\frac{1}{2}$ and of order $\frac{1}{3}$ in an unsymmetrical system.

Chapter 11 is concerned with the so-called "almost periodic oscillation," where the amplitude and phase of the oscillation vary slowly but periodically even in the steady state. However, since the waveform of the oscillation is not usually repeated, the almost periodic oscillation is in general nonperiodic.

The phase-space analysis is used for the present investigation. The solution is assumed in the form of a Fourier series in which the coefficients are assumed to be slowly varying functions of time. These Fourier coefficients constitute the coordinates of a representative point in the phase space. Consequently, a periodic oscillation having constant coefficients is correlated with a singular point in the phase space. An almost periodic oscillation is then represented by a limit cycle. Since the almost periodic oscillation is affected by amplitude and phase modulation, the representative point does not tend to a singular point but keeps on moving along the limit cycle with increasing time. The period required for the representative point to complete one revolution along the limit cycle is not an integral multiple of the period of the external force; the ratio of these periods is in general incommensurable.

Two representative cases of the almost periodic oscillation are studied in Chap. 11. The first is the case in which a harmonic oscillation in a resonant circuit becomes unstable and changes into an almost periodic oscillation. The second case deals with the almost periodic oscillation which develops from a subharmonic oscillation of order $\frac{1}{2}$ in a parametric excitation circuit. A numerical analysis is carried out for these cases; in the second case two distinctive types of limit cycle for different values of the system parameters are obtained.

Part IV, consisting of two chapters, is concerned with forced oscillations in self-oscillatory systems of the negative-resistance type. Chapter 12 deals with entrained oscillations. When no external force is applied, the system produces a self-excited oscillation. Under the impression of a periodic force, the frequency of the self-excited oscillation falls in synchronism with the driving frequency within a certain band of frequencies. This phenomenon of frequency entrainment also occurs when the ratio of the two frequencies is in the neighborhood of an integer (different from unity) or a fraction. Under this condition, the natural frequency of the system is entrained by a frequency which is an integral multiple or submultiple of the driving frequency. In Chap. 12 special attention is directed toward the study of periodic oscillations as caused by frequency entrainment.   The amplitude characteristics of the entrained oscillations are obtained by the method of harmonic balance; the stability of these oscillations is investigated by making use of Hill's equation as a stability criterion.   The regions in which different types of entrained oscillation as well as almost periodic oscillations occur are sought by varying the amplitude and frequency of the external force.   The theoretical results are compared with the solutions obtained by analog-computer analysis and found to be in satisfactory agreement with them.

Chapter 13 deals with almost periodic oscillations which occur in self-oscillatory systems under periodic excitation.   As mentioned in the preceding chapter, the natural frequency of the system is entrained not only by the driving frequency but also by the higher-harmonic and sub-harmonic frequencies of the external force.   Therefore, almost periodic oscillations must be discussed in connection with the entrained oscillations at these frequencies.   Since an entrained oscillation at the higher-harmonic or subharmonic frequency is represented by a sum of the forced and free oscillations having the driving frequency and the entrained frequency, respectively, an almost periodic oscillation which develops from it may also be expressed by the sum of these two, but the amplitude and phase of the free oscillation are allowed to vary slowly with time.

The phase-plane analysis is applied to the study of almost periodic oscillations.   The variables of the phase plane are the time-varying amplitudes of a pair of components of the free oscillation in quadrature.   Consequently, an entrained oscillation is correlated with a singular point and an almost periodic oscillation with a limit cycle in the phase plane.   Almost periodic oscillations investigated in this chapter include those which develop from the entrained oscillations at the harmonic, the higher-harmonic, and the subharmonic frequencies.   Limit cycles representing the almost periodic oscillations are calculated; time-response curves are also obtained by numerical integration along the limit cycles.   This method of analysis is particularly useful in studying the transition between entrained oscillations and almost periodic oscillations.   The generation and extinction of a limit cycle correlated with an almost periodic oscillation are also investigated for a system with nonlinear restoring force.

The text is supplemented by six appendixes.   In Appendix I are given expansions of the Mathieu functions ce_1, se_1, . . . , ce_3, se_3; in Appendix II are given the unstable solutions of Hill's equation computed by making use of Whittaker's method.   The extended form of Hill's equation is the subject of Appendix III.   In Appendix IV the stability condition is compared with one of the conditions derived by Mandelstam and Papalexi [72] for the subharmonic oscillations.   In Appendix V are given complementing remarks concerning integral curves and singular points which are frequently referred to in the analysis in Parts III and IV. Finally, in Appendix VI is shown a circuit diagram of the electronic switch which is used in the experiments, and the sequence of operation is explained in detail.

A set of problems related to the subject matter of each chapter is added at the end of the text.

Part I

Principal Methods
of Nonlinear Analysis

Chapter 1

Analytical Methods

1.1 Introduction

There is usually considerable advantage in finding an analytical solution for a differential equation when that is possible. The analytical solution is obtained in algebraic form without the necessity of introducing numerical values for parameters or initial conditions during the process. Once the solution is obtained, any desired numerical values can be inserted and the entire possible range of solutions explored. Because of this flexibility, it might be well to attempt first to seek a solution of a specified differential equation in analytical form. However, it should be recognized that only a very few equations that arise from actual physical systems are simple enough to allow exact solution. There exist, in general, no methods capable of yielding an exact solution of an arbitrarily selected nonlinear differential equation. Thus, for certain often encountered classes of differential equations, the only methods available are those of various types of approximation procedures.

In order to be able to obtain an analytical solution, all relations needed to describe the physical system must be expressed in mathematical terms. Any empirical relation resulting from experimental measurements, wherewith the data are plotted as a curve between certain quantities, must be converted into the form of an equation. The resulting equations for the system are then analyzed to obtain a solution, which is exhibited as a combination of well-known tabulated mathematical functions.

Let us consider a system of n differential equations of the first order:

$$\frac{dx_1}{dt} = X_1(x_1, x_2, \ldots, x_n, t)$$

$$\frac{dx_2}{dt} = X_2(x_1, x_2, \ldots, x_n, t)$$

$$\cdots \cdots \cdots \cdots \cdots \cdots$$

$$\frac{dx_n}{dt} = X_n(x_1, x_2, \ldots, x_n, t)$$

$$(1.1)$$

where $X_1, \ldots, X_n$ are analytical functions of the unknown variables $x_1, \ldots, x_n$ and the time t.[1] If one views $x_1, \ldots, x_n$ as components of an n-vector $\mathbf{x}$ and $X_1, \ldots, X_n$ as components of an n-vector $\mathbf{X}$, the above system assumes the simpler aspect

$$\frac{d\mathbf{x}}{dt} = \mathbf{X}(\mathbf{x},t) \tag{1.2}$$

It may well happen that $\mathbf{X}$ depends upon $\mathbf{x}$ alone and not upon t. Equation (1.2) then becomes

$$\frac{d\mathbf{x}}{dt} = \mathbf{X}(\mathbf{x}) \tag{1.3}$$

A system of this nature is said to be *autonomous*. In contrast, Eq. (1.2), in which the time t appears explicitly, is referred to as a *nonautonomous system*.

Differential equations having nonlinear terms associated with a small parameter were studied by Poincaré [83, pp. 79–161]. In this case the function $\mathbf{X}(\mathbf{x},t)$ is broken into two parts; thus we write

$$\frac{d\mathbf{x}}{dt} = \mathbf{F}(\mathbf{x},t) + \mu\mathbf{G}(\mathbf{x},t) \tag{1.4}$$

where the nonlinear terms appear in $\mathbf{G}(\mathbf{x},t)$ only. The parameter μ ideally should be a nondimensional number, and its magnitude should be small. Under this condition, solution for Eq. (1.4) can often be found in series form as

$$\mathbf{x} = \mathbf{x}^{(0)}(t) + \mu\mathbf{x}^{(1)}(t) + \mu^2\mathbf{x}^{(2)}(t) + \cdots \tag{1.5}$$

If μ is sufficiently small, the series converges fast enough that merely the first two or three terms give good accuracy. The solution (1.5) is usually so adjusted that, if $\mu \to 0$, the generating solution

$$\mathbf{x} = \mathbf{x}^{(0)}(t)$$

is the exact solution for the linear equation

$$\frac{d\mathbf{x}}{dt} = \mathbf{F}(\mathbf{x},t)$$

The search for the solution (1.5) is the object of the method of small parameters of Poincaré. This method, with its modifications, constitutes the principal subject of this chapter. When the parameter is large, Poincaré's theory ceases to be applicable, and for such cases analytical

---

[1] An equation of order n can readily be reduced to the form of Eqs. (1.1).

methods are largely unexplored.[1] However, it is worth noticing that, from a practical point of view, the method may still be applicable with satisfactory results even for not very small values of the parameter.

1.2 Perturbation Method

One of the important methods for solving nonlinear differential equations is the perturbation method. The use of this method was, in the earlier period, limited to astronomical calculations. But the important contributions of Poincaré and later mathematicians have broadened the applicability of the method to include a more general field of nonlinear mechanics.

As mentioned before, the method is applicable to equations in which a small parameter is associated with nonlinear terms. In application, we develop the desired quantities in powers of the small parameter multiplied by coefficients which are functions of the independent variable; and we then determine the coefficients of the developments one by one, usually by solving a sequence of linear equations. This description of the procedure is, however, perfunctory. By proceeding in this way a serious difficulty may often be encountered in the form of the so-called "secular terms," i.e., terms in Eq. (1.5) which grow up indefinitely when $t \rightarrow \infty$ and thus destroy the convergence of the series solution.[2]

As an example of the appearance of secular terms let us consider the differential equation

$$\frac{d^2x}{dt^2} + x + \mu x^3 = 0 \qquad (1.6)$$

We try to solve this equation with the initial conditions

$$x(0) = A \qquad x'(0) = 0 \qquad ' \equiv \frac{d}{dt} \qquad (1.7)$$

Thus we substitute the power-series solution

$$x(t) = x_0(t) + \mu x_1(t) + \mu^2 x_2(t) + \cdots \qquad (1.8)$$

into Eq. (1.6) and obtain a power series in μ which must vanish identically in μ; that is, the coefficients of the successive powers of μ must vanish

---

[1] A paper by Riabov [92] discusses the applicability limits for the small-parameter method in the theory of nonlinear oscillations.

[2] In certain special cases a solution itself may grow in amplitude without bounds as $t \rightarrow \infty$, and it may be impossible to remove from Eq. (1.5) terms the amplitudes of which increase without bounds. These terms are not referred to as secular terms, since they represent an integral part of the solution.

separately. Equating these coefficients separately to zero gives the following sequence of linear equations:

$$x_0'' + x_0 = 0$$
$$x_1'' + x_1 = -x_0^3 \qquad\qquad (1.9)$$
$$\cdots\cdots\cdots\cdots$$

To determine $x_0, x_1, \ldots$, we have the initial conditions (1.7) which lead to the new conditions

$$x_0(0) = A \qquad x_i(0) = 0$$
$$x_0'(0) = 0 \qquad x_i'(0) = 0 \qquad i = 1, 2, 3, \ldots \qquad (1.10)$$

By virtue of these conditions, the first equation of (1.9) yields

$$x_0 = A \cos t$$

Hence the second equation of (1.9) becomes

$$x_1'' + x_1 = -\tfrac{3}{4}A^3 \cos t - \tfrac{1}{4}A^3 \cos 3t$$

and the solution is

$$x_1 = -\tfrac{3}{8}A^3 t \sin t - \tfrac{1}{32}A^3(\cos t - \cos 3t)$$

Here the first term is a secular term which contains t outside the sign of the trigonometric function.

The appearance of the secular terms in this case may be explained as follows. When $\mu = 0$, the solution is periodic with period 2π. However, owing to the presence of the nonlinear term μx^3 in Eq. (1.6), the solution for $\mu \neq 0$ may not be periodic with the same period. (Actually, it is not, as will be shown shortly.) Since the period of the generating solution $x_0 = A \cos t$ is fixed as 2π, the subsequent terms in Eq. (1.8) must take care of this variation in the period, thus resulting in the appearance of the secular terms. The following simple example will make this statement particularly convincing; namely, we have, for small ϵ, the expansion

$$\sin (\omega + \epsilon)t = \sin \omega t + \epsilon t \cos \omega t - \frac{\epsilon^2 t^2}{2!} \sin \omega t - \frac{\epsilon^3 t^3}{3!} \cos \omega t + \cdots$$

For the elimination of secular terms, precaution must be taken that unknown quantities such as the frequency of a free oscillation and the amplitude of a self-excited oscillation should not be fixed beforehand in the generating solution. In the remainder of this section we describe the practical procedure for obtaining periodic solutions of some representative nonlinear differential equations.

(a) Autonomous Systems

We consider a differential equation of the form

$$\frac{d^2x}{dt^2} + x = \mu f\left(x, \frac{dx}{dt}\right) \tag{1.11}$$

where μ is a nondimensional parameter, which we assume to be small. We also assume that $f(x, dx/dt)$ is a polynomial in x and dx/dt. When $\mu = 0$, the periodic solution of Eq. (1.11) is readily obtained as a linear combination of $\sin t$ and $\cos t$, thus is of period 2π. When $\mu \neq 0$, however, the frequency of the periodic solution becomes unknown; accordingly, it is advantageous to replace the independent variable t by $\tau = \omega t$, where ω is the unknown frequency of the periodic solution. It is clear that now the variable x is of period 2π in τ. Equation (1.11), with τ defined above, becomes[1]

$$\omega^2 \ddot{x} + x = \mu f(x, \omega \dot{x}) \tag{1.12}$$

As mentioned before, the method consists in developing the desired solution $x(\tau)$ in a power series with respect to the small parameter μ, the coefficients in the series being periodic functions of τ. We write, therefore,

$$x(\tau) = x_0(\tau) + \mu x_1(\tau) + \mu^2 x_2(\tau) + \cdots \tag{1.13}$$

the $x_i(\tau)$ being functions of τ of period 2π. In addition to x it is also necessary to develop the unknown quantity ω with respect to μ, that is,

$$\omega = \omega_0 + \mu \omega_1 + \mu^2 \omega_2 + \cdots \tag{1.14}$$

We substitute Eqs. (1.13) and (1.14) into (1.12) and equate the coefficients of the like powers of μ. Then we obtain a sequence of second-order linear differential equations in $x_i(\tau)$, which also involve the unknown quantities ω_i. Since only the periodic solution is under consideration, an origin of τ may be chosen arbitrarily; accordingly, we so choose it that $\dot{x}(\tau) = 0$ at $\tau = 0$. This initial condition and the condition for periodicity of $x(\tau)$ serve to determine the unknown quantities in Eqs. (1.13) and (1.14).

To explain the details of use of the perturbation method, we again take the differential equation

$$\frac{d^2x}{dt^2} + x + \mu x^3 = 0 \tag{1.15}$$

---

[1] Here and throughout this chapter a dot over a quantity denotes differentiation with respect to τ; a second dot denotes a second differentiation.

We replace in Eq. (1.15) the independent variable t by $\tau = \omega t$; then

$$\omega^2 \ddot{x} + x + \mu x^3 = 0 \tag{1.16}$$

Substituting Eqs. (1.13) and (1.14) into (1.16) and equating the coefficients of the like powers of μ leads to the following sequence of linear equations:

$$\mu^0: \qquad \omega_0^2 \ddot{x}_0 + x_0 = 0 \tag{1.17}$$
$$\mu^1: \qquad \omega_0^2 \ddot{x}_1 + x_1 = -2\omega_0\omega_1 \ddot{x}_0 - x_0^3 \tag{1.18}$$
$$\mu^2: \qquad \omega_0^2 \ddot{x}_2 + x_2 = -(2\omega_0\omega_2 + \omega_1^2)\ddot{x}_0 - 2\omega_0\omega_1\ddot{x}_1 - 3x_0^2 x_1 \tag{1.19}$$

. . . .

The initial conditions are given as before [see Eqs. (1.7)]. Since $x(\tau + 2\pi) = x(\tau)$, we have the following conditions to determine the unknown quantities in the above equations; namely,

$$x_i(\tau + 2\pi) = x_i(\tau) \tag{1.20}$$
$$\begin{aligned} x_0(0) &= A \qquad x_{i+1}(0) = 0 \\ \dot{x}_i(0) &= 0 \qquad i = 0, 1, 2, \ldots \end{aligned} \tag{1.21}$$

Solving Eq. (1.17) with use of these conditions gives

$$x_0 = A \cos\tau \tag{1.22}$$
$$\omega_0 = 1 \tag{1.23}$$

The zero-order solution given by Eq. (1.22) is the generating solution. By virtue of Eqs. (1.22) and (1.23), Eq. (1.18) leads to

$$\ddot{x}_1 + x_1 = (2\omega_1 - \tfrac{3}{4}A^2)A \cos\tau - \tfrac{1}{4}A^3 \cos 3\tau \tag{1.24}$$

If the coefficient of $\cos\tau$ were not zero, the solution of Eq. (1.24) would contain a term of $\tau \sin\tau$, that is, a secular term. The periodicity condition for $x_1(\tau)$, therefore, requires that the coefficient of $\cos\tau$ be zero, i.e.,

$$\omega_1 = \tfrac{3}{8}A^2 \tag{1.25}$$

Hence, by virtue of the conditions (1.21), the solution of Eq. (1.24) becomes

$$x_1 = \tfrac{1}{32}A^3(-\cos\tau + \cos 3\tau) \tag{1.26}$$

By proceeding analogously, we obtain

$$\omega_2 = -\tfrac{21}{256}A^4 \tag{1.27}$$
$$x_2 = \tfrac{23}{1024}A^5 \cos\tau - \tfrac{3}{128}A^5 \cos 3\tau + \tfrac{1}{1024}A^5 \cos 5\tau \tag{1.28}$$

From Eqs. (1.13), (1.22), (1.26), and (1.28), the solution of Eq. (1.15), up to and including terms of the second order in μ, is

$$\begin{aligned} x(t) = {}&(A - \tfrac{1}{32}\mu A^3 + \tfrac{23}{1024}\mu^2 A^5) \cos\omega t \\ &+ (\tfrac{1}{32}\mu A^3 - \tfrac{3}{128}\mu^2 A^5) \cos 3\omega t + \tfrac{1}{1024}\mu^2 A^5 \cos 5\omega t + \cdots \end{aligned} \tag{1.29}$$

and, by using Eqs. (1.14), (1.23), (1.25), and (1.27), the frequency ω is given by

$$\omega = 1 + \tfrac{3}{8}\mu A^2 - \tfrac{21}{256}\mu^2 A^4 + \cdots \tag{1.30}$$

One sees that the frequency ω depends on the amplitude A of the oscillation.

As another example, we consider van der Pol's equation

$$\frac{d^2x}{dt^2} - \mu(1 - x^2)\frac{dx}{dt} + x = 0 \tag{1.31}$$

where μ is a small positive quantity. Replacing in Eq. (1.31) the independent variable t by $\tau = \omega t$ as before gives

$$\omega^2 \ddot{x} - \mu\omega(1 - x^2)\dot{x} + x = 0 \tag{1.32}$$

Insertion of Eqs. (1.13) and (1.14) into (1.32) results in a sequence of linear equations:

$$\mu^0: \quad \omega_0^2 \ddot{x}_0 + x_0 = 0 \tag{1.33}$$
$$\mu^1: \quad \omega_0^2 \ddot{x}_1 + x_1 = -2\omega_0\omega_1\ddot{x}_0 + \omega_0(1 - x_0^2)\dot{x}_0 \tag{1.34}$$
$$\mu^2: \quad \omega_0^2 \ddot{x}_2 + x_2 = -(2\omega_0\omega_2 + \omega_1^2)\ddot{x}_0 - 2\omega_0\omega_1\ddot{x}_1$$
$$+ \omega_1(1 - x_0^2)\dot{x}_0 - 2\omega_0 x_0 x_1 \dot{x}_0 + \omega_0(1 - x_0^2)\dot{x}_1 \tag{1.35}$$

· ·

Utilizing $x(\tau + 2\pi) = x(\tau)$ and $\dot{x}(0) = 0$ for determining the unknown quantities in the above equations gives us

$$x_i(\tau + 2\pi) = x_i(\tau) \tag{1.36}$$
and
$$\dot{x}_i(0) = 0 \qquad i = 0, 1, 2, \ldots \tag{1.37}$$

Solving Eq. (1.33) with use of these conditions gives us

$$x_0 = A_0 \cos \tau \tag{1.38}$$
$$\omega_0 = 1 \tag{1.39}$$

where the constant A_0, not yet determined, is fixed in the next step. The zero-order solution given by Eq. (1.38) is the generating solution.

By virtue of Eqs. (1.38) and (1.39), Eq. (1.34) leads to

$$\ddot{x}_1 + x_1 = 2\omega_1 A_0 \cos \tau + A_0\left(\frac{A_0^2}{4} - 1\right)\sin \tau + \frac{A_0^3}{4}\sin 3\tau \tag{1.40}$$

If the coefficients of $\cos \tau$ and $\sin \tau$ were not zero, the solution of Eq. (1.40) would contain terms of the type $\tau \cos \tau$ and $\tau \sin \tau$, that is, the secular terms.

The periodicity condition for $x_1(\tau)$, therefore, requires that the coefficients of $\cos \tau$ and $\sin \tau$ be zero, that is,

$$A_0 = 0$$

or $\qquad \dfrac{A_0{}^2}{4} - 1 = 0 \qquad \text{and} \qquad \omega_1 = 0 \qquad (1.41)$

We note that $A_0 = 0$ provides a solution of zero amplitude (whose significance is discussed in a later section) and proceed with $A_0 = 2$, a solution of the first equation of Eqs. (1.41). Then Eq. (1.38) leads to

$$x_0 = 2 \cos \tau \qquad (1.42)$$

To take $A_0 = -2$ provides no new information, because it gives only a solution of opposite phase, that is, π radians out of phase with the solution in Eq. (1.42). The general solution of Eq. (1.40) may now be written as

$$x_1 = A_1 \cos \tau + B_1 \sin \tau - \tfrac{1}{4} \sin 3\tau \qquad (1.43)$$

The constant B_1 is fixed by the requirement that $\dot{x}_1(0) = 0$; thus

$$B_1 = \tfrac{3}{4} \qquad (1.44)$$

The constant A_1 is determined in the next step.

By virtue of Eqs. (1.39) and (1.41) to (1.44), Eq. (1.35) leads to

$$\ddot{x}_2 + x_2 = (4\omega_2 + \tfrac{1}{4}) \cos \tau + 2A_1 \sin \tau \\ - \tfrac{3}{2} \cos 3\tau + 3A_1 \sin 3\tau + \tfrac{5}{4} \cos 5\tau \qquad (1.45)$$

The periodicity condition for $x_2(\tau)$ yields the following relations:

$$\omega_2 = -\tfrac{1}{16} \qquad A_1 = 0 \qquad (1.46)$$

Hence, from Eqs. (1.43) and (1.44),

$$x_1 = \tfrac{3}{4} \sin \tau - \tfrac{1}{4} \sin 3\tau \qquad (1.47)$$

By virtue of Eqs. (1.46) the general solution of Eq. (1.45) becomes

$$x_2 = A_2 \cos \tau + B_2 \sin \tau + \tfrac{3}{16} \cos 3\tau - \tfrac{5}{96} \cos 5\tau \qquad (1.48)$$

The constant B_2 is determined by use of the requirement that $\dot{x}_2(0) = 0$; thereby we obtain $B_2 = 0$. By proceeding analogously, we may determine the unknown quantities in the right-hand members of Eqs. (1.13) and (1.14). It is found after some calculation that the periodicity condition for $x_3(\tau)$ yields

$$A_2 = -\tfrac{1}{8}$$

Hence,

$$x_2 = -\tfrac{1}{8} \cos \tau + \tfrac{3}{16} \cos 3\tau - \tfrac{5}{96} \cos 5\tau \qquad (1.49)$$

From Eqs. (1.13), (1.42), (1.47), and (1.49), the solution of Eq. (1.31), up to and including terms of the second order in μ, is

$$x = (2 - \tfrac{1}{8}\mu^2) \cos \omega t + \tfrac{3}{4}\mu \sin \omega t + \tfrac{3}{16}\mu^2 \cos 3\omega t$$
$$- \tfrac{1}{4}\mu \sin 3\omega t - \tfrac{5}{96}\mu^2 \cos 5\omega t \quad (1.50)$$

and, by use of Eqs. (1.14), (1.39), (1.41), and (1.46), the frequency ω is given by

$$\omega = 1 - \tfrac{1}{16}\mu^2 \quad (1.51)$$

(b) Nonautonomous Systems

We consider a differential equation of the form

$$\frac{d^2x}{dt^2} + x = \mu f\left(x, \frac{dx}{dt}, t\right) \quad (1.52)$$

where μ is a small parameter and $f(x, dx/dt, t)$ is a function of x, dx/dt, and t. We further assume that $f(x, dx/dt, t)$ is periodic in t with period 2π. We proceed to illustrate use of the perturbation method for obtaining the periodic solution of Eq. (1.52) which has the same frequency as the external force. Equation (1.52) is rewritten as

$$\frac{d^2x}{d\tau^2} + x = \mu f\left(x, \frac{dx}{d\tau}, \tau + \delta\right) \quad (1.53)$$

where $\tau = t - \delta$. Contrary to the previous case of autonomous systems, though the frequency of the desired periodic oscillation is known, the phase angle of the fundamental harmonic of the oscillation cannot be assigned a value arbitrarily, as was done by taking it as zero in Eq. (1.42). Rather, as will be evidenced below, an unknown phase angle δ must be introduced to permit choice of the initial condition such that, as is desired for subsequent analytic convenience,

$$\dot{x}(\tau) = 0 \qquad \text{at} \qquad \tau = 0 \quad (1.54)$$

The perturbation method consists in developing the desired solution $x(\tau)$ in a power series with respect to the small parameter μ, the coefficients in the series being functions of τ with period 2π. In addition to x it is also necessary to develop similarly the unknown quantity δ with respect to μ. Thus a solution of Eq. (1.53) is sought in the form of the series

$$x(\tau) = x_0(\tau) + \mu x_1(\tau) + \mu^2 x_2(\tau) + \cdots \quad (1.55)$$
$$\delta = \delta_0 + \mu \delta_1 + \mu^2 \delta_2 + \cdots \quad (1.56)$$

By proceeding analogously as for the case of autonomous systems, we may determine the unknown quantities in the right-hand members of Eqs. (1.55) and (1.56).

As an example of differential equations of the form (1.52) we consider Duffing's equation

$$\frac{d^2x}{dt^2} + x = \mu(-\alpha x - \beta x^3 + F \cos t) \tag{1.57}$$

Introducing the unknown phase angle δ into Eq. (1.57) gives us

$$\frac{d^2x}{d\tau^2} + x = \mu[-\alpha x - \beta x^3 + F \cos(\tau + \delta)] \tag{1.58}$$

where $\tau = t - \delta$. Substitution of Eqs. (1.55) and (1.56) into (1.58) and collection of terms with like powers of μ gives us the set of differential equations

$$\mu^0: \qquad \ddot{x}_0 + x_0 = 0 \tag{1.59}$$
$$\mu^1: \qquad \ddot{x}_1 + x_1 = -\alpha x_0 - \beta x_0^3 + F \cos(\tau + \delta_0) \tag{1.60}$$
$$\mu^2: \qquad \ddot{x}_2 + x_2 = -\alpha x_1 - 3\beta x_0^2 x_1 - F\delta_1 \sin(\tau + \delta_0) \tag{1.61}$$

. . . .

The unknown quantities in the above equations are to be determined by the conditions

$$x_i(\tau + 2\pi) = x_i(\tau) \tag{1.62}$$
$$\dot{x}_i(0) = 0 \qquad i = 0, 1, 2, \ldots \tag{1.63}$$

Solving Eq. (1.59) with the condition $\dot{x}_0(0) = 0$ gives us

$$x_0(\tau) = A_0 \cos \tau \tag{1.64}$$

Substitution of Eq. (1.64) into (1.60) leads to

$$\ddot{x}_1 + x_1 = -(\alpha A_0 + \tfrac{3}{4}\beta A_0^3 - F \cos \delta_0) \cos \tau - F \sin \delta_0 \sin \tau$$
$$- \tfrac{1}{4}\beta A_0^3 \cos 3\tau \tag{1.65}$$

If the coefficients of $\cos \tau$ and $\sin \tau$ are not zero in the right side of Eq. (1.65), secular terms appear in the solution $x_1(\tau)$. Secular terms are oscillatory terms having amplitudes growing indefinitely with the increase of time τ. The periodicity condition for x_1 requires that these coefficients vanish; accordingly,

$$\alpha A_0 + \tfrac{3}{4}\beta A_0^3 - F \cos \delta_0 = 0 \qquad \sin \delta_0 = 0$$

Hence we obtain

$$\alpha A_0 + \tfrac{3}{4}\beta A_0^3 - F = 0 \qquad \delta_0 = 0 \tag{1.66}$$

The first member of Eqs. (1.66) determines the amplitude A_0. Then, with use of the initial condition $\dot{x}_1(0) = 0$, the general solution of Eq. (1.65) may be written as

$$x_1 = A_1 \cos \tau + \tfrac{1}{32}\beta A_0^3 \cos 3\tau \tag{1.67}$$

where A_1 is a constant of integration to be determined in the following step.

Substitution of Eqs. (1.64) and (1.67) into (1.61) gives the differential equation

$$\ddot{x}_2 + x_2 = -(\alpha A_1 + \tfrac{9}{4}\beta A_0{}^2 A_1 + \tfrac{3}{128}\beta^2 A_0{}^5)\cos\tau - F\delta_1 \sin\tau$$
$$- \tfrac{1}{4}\beta A_0{}^2(3A_1 + \tfrac{1}{8}\alpha A_0 + \tfrac{3}{16}\beta A_0{}^3)\cos 3\tau - \tfrac{3}{128}\beta^2 A_0{}^5 \cos 5\tau \quad (1.68)$$

In order that secular terms do not appear in the solution $x_2(\tau)$, we equate the coefficients of $\cos\tau$ and $\sin\tau$ to zero. Thereby we obtain

$$A_1 = -\frac{3\beta^2 A_0{}^5}{128(\alpha + \tfrac{9}{4}\beta A_0{}^2)} \qquad \delta_1 = 0 \qquad (1.69)$$

The first member of Eqs. (1.69) fixes the coefficient A_1 in Eq. (1.67). By virtue of Eqs. (1.69) the general solution of Eq. (1.68) may be written as

$$x_2(\tau) = A_2 \cos\tau + \tfrac{1}{32}\beta A_0{}^2(3A_1 + \tfrac{1}{8}\alpha A_0 + \tfrac{3}{16}\beta A_0{}^3)\cos 3\tau$$
$$+ \tfrac{3}{3072}\beta^2 A_0{}^5 \cos 5\tau \quad (1.70)$$

where A_2 is a constant of integration.

By proceeding analogously, one may determine $x_3(\tau)$, $x_4(\tau)$, . . . and δ_3, δ_4, . . . successively. Use of the periodicity condition for $x_3(\tau)$ fixes the amplitude A_2 and the phase δ_2 as

$$A_2 = \frac{-3\beta A_0(\alpha\beta A_0{}^4 + 2\beta^2 A_0{}^6 + 40\beta A_0{}^3 A_1 + 768 A_1{}^2)}{1024(\alpha + \tfrac{9}{4}\beta A_0{}^2)} \qquad \delta_2 = 0 \quad (1.71)$$

When we summarize the above results, the solution $x(t)$, up to and including terms of the second order in μ, is

$$x(t) = (A_0 + \mu A_1 + \mu^2 A_2)\cos t$$
$$+ \frac{\mu}{32}\beta A_0{}^2\left[A_0 + \frac{\mu}{16}(2\alpha A_0 + 3\beta A_0{}^3 + 48 A_1)\right]\cos 3t$$
$$+ \frac{3\mu^2}{3072}\beta^2 A_0{}^5 \cos 5t \quad (1.72)$$

where the amplitudes A_0, A_1, and A_2 are given by Eqs. (1.66), (1.69), and (1.71), respectively.

The harmonic solution of Duffing's equation with a term for dissipation,

$$\frac{d^2 x}{dt^2} + x = \mu\left(-\alpha x - \beta x^3 - k\frac{dx}{dt} + F\cos t\right) \quad (1.73)$$

can be obtained in much the same way. Equation (1.73) may be rewritten as

$$\ddot{x} + x = \mu[-\alpha x - \beta x^3 - k\dot{x} + F\cos(\tau + \delta)] \quad (1.74)$$

where $\tau = t - \delta$. The solution $x(\tau)$ and the phase angle δ are expanded as power series in μ, that is,

$$x(\tau) = x_0(\tau) + \mu x_1(\tau) + \mu^2 x_2(\tau) + \cdots \qquad (1.75)$$

$$\delta = \delta_0 + \mu \delta_1 + \mu^2 \delta_2 + \cdots \qquad (1.76)$$

Substitution of Eqs. (1.75) and (1.76) into Eq. (1.74) and collection of like powers of μ gives us a set of differential equations with respect to the unknown quantities in the right sides of Eqs. (1.75) and (1.76). These quantities are to be determined by use of the conditions

$$x_i(\tau + 2\pi) = x_i(\tau) \qquad (1.77)$$

$$\dot{x}_i(0) = 0 \qquad i = 0, 1, 2, \ldots \qquad (1.78)$$

A procedure analogous to that for the preceding case gives the solution for the first approximation, that is,

$$x_0(\tau) = A_0 \cos \tau \qquad (1.79)$$

The amplitude A_0 and the phase angle δ_0 are determined by use of the periodicity condition for $x_1(\tau)$, thus,

$$\alpha A_0 + \tfrac{3}{4}\beta A_0^3 - F \cos \delta_0 = 0 \qquad k A_0 - F \sin \delta_0 = 0$$

from which we obtain

$$[(\alpha + \tfrac{3}{4}\beta A_0^2)^2 + k^2]A_0^2 = F^2$$

$$\cos \delta_0 = \left(\alpha + \frac{3}{4}\beta A_0^2\right)\frac{A_0}{F} \qquad \sin \delta_0 = k\frac{A_0}{F} \qquad (1.80)$$

The solution $x_1(\tau)$, which provides the correction term of order μ, is found to be

$$x_1(\tau) = A_1 \cos \tau + \tfrac{1}{32}\beta A_0^3 \cos 3\tau \qquad (1.81)$$

The amplitude A_1 and the phase angle δ_1 are determined by use of the periodicity condition for $x_2(\tau)$, thus,

$$A_1 = -\frac{3\beta^2 A_0^5}{128(\alpha + \tfrac{9}{4}\beta A_0^2 + k \tan \delta_0)}$$

$$\delta_1 = -\frac{3k\beta^2 A_0^5}{128(\alpha + \tfrac{9}{4}\beta A_0^2 + k \tan \delta_0)F \cos \delta_0} \qquad (1.82)$$

By summarizing the above results, the solution $x(t)$, up to and including terms of the first order in μ, is found to be

$$x(t) = (A_0 + \mu A_1) \cos (t - \delta_0 - \mu \delta_1)$$

$$+ \frac{\mu}{32}\beta A_0^3 \cos 3(t - \delta_0 - \mu \delta_1) \qquad (1.83)$$

where A_0, δ_0 and A_1, δ_1 are given by Eqs. (1.80) and (1.82), respectively.

1.3 Iteration Method

A method of solving nonlinear differential equations based on the process of successive iteration is called the *iteration method*. Iteration may be performed in a number of ways, but the basic procedure is this: first, solve the equation with certain terms neglected; second, insert the resulting solution in the terms at first neglected and obtain a second solution, of improved accuracy.

Let us consider, as an example, obtaining the harmonic solution of Duffing's equation

$$\ddot{x} + x = \mu(-\alpha x - \beta x^3 + F \cos \tau) \tag{1.84}$$

where μ is a small parameter. We start with the solution[1]

$$x_0 = A_0 \cos \tau \tag{1.85}$$

as a first approximation. Since this solution is obtained by ignoring the right side of Eq. (1.84), the difference between x_0 and the exact solution x should be of the first order in μ.

Inserting x_0 into the right side of Eq. (1.84) gives us the differential equation whose solution furnishes the second approximation, namely,

$$\ddot{x}_1 + x_1 = \mu(-\alpha A_0 - \tfrac{3}{4}\beta A_0^3 + F) \cos \tau - \tfrac{1}{4}\mu\beta A_0^3 \cos 3\tau \tag{1.86}$$

Since the right side of this equation may differ from that of Eq. (1.84) by an amount of order μ^2, one may expect to find a second approximation x_1 that is correct up to the first order in μ. The periodicity condition for x_1 requires that no secular terms should appear in the solution x_1; hence,

$$\alpha A_0 + \tfrac{3}{4}\beta A_0^3 - F = 0 \tag{1.87}$$

which fixes the amplitude A_0. By using Eq. (1.87), we solve Eq. (1.86) and obtain

$$x_1 = A_1 \cos \tau + \tfrac{1}{32}\mu\beta A_0^3 \cos 3\tau \tag{1.88}$$

where the amplitude A_1 may be expected to differ from A_0 only by an amount of the first order in μ. Inserting Eq. (1.88) into the right side of Eq. (1.84) gives us

$$\begin{aligned}
\ddot{x}_2 + x_2 = {}& \mu(-\alpha A_1 - \tfrac{3}{4}\beta A_1^3 - \tfrac{3}{128}\mu\beta^2 A_0^3 A_1^2 + F) \cos \tau \\
& - \tfrac{1}{4}\mu\beta[A_1^3 + \tfrac{1}{16}\mu(2\alpha + 3\beta A_1^2)A_0^3] \cos 3\tau \\
& - \tfrac{3}{128}\mu^2\beta^2 A_0^3 A_1^2 \cos 5\tau
\end{aligned} \tag{1.89}$$

Terms of order higher than μ^2 are omitted in the right side of this equation. Since the right side of Eq. (1.89) may differ from that of Eq. (1.84)

---

[1] A term $B_0 \sin \tau$ may be added to the solution x_0; if so, however, B_0 will be evaluated as zero in the next step of iteration.

by an amount of order μ^3, the third approximation x_2 must be a correct solution up to the order of μ^2. The periodicity condition for x_2 requires that

$$\alpha A_1 + \tfrac{3}{4}\beta A_1^3 + \tfrac{3}{128}\mu\beta^2 A_0^3 A_1^2 - F = 0 \qquad (1.90)$$

Bearing in mind that the difference between A_0 and A_1 is of the first order in μ, we solve Eq. (1.90) for A_1 and obtain

$$A_1 = A_0 - \frac{3\mu\beta^2 A_0^5}{128(\alpha + \tfrac{9}{4}\beta A_0^2)} \qquad (1.91)$$

Therefore, we write

$$x_1 = \left[A_0 - \frac{3\mu\beta^2 A_0^5}{128(\alpha + \tfrac{9}{4}\beta A_0^2)} \right] \cos\tau + \frac{1}{32}\mu\beta A_0^3 \cos 3\tau$$

where A_0 is obtained by solving Eq. (1.87). The result agrees, as it should, up to the first order in μ with the solution obtained by the perturbation method [Eq. (1.72)].

1.4 Averaging Method

(a) Autonomous Systems

We consider the differential equation

$$\ddot{x} + x = \mu f(x,\dot{x}) \qquad (1.92)$$

or equivalently,

$$\dot{x} = y \qquad \dot{y} + x = \mu f(x,y) \qquad (1.93)$$

where μ is a dimensionless parameter, which we assume to be small. We also assume that $f(x,y)$ is a polynomial in x and y. The behavior of the system is described by the movement of a representative point $(x(\tau),y(\tau))$ along the solution curves of Eq. (1.93) in the xy plane. These solution curves are called *trajectories of the representative point*. For $\mu = 0$, Eqs. (1.93) become a harmonic oscillator; so the trajectory is a circle with its center at the origin. The representative point rotates along the circle in the clockwise direction with unity angular velocity.

We now consider a new coordinate system (a,b) in the phase plane, which rotates together with the representative point with unity angular velocity. If μ is not zero but is small, the coordinates $(a(\tau),b(\tau))$ of the representative point vary slowly with τ. As one readily sees in Fig. 1.1,

$$a(\tau) = x(\tau) \cos\tau - y(\tau) \sin\tau \qquad b(\tau) = x(\tau) \sin\tau + y(\tau) \cos\tau$$

so that

$$x(\tau) = a(\tau) \cos\tau + b(\tau) \sin\tau \qquad y(\tau) = -a(\tau) \sin\tau + b(\tau) \cos\tau \qquad (1.94)$$

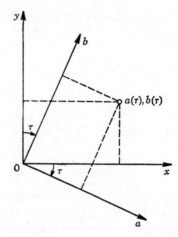

FIGURE 1.1 Transformation of the coordinates.

Inserting Eqs. (1.94) into (1.93) yields

$$\dot{a} \cos \tau + \dot{b} \sin \tau = 0$$

and

$$-\dot{a} \sin \tau + \dot{b} \cos \tau = \mu f(a \cos \tau + b \sin \tau, -a \sin \tau + b \cos \tau) \tag{1.95}$$

Solving Eqs. (1.95) for the derivatives $\dot{a}$ and $\dot{b}$ gives us

$$\dot{a} = -\mu f(a \cos \tau + b \sin \tau, -a \sin \tau + b \cos \tau) \sin \tau$$
$$\dot{b} = \mu f(a \cos \tau + b \sin \tau, -a \sin \tau + b \cos \tau) \cos \tau \tag{1.96}$$

Expanding the right sides into Fourier series, treating a and b as though constant, gives us

$$\dot{a} = \mu \left(\frac{A_0}{2} + A_{s1} \sin \tau + A_{c1} \cos \tau + A_{s2} \sin 2\tau \right.$$
$$\left. + A_{c2} \cos 2\tau + \cdots \right)$$
$$\dot{b} = \mu \left(\frac{B_0}{2} + B_{s1} \sin \tau + B_{c1} \cos \tau + B_{s2} \sin 2\tau \right. \tag{1.97}$$
$$\left. + B_{c2} \cos 2\tau + \cdots \right)$$

where the A's and B's are the corresponding Fourier coefficients of the functions

$$-f(a \cos \tau + b \sin \tau, -a \sin \tau + b \cos \tau) \sin \tau$$
$$f(a \cos \tau + b \sin \tau, -a \sin \tau + b \cos \tau) \cos \tau$$

From the form of the right sides of Eqs. (1.97) it is seen that both $\dot{a}$ and $\dot{b}$ are periodic functions of time. However, owing to the presence of the small parameter μ, both $a(\tau)$ and $b(\tau)$ are functions which vary slowly with τ. It may therefore be considered that $a(\tau)$ and $b(\tau)$ remain approximately constant during one period 2π.   If so, we can write Eqs. (1.96), to a first approximation, as

$$\dot{a} = -\frac{\mu}{2\pi} \int_0^{2\pi} f(a \cos \tau + b \sin \tau, -a \sin \tau + b \cos \tau) \sin \tau \, d\tau$$
$$\dot{b} = \frac{\mu}{2\pi} \int_0^{2\pi} f(a \cos \tau + b \sin \tau, -a \sin \tau + b \cos \tau) \cos \tau \, d\tau$$
$$\text{(1.98)}$$

From Eqs. (1.97) we have

$$\dot{a} = \mu \frac{A_0(a,b)}{2} \qquad \dot{b} = \mu \frac{B_0(a,b)}{2} \qquad \text{(1.99)}$$

If these equations are compared with the exact equations (1.97), it is seen that the equations of the first approximation are obtained from the exact equations by averaging the latter equations over the period 2π, thus retaining only the first terms and eliminating all the other terms of the Fourier series in the right sides of Eqs. (1.97).

The system of Eqs. (1.98) has the advantage that, when transformed to polar coordinates, the variables can be separated.   By putting

$$a = r \cos \theta \qquad b = r \sin \theta \qquad \text{(1.100)}$$

we obtain from Eqs. (1.98)

$$\dot{a} = \dot{r} \cos \theta - r \sin \theta \, \dot{\theta} = -\frac{\mu}{2\pi} \int_0^{2\pi} f[r \cos (\theta - \tau), r \sin (\theta - \tau)] \sin \tau \, d\tau$$
$$\dot{b} = \dot{r} \sin \theta + r \cos \theta \, \dot{\theta} = \frac{\mu}{2\pi} \int_0^{2\pi} f[r \cos (\theta - \tau), r \sin (\theta - \tau)] \cos \tau \, d\tau$$

Since $r(\tau)$ and $\theta(\tau)$ also remain approximately constant during one period 2π, multiplying the first equation by $\cos \theta$, multiplying the second by $\sin \theta$, and adding gives us

$$\dot{r} = -\frac{\mu}{2\pi} \int_0^{2\pi} f(r \cos u, -r \sin u) \sin u \, du$$

Similarly, we obtain   (1.101)

$$\dot{\theta} = \frac{\mu}{2\pi r} \int_0^{2\pi} f(r \cos u, -r \sin u) \cos u \, du$$

where $u = \tau - \theta$.

(b) *Nonautonomous Systems*

We consider the differential equation

$$\ddot{x} + x = \mu f(x, \dot{x}, \tau) \qquad \text{(1.102)}$$

where the function $f(x, \dot{x}, \tau)$ contains the time τ explicitly and we assume that $f(x, \dot{x}, \tau + 2\pi) = f(x,\dot{x},\tau)$. By introducing new variables $a(\tau)$ and $b(\tau)$ as before, we obtain

$$x(\tau) = a(\tau) \cos \tau + b(\tau) \sin \tau \qquad y(\tau) = -a(\tau) \sin \tau + b(\tau) \cos \tau \quad (1.103)$$

Since $y(\tau) = \dot{x}(\tau)$ by definition, we obtain

$$\begin{aligned} \dot{a} \cos \tau + \dot{b} \sin \tau &= 0 \\ -\dot{a} \sin \tau + \dot{b} \cos \tau &= \mu f(a \cos \tau + b \sin \tau, -a \sin \tau + b \cos \tau, \tau) \end{aligned} \quad (1.104)$$

Solving Eqs. (1.104) for the derivatives $\dot{a}$ and $\dot{b}$ gives us

$$\begin{aligned} \dot{a} &= -\mu f(a \cos \tau + b \sin \tau, -a \sin \tau + b \cos \tau, \tau) \sin \tau \\ \dot{b} &= \mu f(a \cos \tau + b \sin \tau, -a \sin \tau + b \cos \tau, \tau) \cos \tau \end{aligned} \quad (1.105)$$

Since $\dot{a}$ and $\dot{b}$ are slowly varying functions of the time τ, we assume that they are constant during one period 2π. Then $f(a \cos \tau + b \sin \tau, -a \sin \tau + b \cos \tau, \tau)$ is periodic, of period 2π. Therefore, expanding the right sides of Eqs. (1.105) into Fourier series leads to

$$\dot{a} = \mu \left(\frac{A_0}{2} + A_{s1} \sin \tau + A_{c1} \cos \tau + A_{s2} \sin 2\tau \right.$$
$$\left. + A_{c2} \cos 2\tau + \cdots \right)$$
$$\dot{b} = \mu \left(\frac{B_0}{2} + B_{s1} \sin \tau + B_{c1} \cos \tau + B_{s2} \sin 2\tau \right. \qquad (1.106)$$
$$\left. + B_{c2} \cos 2\tau + \cdots \right)$$

where the A's and B's are the corresponding Fourier coefficients of the functions

$$-f(a \cos \tau + b \sin \tau, -a \sin \tau + b \cos \tau, \tau) \sin \tau$$
$$f(a \cos \tau + b \sin \tau, -a \sin \tau + b \cos \tau, \tau) \cos \tau$$

We neglect the oscillatory terms in the right sides of Eqs. (1.106) and obtain the relations to determine $\dot{a}$ and $\dot{b}$ to a first approximation, namely,

$$\dot{a} = -\frac{\mu}{2\pi} \int_0^{2\pi} f(a \cos \tau + b \sin \tau, -a \sin \tau + b \cos \tau, \tau) \sin \tau \, d\tau$$
$$\dot{b} = \frac{\mu}{2\pi} \int_0^{2\pi} f(a \cos \tau + b \sin \tau, -a \sin \tau + b \cos \tau, \tau) \cos \tau \, d\tau \qquad (1.107)$$

Equations (1.107) were obtained rather simply by van der Pol. He set $x(\tau) = a(\tau) \cos \tau + b(\tau) \sin \tau$, inserted it into Eq. (1.102), and thus obtained

$$\ddot{a} \cos \tau - 2\dot{a} \sin \tau + \ddot{b} \sin \tau + 2\dot{b} \cos \tau$$
$$= \mu f(a \cos \tau + b \sin \tau, -a \sin \tau + b \cos \tau + \dot{a} \cos \tau + \dot{b} \sin \tau, \tau)$$

Following his method, the terms containing $\ddot{a}$, $\ddot{b}$, $\mu\dot{a}$, and $\mu\dot{b}$ are neglected because $a(\tau)$ and $b(\tau)$ are assumed to be slowly varying functions of τ. Expanding f into a Fourier series and equating the coefficients of $\sin \tau$ and $\cos \tau$ gives us Eqs. (1.107).

The averaging method is applicable not only to the study of periodic oscillations but also to the study of transient oscillations where the amplitude and phase of the oscillations vary slowly with time.

1.5 Principle of Harmonic Balance

This is a method of widest utility for obtaining a periodic solution of a nonlinear differential equation. We first explain this method by making use of the averaging method described in the preceding section. The harmonic solution of Eq. (1.102) may be written, to a first approximation, as

$$x(\tau) = a \cos \tau + b \sin \tau \qquad (1.108)$$

where the amplitudes a and b are expressed as constants, since we are concerned with the periodic solution. Then Eqs. (1.107) become

$$\int_0^{2\pi} f(a \cos \tau + b \sin \tau, -a \sin \tau + b \cos \tau, \tau) \sin \tau \, d\tau = 0$$
$$\int_0^{2\pi} f(a \cos \tau + b \sin \tau, -a \sin \tau + b \cos \tau, \tau) \cos \tau \, d\tau = 0 \qquad (1.109)$$

One will readily see that the above relations are immediately obtained by inserting Eq. (1.108) into Eq. (1.102) and equating the coefficients of $\sin \tau$ and $\cos \tau$ separately to zero. This procedure is the basic principle of harmonic balance.[1]

The periodic solution of a higher approximation may be obtained in the following manner. The periodic solution is first expanded into Fourier series with unknown coefficients. The assumed solution is then inserted into the original equation, and the sine and cosine terms of the respective frequencies are set to zero separately. By solving the simultaneous equations thus obtained, one may fix the unknown coefficients of the assumed solution. In assuming the periodic solution, only terms of the harmonic frequency and a few additional terms of different frequencies (usually subharmonic or higher-harmonic frequency) are considered, and these because of their prime importance. Terms of frequency other than those are certain to be present also, but they may tolerably be omitted in most cases. A closer approximation may be

---

[1] The describing-function method [54] may be regarded as an application of the principle of harmonic balance. The method is used widely for the analysis of nonlinear control systems [17].

obtained if more terms of the Fourier series are taken into account; however, numerical computation will become too unwieldy. We describe a practical method of improving the approximation resulting from use of the principle of harmonic balance.

Let us consider a harmonic solution of Duffing's equation

$$\frac{d^2x}{d\tau^2} + k\frac{dx}{d\tau} + \alpha x + \beta x^3 = F\cos\tau \tag{1.110}$$

We assume the solution to a first approximation as

$$x_0(\tau) = A_{10}\cos\tau + B_{10}\sin\tau \tag{1.111}$$

and seek a solution of higher accuracy which does not substantially differ from the first approximation.[1] Substitution of Eq. (1.111) into (1.110) gives us

$$[-(1-\alpha-\tfrac{3}{4}\beta R_{10}^2)A_{10} + kB_{10}]\cos\tau$$
$$+ [-(1-\alpha-\tfrac{3}{4}\beta R_{10}^2)B_{10} - kA_{10}]\sin\tau$$
$$+ \tfrac{1}{4}\beta(A_{10}^2 - 3B_{10}^2)A_{10}\cos 3\tau$$
$$+ \tfrac{1}{4}\beta(3A_{10}^2 - B_{10}^2)B_{10}\sin 3\tau = F\cos\tau \tag{1.112}$$

where $R_{10}^2 = A_{10}^2 + B_{10}^2$. Since the first approximation is concerned only with the terms of harmonic frequency, we ignore the third-harmonic components and equate the coefficients of $\cos\tau$ and $\sin\tau$ on both sides, that is,

$$-(1-\alpha-\tfrac{3}{4}\beta R_{10}^2)A_{10} + kB_{10} = F$$
$$kA_{10} + (1-\alpha-\tfrac{3}{4}\beta R_{10}^2)B_{10} = 0 \tag{1.113}$$

from which we obtain

$$A_{10} = -\left(1-\alpha-\frac{3}{4}\beta R_{10}^2\right)\frac{R_{10}^2}{F} \qquad B_{10} = k\frac{R_{10}^2}{F}$$

where
$$[(1-\alpha-\tfrac{3}{4}\beta R_{10}^2)^2 + k^2]R_{10}^2 = F^2 \tag{1.114}$$

The second approximation is assumed in the form

$$x(\tau) = (A_{10} + \epsilon A_{11})\cos\tau + (B_{10} + \epsilon B_{11})\sin\tau + \epsilon A_{31}\cos 3\tau + \epsilon B_{31}\sin 3\tau \tag{1.115}$$

It is to be noted that inclusion of the small parameter ϵ in the correction terms shows the order of magnitude among the coefficients. We substitute Eq. (1.115) into Eq. (1.110) and equate the coefficients of $\cos\tau$,

---

[1] Equation (1.110) may have other periodic solutions in which higher-harmonic or subharmonic components are predominant. In such cases these components should be included in the solution (1.111).

$\sin \tau$, $\cos 3\tau$, and $\sin 3\tau$ separately to zero. Neglecting terms of order higher than the first in ϵ leads to

$$(1 - \alpha - 3l - n)\epsilon A_{11} - (k + m)\epsilon B_{11} - (l - n)\epsilon A_{31} - m\epsilon B_{31}$$
$$= 0$$

$$(k - m)\epsilon A_{11} + (1 - \alpha - l - 3n)\epsilon B_{11} + m\epsilon A_{31} - (l - n)\epsilon B_{31}$$
$$= 0$$

$$-(l - n)\epsilon A_{11} + m\epsilon B_{11} + (9 - \alpha - 2l - 2n)\epsilon A_{31} - 3k\epsilon B_{31} \qquad (1.116)$$
$$= (\tfrac{1}{3}l - n)A_{10}$$

$$-m\epsilon A_{11} - (l - n)\epsilon B_{11} + 3k\epsilon A_{31} + (9 - \alpha - 2l - 2n)\epsilon B_{31}$$
$$= (l - \tfrac{1}{3}n)B_{10}$$

where $\quad l = \tfrac{3}{4}\beta A_{10}{}^2 \qquad m = \tfrac{3}{2}\beta A_{10}B_{10} \qquad$ and $\qquad n = \tfrac{3}{4}\beta B_{10}{}^2$

Solving the simultaneous linear equations (1.116) gives us the coefficients ϵA_{11}, ϵB_{11}, ϵA_{31}, and ϵB_{31} of the correction terms in Eq. (1.115). The method of improving the approximation described above is particularly useful when the amplitude of each harmonic component decreases with increasing order of the harmonics.

1.6 Numerical Examples of the Solution for Duffing's Equation

The analytical methods described in the preceding sections are considered to be legitimate only for solving differential equations with small nonlinearity. By use of those methods, an approximate solution is found as a power series with terms involving the small parameter μ raised to successively higher powers. When the parameter μ becomes large, the successive approximation will no longer be reasonable. However, the methods are still applicable to some extent under such a condition.

In this section we consider Duffing's equation with particular values of the parameters, that is,

$$\frac{d^2x}{d\tau^2} + x^3 = 0.2 \cos \tau \qquad (1.117)$$

and
$$\frac{d^2x}{d\tau^2} + 0.2\frac{dx}{d\tau} + x^3 = 0.3 \cos \tau \qquad (1.118)$$

These equations may be obtained by equating the small parameter μ equal to unity in the preceding sections [Eqs. (1.57) and (1.73)]. The harmonic solution of Eq. (1.117) may be written as

$$x(\tau) = A_1 \cos \tau + A_3 \cos 3\tau + A_5 \cos 5\tau + \cdots \qquad (1.119)$$

The coefficients of the series, as determined by both the perturbation method and the method of harmonic balance, are as shown in Tables 1.1 and 1.2.

Table 1.1   Harmonic Solutions, Obtained by the
Perturbation Method, for Eq. (1.117)

Harmonic solution	Order of approximation	Approximate solution		
		A_1	A_3	A_5
1	first	-0.207		
	second	-0.207	-0.000	
	third	-0.207	-0.000	0.000
2	first	1.244		
	second	1.216	0.060	
	third	1.211	0.066	0.003
3	first	-1.037		
	second	-1.017	-0.035	
	third	-1.016	-0.036	-0.001

Table 1.2   Harmonic Solutions, Obtained by the Method of
Harmonic Balance, for Eq. (1.117)

Harmonic solution	Order of approximation	Approximate solution		
		A_1	A_3	A_5
1	first	-0.207		
	second	-0.207	-0.000	
	third	-0.207	-0.000	0.000
2	first	1.244		
	second	1.213	0.067	
	third	1.212	0.066	0.003
3	first	-1.037		
	second	-1.017	-0.036	
	third	-1.016	-0.035	-0.001

There are three kinds of harmonic oscillation in this particular case. The exact solutions, correct to four decimal places, are obtained by making use of a digital computer. They are

$$x_1(\tau) = -0.2066 \cos \tau - 0.0003 \cos 3\tau - 0.0000 \cos 5\tau - 0.0000 \cos 7\tau + \cdots$$

$$x_2(\tau) = 1.2103 \cos \tau + 0.0658 \cos 3\tau + 0.0035 \cos 5\tau + 0.0001 \cos 7\tau + \cdots$$

$$x_3(\tau) = -1.0161 \cos \tau - 0.0352 \cos 3\tau - 0.0012 \cos 5\tau - 0.0000 \cos 7\tau + \cdots$$

The third approximations in Tables 1.1 and 1.2 show satisfactory agreement with the above solutions.

As the second example, the harmonic solution of Eq. (1.118), that is,

$$x(\tau) = A_1 \cos \tau + B_1 \sin \tau + A_3 \cos 3\tau + B_3 \sin 3\tau + \cdots$$

is sought. The coefficients of the solution are as shown in Tables 1.3 and 1.4.

The exact solutions, with coefficients correct to four decimal places, are

$$x_1(\tau) = -0.3101 \cos \tau + 0.0670 \sin \tau - 0.0007 \cos 3\tau + 0.0004 \sin 3\tau + \cdots$$

$$x_2(\tau) = 0.6864 \cos \tau + 0.9841 \sin \tau - 0.0597 \cos 3\tau + 0.0214 \sin 3\tau + \cdots$$

$$x_3(\tau) = -0.7404 \cos \tau + 0.6768 \sin \tau + 0.0223 \cos 3\tau + 0.0231 \sin 3\tau + \cdots$$

The second approximations in Tables 1.3 and 1.4 agree well with these solutions.

Table 1.3   Harmonic Solutions, Obtained by the Perturbation Method, for Eq. (1.118)

Harmonic solution	Order of approximation	Approximate solution			
		A_1	B_1	A_3	B_3
1	first	−0.310	0.067		
	second	−0.310	0.067	−0.001	0.001
2	first	0.703	1.012		
	second	0.717	0.972	−0.055	0.019
3	first	−0.748	0.699		
	second	−0.745	0.669	0.020	0.027

Table 1.4   Harmonic Solutions, Obtained by the Method of Harmonic Balance, for Eq. (1.118)

Harmonic solution	Order of approximation	Approximate solution			
		A_1	B_1	A_3	B_3
1	first	−0.310	0.067		
	second	−0.310	0.067	−0.001	0.001
2	first	0.703	1.012		
	second	0.684	0.988	−0.061	0.021
3	first	−0.748	0.699		
	second	−0.744	0.671	0.022	0.026

Topological Methods and Graphical Solutions

2.1 Introduction

The topological method of analysis is one of the important means of investigating various phenomena of nonlinear oscillations, and it is applicable to the study of autonomous systems. By this method solutions of differential equations are sought not as explicit functions of the time, but as solution (or integral) curves in a phase space or, more generally, in the state space. Considerable insight into the qualitative aspects of the solution, and some quantitative information as well, can be obtained through a study of integral curves. If, however, we rely on graphical methods for the representation of solutions, the applicability of the method is usually confined to systems of lower orders.

An equilibrium condition in which the variables of a system are at rest is correlated with a singular point[1] in the state space; a periodic solution in which the variables are undergoing periodic change is correlated with a limit cycle. Hence, an integral curve which tends to such a singularity or limit cycle may represent a transient state of the system. This chapter deals with general properties of singular points and limit cycles and also describes several methods of graphical solution of nonlinear differential equations.

The theory of mapping is also useful for studying nonlinear differential equations. Fixed points and invariant curves under iteration of the mapping have properties similar to those of the singular points and integral curves of differential equations. Investigation along this line will be given in Chap. 10.

2.2 Integral Curves and Singular Points in a State Plane

In this section we shall be concerned with an autonomous system of two first-order differential equations, that is,

$$\frac{dx}{d\tau} = X(x,y) \qquad \frac{dy}{d\tau} = Y(x,y) \tag{2.1}$$

---

[1] Technical terms are defined in the following section.

where $X(x,y)$ and $Y(x,y)$ are polynomials of x and y. It is possible to eliminate $d\tau$ between the equations and write

$$\frac{dy}{dx} = \frac{Y(x,y)}{X(x,y)} \tag{2.2}$$

As mentioned before, we seek the integral curves of Eq. (2.2) on the xy plane, that is, on the state plane of the variables x and y. If a relation exists between the variables such that $dx/d\tau = y$, the state plane is referred to as the phase plane. Once the integral curves of Eq. (2.2) are obtained on the state plane, it is not difficult to find the solutions $x(\tau)$ and $y(\tau)$ of Eqs. (2.1). Thus the behavior of the system may be described by the movement of the representative point $(x(\tau),y(\tau))$ along the integral curves of Eq. (2.2). These integral curves are called *trajectories of the representative point.*

Any point (x,y) for which the two functions $X(x,y)$ and $Y(x,y)$ do not vanish simultaneously is called an *ordinary point* with respect to Eq. (2.2). A point (x_0,y_0) for which $X(x_0,y_0) = Y(x_0,y_0) = 0$ is called a *singular point.* Physically, a singular point represents an equilibrium point, since both $x(\tau)$ and $y(\tau)$ are constant under such a condition.

We consider a system of differential equations of the form

$$\frac{dx}{d\tau} = a_1x + a_2y + X_2(x,y) \qquad \frac{dy}{d\tau} = b_1x + b_2y + Y_2(x,y) \tag{2.3}$$

where X_2, Y_2 are polynomials containing terms of order higher than the first in x and y. Since the right-hand sides of Eqs. (2.3) vanish for $x = y = 0$, the origin is a singular point.[1] We shall be interested for the time being in the integral curves in the neighborhood of the origin. Then, instead of (2.3), we may generally deal with the linear equations

$$\frac{dx}{d\tau} = a_1x + a_2y \qquad \frac{dy}{d\tau} = b_1x + b_2y \tag{2.4}$$

from which, by eliminating τ, we obtain

$$\frac{dy}{dx} = \frac{b_1x + b_2y}{a_1x + a_2y} \tag{2.5}$$

Putting

$$x = Ae^{\lambda\tau} \qquad y = Be^{\lambda\tau} \tag{2.6}$$

---

[1] Notice that we can always make a change of the variables that brings a singular point (x_0,y_0) to the origin, and thus gives the differential equations the form (2.3) in the neighborhood of the singular point.

and inserting these solutions into Eqs. (2.4) gives us

$$(a_1 - \lambda)A + a_2 B = 0 \qquad b_1 A + (b_2 - \lambda)B = 0 \qquad (2.7)$$

Nontrivial solutions for A and B exist only when the λ's are the roots of the characteristic equation

$$\begin{vmatrix} a_1 - \lambda & a_2 \\ b_1 & b_2 - \lambda \end{vmatrix} = 0 \qquad (2.8)$$

Solving this quadratic equation gives

$$\lambda_{1,2} = \frac{a_1 + b_2 \pm \sqrt{(a_1 - b_2)^2 + 4a_2 b_1}}{2} \qquad (2.9)$$

We consider the case where the roots λ_i $(i = 1, 2)$ are not zero and are distinct. For a particular value of λ_i the ratio B_i/A_i is determined from Eqs. (2.7), namely,

$$\frac{B_i}{A_i} = \frac{a_1 - \lambda_i}{-a_2} = \frac{-b_1}{b_2 - \lambda_i} \qquad (2.10)$$

General solutions of Eqs. (2.4) may be written as

$$x = A_1 e^{\lambda_1 \tau} + A_2 e^{\lambda_2 \tau} \qquad y = B_1 e^{\lambda_1 \tau} + B_2 e^{\lambda_2 \tau} \qquad (2.11)$$

Since the ratio A_i/B_i is fixed by Eqs. (2.10), two arbitrary constants can be chosen in Eqs. (2.11).

(a) Classification of Singular Points

Poincaré [84, p. 14] classified the types of singular points according to the character of the integral curves near the singular points, that is, according to the nature of the characteristic roots λ. They are as follows:

1. The singularity is a nodal point (or simply a node) if the characteristic roots are both real and of the same sign, so that

$$(a_1 - b_2)^2 + 4a_2 b_1 \geqq 0 \qquad \text{and} \qquad a_1 b_2 - a_2 b_1 > 0 \qquad (2.12)$$

2. The singularity is a saddle point if the two roots are real but of opposite sign, so that

$$(a_1 - b_2)^2 + 4a_2 b_1 > 0 \qquad \text{and} \qquad a_1 b_2 - a_2 b_1 < 0 \qquad (2.13)$$

3. The singularity is a focal point (or simply a focus) if the two roots are complex conjugates, so that

$$(a_1 - b_2)^2 + 4a_2 b_1 < 0 \qquad (2.14)$$

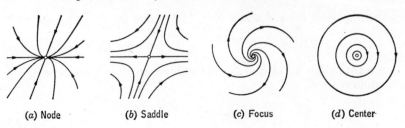

(a) Node   (b) Saddle   (c) Focus   (d) Center

FIGURE 2.1   Types of singular points in the state plane.

4. The singularity is either a center or a focal point[1] if the two roots are imaginary, so that

$$(a_1 - b_2)^2 + 4a_2b_1 < 0 \qquad \text{and} \qquad a_1 + b_2 = 0 \qquad (2.15)$$

In this case it is impossible to distinguish between these two singularities on the basis of the linear equations (2.4).[2]

We also infer a singularity to be stable or unstable according as a representative point on any integral curve moves into the said singularity or not with increasing τ, that is, according as the real part of λ is negative or positive.

We illustrate the types of singularities in what follows. Figure 2.1a gives an example of a nodal point. In this example, the singular point is stable, since a representative point $(x(\tau),y(\tau))$ moves on the integral curves in the direction of the arrows as τ increases and ultimately leads to the nodal point. The saddle point in Fig. 2.1b is a singularity which is terminated by four trajectories forming two distinct integral curves. Two of these trajectories approach the saddle point with increasing τ, while two others move away from it with increasing τ, so that the saddle point is intrinsically unstable. We also see that, between these trajectories, there exist four regions containing continua of hyperbolically shaped integral curves which do not approach the singularity. Figure 2.1c gives an example of an unstable focal point from which the trajectories start without definite direction. Finally, Fig. 2.1d shows a center which is surrounded by a continuum of closed curves such that none approaches it.

The direction of integral curves at a singularity of the node or saddle type may be found by the following procedure. We first assume an

---

[1] The term *spiral point* is also used for a *focal point*, and the term *vortex point* for a *center*.

[2] Referring to the original Eqs. (2.3), Poincaré [84, p. 95] gave a criterion to distinguish between these singularities (Sec. V.2).

initial condition such that $A_1 = 0$ or $A_2 = 0$ in Eqs. (2.11); then

$$\frac{y}{x} = \frac{B_2}{A_2} \quad \text{or} \quad \frac{y}{x} = \frac{B_1}{A_1}$$

These equations represent straight lines which pass through the singularity and thus show the direction of the integral curves at the singularity. By denoting y/x by μ, we obtain, through use of (2.9) and (2.10),

$$\mu_{1,2} = \frac{-(a_1 - b_2) \pm \sqrt{(a_1 - b_2)^2 + 4a_2 b_1}}{2a_2} \tag{2.16}$$

Thus, as it should be, the direction of integral curves is definite when μ_1 and μ_2 are real, that is, when the singularity is either a node or a saddle.

(b) Canonical Form of the Differential Equations

By means of a linear transformation, Eqs. (2.4) can be converted into a canonical form. By multiplying the first equation by l, the second by m, and adding, we obtain

$$l \frac{dx}{d\tau} + m \frac{dy}{d\tau} = (a_1 l + b_1 m)x + (a_2 l + b_2 m)y \tag{2.17}$$

where l and m are constants to be fixed shortly. We choose a constant λ such that

$$\begin{array}{ll} a_1 l + b_1 m = \lambda l & \qquad (a_1 - \lambda)l + b_1 m = 0 \\ \text{or} & \\ a_2 l + b_2 m = \lambda m & \qquad a_2 l + (b_2 - \lambda)m = 0 \end{array} \tag{2.18}$$

Nontrivial solutions for l and m exist only when λ is the root of the quadratic equation

$$\begin{vmatrix} a_1 - \lambda & b_1 \\ a_2 & b_2 - \lambda \end{vmatrix} = 0 \tag{2.19}$$

Corresponding to the roots λ_1 and λ_2 (which we assume to be distinct)[1] of Eq. (2.19), the ratio l_1/m_1 and l_2/m_2 may be fixed from Eqs. (2.18). By putting

$$u = l_1 x + m_1 y \qquad v = l_2 x + m_2 y \tag{2.20}$$

[1] If the characteristic equation (2.19) has a double root $\lambda = \lambda_1 = \lambda_2$, we generally obtain, instead of Eqs. (2.21), the following canonical form:

$$\frac{du}{d\tau} = \lambda u \qquad \frac{dv}{d\tau} = u + \lambda v$$

For further details, see Ref. 16, pp. 371–373.

we obtain, through use of (2.17) and (2.18),

$$\frac{du}{d\tau} = l_1 \frac{dx}{d\tau} + m_1 \frac{dy}{d\tau} = \lambda_1(l_1 x + m_1 y) = \lambda_1 u \qquad \frac{dv}{d\tau} = \lambda_2 v \qquad (2.21)$$

Thus one sees that, by making use of the linear transformation (2.20), Eqs. (2.4) are converted into the canonical form (2.21). Equation (2.19) is the same as (2.8), that is, the characteristic equation of the system (2.4).

When the characteristic roots λ_1 and λ_2 are conjugate complex, u and v are also conjugate complex. Putting

$$\lambda_1 = \alpha + j\beta \qquad u = \xi + j\eta$$
$$\lambda_2 = \alpha - j\beta \qquad v = \xi - j\eta$$

and substituting into (2.21), we obtain

$$\frac{d\xi}{d\tau} = \alpha\xi - \beta\eta \qquad \frac{d\eta}{d\tau} = \beta\xi + \alpha\eta \qquad (2.22)$$

Equations (2.22) are easily solved by introducing polar coordinates $\xi = \rho \cos \phi$, $\eta = \rho \sin \phi$. In these variables the equations reduce to

$$\frac{d\rho}{d\tau} = \alpha\rho \qquad \frac{d\phi}{d\tau} = \beta \qquad (2.23)$$

and the integral curves are given by $\rho = Ce^{(\alpha/\beta)\phi}$, C being a constant of integration. They are thus logarithmic spirals.

(c) Integration of Eq. (2.5)

We may easily integrate Eq. (2.5) by a well-known procedure for homogeneous equations. Thus we obtain

$$(y - \mu_1 x)^{\lambda_1}(y - \mu_2 x)^{-\lambda_2} = \text{const} \qquad (2.24)$$

where $\lambda_{1,2}$ and $\mu_{1,2}$ are given by Eqs. (2.9) and (2.16), respectively.

If, in particular, $(a_1 - b_2)^2 + 4a_2 b_1 = 0$ in Eq. (2.5), we have $\lambda_1 = \lambda_2$ and $\mu_1 = \mu_2$. In this case the solution becomes

$$\log |(a_1 - b_2)x + 2a_2 y| = \frac{(a_1 + b_2)x}{(a_1 - b_2)x + 2a_2 y} + \text{const} \qquad (2.25)$$

If $a_2 = 0$, the solution is

$$[b_1 x + (b_2 - a_1)y]x^{-b_2/a_1} = \text{const} \qquad (2.26)$$

In the case where the singularity is a focal point, λ and μ in (2.24) are complex numbers, and it may be expedient to express the integral curves in polar coordinates, that is,

$$x = r \cos \theta \qquad y = r \sin \theta$$

Then Eq. (2.5) becomes

$$\frac{1}{r}\frac{dr}{d\theta} = \frac{a_1 \cos^2 \theta + (a_2 + b_1) \sin \theta \cos \theta + b_2 \sin^2 \theta}{b_1 \cos^2 \theta + (b_2 - a_1) \sin \theta \cos \theta - a_2 \sin^2 \theta} \tag{2.27}$$

By integrating this, we obtain [109]

$$r = \frac{C \cdot e^{M \tan^{-1} G}}{\sqrt{b_1 \cos^2 \theta + (b_2 - a_1) \sin \theta \cos \theta - a_2 \sin^2 \theta}}$$

where
$$M = \frac{a_1 + b_2}{\sqrt{-D}} \qquad G = \frac{-2a_2 \tan \theta - a_1 + b_2}{\sqrt{-D}} \tag{2.28}$$

$$D = (a_1 - b_2)^2 + 4a_2 b_1 \ (<0)$$

C being an arbitrary constant.

The integral curves spiral around the origin, and for each revolution the change in r is determined by

$$r(\theta + 2\pi) = e^{\sigma \cdot 2\pi M} \cdot r(\theta) \tag{2.29}$$

where σ is 1 or -1 according as $a_2 < 0$ or $a_2 > 0$. Further, from (2.4) we obtain

$$\frac{d\theta}{d\tau} = \frac{1}{2}[(b_1 - a_2) + (a_2 + b_1) \cos 2\theta + (b_2 - a_1) \sin 2\theta]$$

Since the singularity is of the focus type, we have

$$(a_1 - b_2)^2 + 4a_2 b_1 < 0 \qquad \text{or} \qquad (b_1 - a_2)^2 > (a_2 + b_1)^2 + (b_2 - a_1)^2$$

so that $d\theta/d\tau$ has the same sign as $b_1 - a_2$. Since a_2 and b_1 are of opposite signs, it is concluded that

1: $\qquad$ If $a_2 > 0$ (or $b_1 < 0$) $\qquad$ then $\qquad \dfrac{d\theta}{d\tau} < 0$

2: $\qquad$ If $a_2 < 0$ (or $b_1 > 0$) $\qquad$ then $\qquad \dfrac{d\theta}{d\tau} > 0$
$$\tag{2.30}$$

Thus it follows that when $a_2 > 0$, a representative point (r, θ) moves, with the increase of τ, along the integral curve in the clockwise direction and tends to the singularity provided that $e^{-2\pi M} > 1$ or $a_1 + b_2 < 0$. In this case the singular point is stable. Other possible cases may be treated in like manner.

(d) Singular Points in Special Cases; Singular Points of Higher Order

Following the classification as mentioned in Sec. 2.2a, the type of a singular point will be definite when the characteristic roots λ_1 and λ_2 are neither zero nor imaginary and when they are distinct. Now in this part we shall see some special cases which have not been referred to.

To begin with, we consider the case in which $(a_1 - b_2)^2 + 4a_2 b_1 = 0$.

The roots λ_1 and λ_2 are then real and equal. Although the detailed discussion is omitted here,[1] one may conclude that the corresponding singularity is a nodal point. An example for such a case will be given in Sec. 8.6a.

It is worth noticing that the condition $a_1 + b_2 = 0$ is not sufficient to distinguish between a center and a focus in the case where $(a_1 - b_2)^2 + 4a_2b_1 < 0$. In this case the roots λ_1 and λ_2 are imaginary and of opposite sign. Following the analysis due to Poincaré [84, p. 95], a criterion for distinguishing these two types of singularity is given in Sec. V.2. An application of this criterion will appear in Sec. 13.4b.

Thus far we have dealt with the singularity which is simple. A singular point for which either one or both roots of the characteristic equation are zero is referred to as a *singularity of a higher order*. We have at least one zero root if one of the following conditions is satisfied; namely,

$$a_1 = a_2 = 0 \qquad b_1 = b_2 = 0 \qquad \text{or} \qquad \frac{a_1}{a_2} = \frac{b_1}{b_2} \qquad (2.31)$$

When $a_1 = a_2 = 0$ (or $b_1 = b_2 = 0$), the curve $X(x,y) = 0$ [or $Y(x,y) = 0$] has the origin (i.e., the singularity) for multiple point; when $a_1/a_2 = b_1/b_2$, the curves $X(x,y) = 0$ and $Y(x,y) = 0$ have a common tangent at the origin. Under such circumstances the type of a singular point is dependent upon $X_2(x,y)$ and $Y_2(x,y)$ in Eqs. (2.3), which contain terms of order higher than the first in x and y.

By following the method of analysis due to Bendixson, the singularities of higher order may be investigated; an approximate analytical expression of integral curves near the singularity may also be obtained. One may expect (though not always) the coalescence of singularities when one of the characteristic roots is zero. Examples for such cases will be treated in Secs. 8.6b, 13.3b, 13.4a, and 13.6b, where the singularities of the node-saddle type are obtained. The case when both roots of the characteristic equation are zero and the singularity is of the cusp type is treated in Sec. 8.5b.

(e) Limit Cycles

Thus far we have dealt with the singular points and integral curves of a system of differential equations. An integral curve usually terminates in a stable singularity of the node or focus type. When the singularity is a center, the integral curves form a continuum of concentric loops around a center such that none approaches it. This occurs physically when we deal with conservative systems. In certain nonconserva-

---

[1] A rigorous treatment concerning the nature of a singularity in this case is given by Bendixson [6, p. 50].

tive systems, however, we may find closed trajectories or limit cycles toward which the neighboring trajectories spiral on both sides.

For a given nonlinear system governed by

$$\frac{dx}{d\tau} = X(x,y) \qquad \frac{dy}{d\tau} = Y(x,y) \qquad (2.32)$$

the problem of establishing the existence of a limit cycle is generally very difficult. The method of contact curves introduced by Poincaré is sometimes useful in locating possible limit cycles. For this purpose we first consider a family of concentric circles with center at the singular point of Eqs. (2.32) and then determine on the xy plane the locus of the points where these circles are tangent to the integral curves of Eqs. (2.32). This locus is the contact curve. We assume that the contact loci lie in a bounded region of the xy plane. If a limit cycle occurs at all, it must of necessity lie in a ring domain with center at the singular point whose boundaries are the innermost and outermost circles of radii r_{min} and r_{max}, respectively, which touch the contact curve.

Thus far we have considered a family of concentric circles which are in contact with the integral curves of Eqs. (2.32). One may obtain a closer location of limit cycles provided an appropriate form of closed curves instead of concentric circles is chosen.

It is sometimes possible to know that no limit cycle exists, and the Bendixson criterion [6, p. 78], which establishes a condition for the non-existence of closed trajectories, is useful in some cases. Bendixson's criterion is as follows: If the expression $\partial X/\partial x + \partial Y/\partial y$ does not change its sign within a domain D of the xy plane, no closed trajectories can exist in that domain.

The proof is immediate. Suppose a closed curve such as γ existed in D. By Gauss' well-known theorem, if D' is the domain bounded by γ, then

$$\iint_{D'} \left(\frac{\partial X}{\partial x} + \frac{\partial Y}{\partial y} \right) dx\, dy = \oint_{\gamma} (X\, dy - Y\, dx) \neq 0 \qquad (2.33)$$

However, along the path γ Eqs. (2.32) hold, and so $X\, dy - Y\, dx = 0$. Hence the simple integral is zero. This contradiction proves the criterion.

NUMERICAL EXAMPLE

We consider the differential equation

$$\frac{dy}{dx} = \frac{x + y(x^2 + xy + y^2 - 1)}{x(x^2 + xy + y^2 - 1) - y} \qquad (2.34)$$

Since the origin is a singular point, we consider a family of concentric

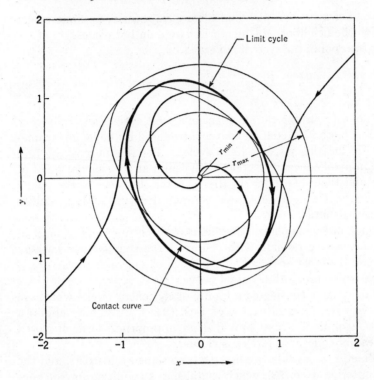

FIGURE 2.2   Integral curves of $\dfrac{dy}{dx} = \dfrac{x + y(x^2 + xy + y^2 - 1)}{x(x^2 + xy + y^2 - 1) - y}$ and
the ring domain in which a limit cycle exists.

circles around the origin, that is,

$$x^2 + y^2 = \text{const}$$

from which we obtain

$$\frac{dy}{dx} = -\frac{x}{y} \tag{2.35}$$

The contact curve is given by

$$\frac{x + y(x^2 + xy + y^2 - 1)}{x(x^2 + xy + y^2 - 1) - y} = -\frac{x}{y}$$

or $$(x^2 + xy + y^2 - 1)(x^2 + y^2) = 0 \tag{2.36}$$

By the introduction of the polar coordinates $x = r \cos \theta$ and $y = r \sin \theta$,
Eq. (2.36) is transformed to

$$[(1 + 0.5 \sin 2\theta)r^2 - 1]r^2 = 0$$

Aside from the point $r = 0$, the contact curve is given by

$$r^2 = \frac{1}{1 + 0.5 \sin 2\theta} \tag{2.37}$$

from which we obtain

$$r_{\min} = \sqrt{\tfrac{2}{3}} \qquad \text{and} \qquad r_{\max} = \sqrt{2}$$

Consequently, we know that a ring domain centered at the origin exists and contains all possible limit cycles; its boundaries are the radii $r_{\min}$ and $r_{\max}$ of the smallest and largest circles which touch the contact curves as given by Eq. (2.37). Figure 2.2 shows this ring domain. Also plotted in the figure are the integral curves and the limit cycle of Eq. (2.34). These curves are obtained first by transforming Eq. (2.34) into

$$\frac{dr}{r\, d\theta} = (1 + 0.5 \sin 2\theta)r^2 - 1$$

and then by making use of the iso-$(r\, d\theta/dr)$ curves (Sec. 2.4). Clearly, the contact curve is given by one of these curves for which $dr/(r\, d\theta) = 0$.

2.3 Integral Curves and Singular Points in a State Space

We consider a set of differential equations of the first order,

$$\begin{aligned}
\frac{dx}{d\tau} &= a_1x + a_2y + a_3z + X_2(x,y,z) \\
\frac{dy}{d\tau} &= b_1x + b_2y + b_3z + Y_2(x,y,z) \\
\frac{dz}{d\tau} &= c_1x + c_2y + c_3z + Z_2(x,y,z)
\end{aligned} \tag{2.38}$$

where X_2, Y_2, and Z_2 are polynomials containing terms of order higher than the first in x, y, and z. Since the right sides of Eqs. (2.38) vanish for $x = y = z = 0$, the origin is a singular point. In the neighborhood of the singular point we may consider the following set of linear equations

$$\begin{aligned}
\frac{dx}{d\tau} &= a_1x + a_2y + a_3z \\
\frac{dy}{d\tau} &= b_1x + b_2y + b_3z \\
\frac{dz}{d\tau} &= c_1x + c_2y + c_3z
\end{aligned} \tag{2.39}$$

Putting

$$x = Ae^{\lambda\tau} \qquad y = Be^{\lambda\tau} \qquad z = Ce^{\lambda\tau} \tag{2.40}$$

and inserting these solutions into (2.39), we obtain

$$
\begin{aligned}
(a_1 - \lambda)A + a_2B + a_3C &= 0 \\
b_1A + (b_2 - \lambda)B + b_3C &= 0 \\
c_1A + c_2B + (c_3 - \lambda)C &= 0
\end{aligned}
\tag{2.41}
$$

Nontrivial solutions for A, B, and C exist only when λ's are the roots of the characteristic equation

$$
\begin{vmatrix}
a_1 - \lambda & a_2 & a_3 \\
b_1 & b_2 - \lambda & b_3 \\
c_1 & c_2 & c_3 - \lambda
\end{vmatrix} = 0
\tag{2.42}
$$

We consider the case where the roots λ_i $(i = 1, 2, 3)$ are not zero and are distinct.   For a particular value of λ_i the ratios $A_i : B_i : C_i$ are determined from Eqs. (2.41); thus we obtain

$$
\frac{A_i}{D_{11}(\lambda_i)} = \frac{B_i}{D_{12}(\lambda_i)} = \frac{C_i}{D_{13}(\lambda_i)}
\tag{2.43}
$$

where $D_{jk}(\lambda_i)$ is the cofactor of the element in the jth row and the kth column.   General solutions of Eqs. (2.39) may be written as

$$
\begin{aligned}
x &= A_1 e^{\lambda_1 \tau} + A_2 e^{\lambda_2 \tau} + A_3 e^{\lambda_3 \tau} \\
y &= B_1 e^{\lambda_1 \tau} + B_2 e^{\lambda_2 \tau} + B_3 e^{\lambda_3 \tau} \\
z &= C_1 e^{\lambda_1 \tau} + C_2 e^{\lambda_2 \tau} + C_3 e^{\lambda_3 \tau}
\end{aligned}
\tag{2.44}
$$

It is noted that, since the ratios $A_i : B_i : C_i$ are fixed by Eqs. (2.43), three arbitrary constants can be chosen in Eqs. (2.44).

(a) Classification of Singular Points

Following Poincaré [84, p. 167] the types of singular points of Eqs. (2.39) may be classified according to the nature of the roots λ of Eq. (2.42).

1. If all the roots are real and of the same sign, the singularity is a nodal point.

2. If all the roots are real, but not of the same sign, the singularity is a saddle point.

3. If the roots consist of one real root and two conjugate complex roots the sum of which has the same sign as the real root, the singularity is a focal point.

4. If the roots consist of one real root and two conjugate complex roots the sum of which has a sign opposite to that of the real root, the singularity is a saddle-focus.

5. If the two conjugate complex roots are imaginary, the singularity is a center, a focus, or a saddle-focus.

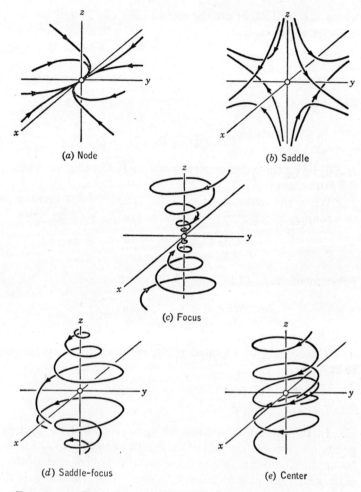

FIGURE 2.3   Types of singular points in the state space.

The types of singular points in the state space are shown in Fig. 2.3. If we proceed in the same manner as in Sec. 2.2a, the direction of integral curves at the singularity is, for real values of λ, given by

$$x:y:z = A_i:B_i:C_i = D_{11}(\lambda_i):D_{12}(\lambda_i):D_{13}(\lambda_i) \tag{2.45}$$

(b) Canonical Form of the Differential Equations

By a procedure analogous to that for the two-dimensional case of Sec. 2.2b, Eqs. (2.39) can be transformed to a canonical form

$$\frac{du}{d\tau} = \lambda_1 u \qquad \frac{dv}{d\tau} = \lambda_2 v \qquad \frac{dw}{d\tau} = \lambda_3 w \tag{2.46}$$

where λ_i $(i = 1, 2, 3)$ are the roots of Eq. (2.42) and the new variables u, v, and w are defined by

$$
\begin{aligned}
u &= l_1 x + m_1 y + n_1 z \\
v &= l_2 x + m_2 y + n_2 z \\
w &= l_3 x + m_3 y + n_3 z
\end{aligned}
\tag{2.47}
$$

The ratios $l_i : m_i : n_i$ $(i = 1, 2, 3)$ are determined by

$$
\frac{l_i}{D_{11}(\lambda_i)} = \frac{m_i}{D_{21}(\lambda_i)} = \frac{n_i}{D_{31}(\lambda_i)}
\tag{2.48}
$$

$D_{jk}(\lambda_i)$ being the cofactor corresponding to the element at the jth row and the kth column.

When the roots λ of Eq. (2.42) consist of one real root, say, λ_1, and two conjugate complex roots, say, λ_2 and λ_3, we may write

$$
\begin{aligned}
\lambda_2 &= \alpha + j\beta & v &= \xi + j\eta \\
\lambda_3 &= \alpha - j\beta & w &= \xi - j\eta
\end{aligned}
\tag{2.49}
$$

Substituting Eqs. (2.49) into (2.46) gives us

$$
\frac{d\xi}{d\tau} = \alpha\xi - \beta\eta \qquad \frac{d\eta}{d\tau} = \beta\xi + \alpha\eta
\tag{2.50}
$$

which are readily integrated by introducing polar coordinates [cf. Eqs. (2.22)].

2.4 Isocline Method

In the preceding sections we have discussed the nature of singular points and limit cycles of the differential equations (2.3) and (2.38). So far as the singular points of these systems are concerned, the linear parts of the equations play an important role. Integral curves of the equations located away from the singularities are affected by the presence of higher-order terms of the variables. Since the general solutions of these equations in analytical form are obtained with difficulty, graphical methods of solution are commonly used.

The basic graphical method is that known as the *isocline method*. We shall apply this method to a system governed by

$$
\frac{dx}{d\tau} = X(x,y) \qquad \frac{dy}{d\tau} = Y(x,y)
\tag{2.51}
$$

and seek the integral curves of

$$
\frac{dy}{dx} = \frac{Y(x,y)}{X(x,y)}
\tag{2.52}
$$

which is derived from Eqs. (2.51) by eliminating the independent variable τ. The functions $X(x,y)$ and $Y(x,y)$ may be nonlinear functions of the variables x and y. Suppose that

$$\frac{dy}{dx} = \frac{Y(x,y)}{X(x,y)} \equiv F(x,y) = c = \text{const} \qquad (2.53)$$

This equation clearly defines a curve in the xy plane along which the slope dy/dx of the integral curves remains constant. Such a curve is called an *isocline*. It is apparent that the slope of integral curves at the singularity is not uniquely determined and that the isoclines will intersect at such a point.

The isocline method makes it possible to explore the field of integral curves graphically without solving the differential equation. Once the curve $F(x,y) = c$ is traced, one draws along it small line segments having the prescribed slope $dy/dx = c$. One repeats the procedure for other values of c so that finally one obtains a series of curves

$$F(x,y) = c, c_1, c_2, \ldots$$

with the corresponding slopes drawn along these curves. These slopes then determine the field of directions of tangents to the integral curves in a certain region of the xy plane.

Starting from a point (x_0,y_0), a continuous curve can be traced by always following the direction of line elements of the field. The curve so obtained is clearly the integral curve passing through the initial point (x_0,y_0). The method is very valuable when the explicit form of the solution of the differential equation is not known. It was applied, for instance, by van der Pol in his early studies of the equation

$$\frac{d^2x}{d\tau^2} - \mu(1 - x^2)\frac{dx}{d\tau} + x = 0 \qquad \mu > 0 \qquad (2.54)$$

This equation can be reduced to the system

$$\frac{dx}{d\tau} = y \qquad \frac{dy}{d\tau} = \mu(1 - x^2)y - x$$

the trajectories of which are given by

$$\frac{dy}{dx} = \frac{\mu(1 - x^2)y - x}{y} \qquad (2.55)$$

The only singularity of Eq. (2.55) is clearly the origin, $x = y = 0$; hence the method is applicable everywhere except at that point. Equation (2.53) of the isoclines in this case is

$$\mu(1 - x^2)y - x = cy \qquad (2.56)$$

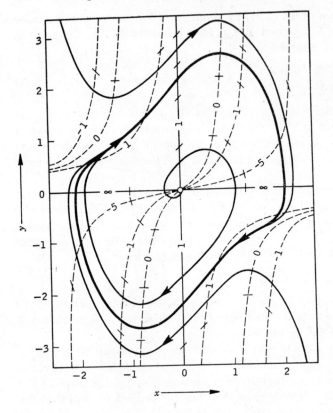

FIGURE 2.4   Solution of van der Pol's equation by the isocline method.

For a fixed μ and for a number of different values of the slope c, a series of curves (2.56) along which the slope of the trajectories is constant is obtained.

Figure 2.4 shows this construction of van der Pol [85] for $\mu = 1$ in Eq. (2.54), which is self-explanatory.   This graphical construction makes it possible to establish the existence of a stable limit cycle to which the spirals near the origin as well as those far away from the origin tend with increasing τ.   Once the integral curve is known, the time-response curve of the variable x in Eq. (2.54) may be found by the numerical integration of

$$\tau = \int \frac{dx}{y} \tag{2.57}$$

It is sometimes useful to transform the system (2.51) into polar coordinates, so that one may lessen the labor in drawing the integral

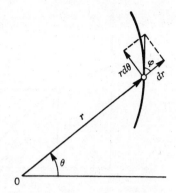

FIGURE 2.5   Integral curve in polar coordinates.

curves.   Let the system be described by

$$\frac{dr}{d\tau} = R(r,\theta) \qquad \frac{d\theta}{d\tau} = \Theta(r,\theta) \tag{2.58}$$

Then the isocline with respect to the radius vector is defined by

$$\frac{r\,d\theta}{dr} = \frac{r\Theta(r,\theta)}{R(r,\theta)} = \tan \phi = \text{const} \tag{2.59}$$

where ϕ is the angle with which the integral curve traverses the radius vector (Fig. 2.5).   This modification of the isocline method will expediently be used in Chap. 13 for the phase-plane analysis of subharmonic oscillations.

2.5 Liénard's Method

A method due to Liénard [62] is applicable to certain special types of second-order differential equations.[1]   The method is most conveniently used to deal with what are called *self-excited oscillations,* but it is also applicable in other cases.   The special cases in question are those in which the damping term is nonlinear but the restoring force is linear in x, and the differential equation is given by

$$\frac{d^2x}{d\tau^2} + f\left(\frac{dx}{d\tau}\right) + x = 0 \tag{2.60}$$

or

$$\frac{dv}{dx} = -\frac{f(v) + x}{v} \tag{2.61}$$

where v is the derivative of x with respect to the time τ.

[1] A generalization of this method is given by Le Corbeiller [58].

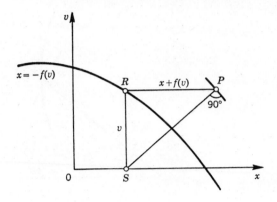

FIGURE 2.6   Liénard's method for determining the
field direction.

Liénard's method of graphical construction is indicated in Fig. 2.6.
Its purpose is to obtain the direction of a solution curve at any point in
the xv plane.   The procedure is as follows:

1. The curve $x = -f(v)$ is first plotted in the xv plane.
2. To determine the field direction at any point $P(x,v)$, a line is
drawn from P parallel to the x axis until it cuts the curve $x = -f(v)$ at R.
3. From R a perpendicular is dropped to the x axis at S; the field
direction at P is then orthogonal to the line SP.

That the construction yields the correct field direction is seen at once
from Eq. (2.61) and the fact that the slope of the line SP is $v/[x + f(v)]$.
It is also readily seen that the field direction of any other point, say, P',
in the xv plane is orthogonal to the line SP' provided the ordinate of P' is
equal to RS.

Differentiating Eq. (2.60) with respect to τ gives

$$\frac{d^2v}{d\tau^2} + \frac{d}{dv}[f(v)]\frac{dv}{d\tau} + v = 0 \qquad (2.62)$$

Van der Pol's equation

$$\frac{d^2v}{d\tau^2} - \mu(1 - v^2)\frac{dv}{d\tau} + v = 0 \qquad \mu > 0$$

is an example which takes the form of Eq. (2.62).   Equation (2.62) can
be written as

$$\frac{dw}{dv} + \frac{d}{dv}[f(v)] + \frac{v}{w} = 0 \qquad (2.63)$$

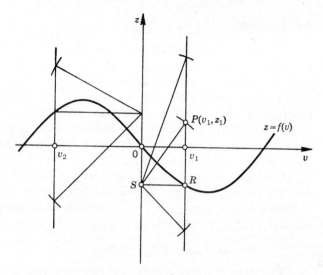

FIGURE 2.7   Liénard's method applied to the solution of Eq. (2.64).

where w is the derivative of v with respect to the time τ.   Upon introducing a new variable $z = w + f(v)$, we replace Eq. (2.63) by

$$\frac{dz}{dv} + \frac{v}{z - f(v)} = 0 \tag{2.64}$$

This equation takes the same form as Eq. (2.61), so one may apply the Liénard construction to obtain the solution curve.   The procedure is as follows.   The curve $z = f(v)$ is plotted in the vz plane of Fig. 2.7.   In order to determine the direction of the solution curve at $P(v_1,z_1)$, the line $v = v_1$ is drawn to cut the curve $z = f(v)$ at R.   From R a line parallel to the v axis is drawn to cut the z axis at S.   The field direction at P is then orthogonal to the line SP.   It is obvious that the normals to the solution curves of Eq. (2.64), for $v = v_1$, all pass through the same point S, whose coordinates are $(0,f(v_1))$.   Hence we obtain the elements of solution curves along the line $v = v_1$ by taking on the z axis a point S whose ordinate is $f(v_1)$ and by describing with S as center a series of circular arcs, as illustrated in Fig. 2.7.   By taking other points $v = v_2, v_3, \ldots$ on the v axis and by repeating the same procedure, additional elements of solution curves are obtained.   By having a field of line elements, and by starting from an initial point (v_0,z_0), a continuous curve can be traced following the line elements so traced; this curve will be clearly a solution curve of Eq. (2.64).   Finally, since $w = z - f(v)$, the phase trajectories in the vw plane are easily obtained from the solution curves of Fig. 2.7.

2.6 Delta Method

In applying the isocline method or the Liénard method to the graphical solution of differential equations, the entire phase plane must be filled with line segments fixing the direction of solution curves. If only a single solution curve is needed, only a few of these line segments are actually put to use. A technique of graphical construction known as the *delta method* [13, 49] is a more straightforward method that will provide the desired solution, since only information related directly to the solution curve is obtained by this method.

The delta method of solving second-order differential equations is described in this section. The *double-delta method*, a modification of this method which is applicable to the solution of differential equations of a more general type, will also be presented.

(a) Development of the Method

The delta method is applicable to the solution of differential equations of the form

$$\frac{d^2x}{dt^2} + f\left(\frac{dx}{dt}, x, t\right) = 0 \tag{2.65}$$

where the function $f(dx/dt,x,t)$ is continuous and single-valued but may be nonlinear. In applying this method, the equation is rewritten by adding and subtracting a term $\omega_0^2 x$ to give

$$\frac{d^2x}{dt^2} + \omega_0^2 x + f\left(\frac{dx}{dt}, x, t\right) - \omega_0^2 x = 0 \tag{2.66}$$

The term $\omega_0^2 x$ can be separated out of the term $f(dx/dt,x,t)$; if it cannot be, Eq. (2.66) is of a fictitious nature. The constant ω_0 can be determined by the form of Eq. (2.65), or it may have to be chosen from other information. By introducing new variables defined by

$$\tau = \omega_0 t \quad \text{and} \quad v = \frac{dx}{d\tau} \tag{2.67}$$

we may write Eq. (2.66) as

$$\frac{dv}{dx} = -\frac{x + \delta(v,x,\tau)}{v} \tag{2.68}$$

where

$$\delta(v,x,\tau) = \frac{1}{\omega_0^2} f\left(\omega_0 v, x, \frac{\tau}{\omega_0}\right) - x \tag{2.69}$$

The function $\delta(v,x,\tau)$ depends, in general, upon all three variables v, x, and τ, but for small changes in these variables it remains practically constant. This is the basic assumption of the method. If δ is constant, the variables

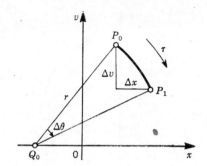

FIGURE 2.8   Short arc of the solution curve constructed by the delta method.

of Eq. (2.68) can be separated and integrated to give

$$(x + \delta)^2 + v^2 = \text{const} = r^2 \tag{2.70}$$

This is the equation of a circle of radius r centered at the point ($x = -\delta$, $v = 0$). Therefore, δ is the displacement of the center of the circle in the negative direction of the x axis. This displacement δ gives the method its name. Thus, for a small increment of τ, the solution curve may be approximated by a small arc of this circle.

The delta method is most immediately applicable to equations with oscillatory solutions. The constant ω_0 in Eq. (2.67) may preferably be chosen equal to the frequency of the oscillation, or more generally, ω_0 should be so chosen that the change in $\delta(v,x,\tau)$ is as small as possible during the process of graphical computation. Figure 2.8 shows the graphical construction of this method. The procedure is as follows:

1. Locate the initial point $P_0(x_0,v_0)$ at $\tau = \tau_0$ in the xv plane.
2. By using Eq. (2.69), calculate the initial value of δ and fix the point $Q_0(-\delta,0)$ on the x axis.
3. Starting from P_0 draw a circular arc with its center at Q_0. The arc P_0P_1 represents a portion of the solution curve. The arc must be short enough that the change in δ is small.
4. Repetition of steps 1 to 3 yields the solution curve as the continuation of small arcs centered on the x axis.

By making use of Taylor's expansion for the increments of x and v, one may obtain the general expression for the local error, i.e., the error committed at each step, by

$$\epsilon_x = \frac{1}{6}\left(\frac{d\delta}{d\tau}\right)_0 (\Delta\theta)^3 + 0_4(\Delta\theta)$$

$$\epsilon_v = \frac{1}{2}\left(\frac{d\delta}{d\tau}\right)_0 (\Delta\theta)^2 + \frac{1}{6}\left(\frac{d^2\delta}{d\tau^2}\right)_0 (\Delta\theta)^3 + 0_4(\Delta\theta) \tag{2.71}$$

where $(d\delta/d\tau)_0$ and $(d^2\delta/d\tau^2)_0$ denote the values of $d\delta/d\tau$ and $d^2\delta/d\tau^2$ at

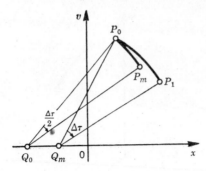

FIGURE 2.9   Modified procedure of the delta method.

$\tau = \tau_0$ and $\Delta\theta$ is the incremental angle of the radius vector for the individual circular arc.

The increment in time τ may readily be found; an explicit expression for time can be obtained from the relation

$$d\tau = \frac{dx}{v} = d\theta \qquad (2.72)$$

so that $\Delta\theta$ in Eqs. (2.71) may be replaced by $\Delta\tau$, which is the time increment corresponding to the individual circular arc.

(b) Higher Approximation

In the foregoing graphical solution the value of δ calculated at the beginning of each step is used throughout that interval. Clearly, the approximation will be improved if the average value of δ during that interval is used instead of the initial value. Figure 2.9 shows the graphical construction of a higher approximation which takes care of this consideration. The procedure is as follows:

1. Locate the initial point $P_0(x_0, v_0)$ at $\tau = \tau_0$.
2. By using Eq. (2.69), calculate the initial value of δ and fix the point $Q_0(-\delta, 0)$ as before.
3. Draw the circular arc P_0P_m with its center at Q_0, the incremental angle being chosen equal to $\Delta\tau/2$.
4. Again calculate $\delta = \delta_m$ by using the intermediate values x_m, v_m, and $\tau_0 + \Delta\tau/2$. Locate $Q_m(-\delta_m, 0)$ on the x axis.
5. Draw the circular arc P_0P_1 with its center at Q_m, the incremental angle being equal to $\Delta\tau$. The arc P_0P_1 represents a portion of the solution curve.

The local errors are estimated to be

$$\epsilon_x = -\frac{1}{12}\left(\frac{d\delta}{d\tau}\right)_0 (\Delta\tau)^3 + 0_4(\Delta\tau) \qquad \epsilon_v = \frac{1}{24}\left(\frac{d^2\delta}{d\tau^2}\right)_0 (\Delta\tau)^3 + 0_4(\Delta\tau) \qquad (2.73)$$

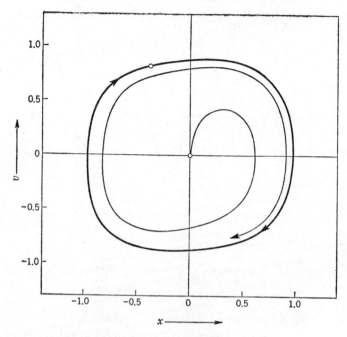

FIGURE 2.10   Phase-plane diagram for Eqs. (2.75).

These equations, in comparison with Eqs. (2.71), show the improvement of accuracy.

NUMERICAL EXAMPLE

We consider as an example a specific instance of Duffing's equation,

$$\frac{d^2x}{dt^2} + 0.7\frac{dx}{dt} + x^3 = 0.75\cos t \qquad (2.74)$$

The equivalent δ form of this equation is

$$\frac{dv}{dx} = -\frac{x + \delta(v,x,\tau)}{v}$$

where $\delta(v,x,\tau) = -x + x^3 + 0.7v - 0.75\cos\tau$ (2.75)

$$v = \frac{dx}{d\tau} \qquad \tau = t$$

In Fig. 2.10 is shown a solution curve of Eqs. (2.75) which has started from the initial conditions $x = 0$ and $v = 0$ at $\tau = 0$.   By using the relation (2.72), the trajectory in the xv plane may readily be converted to the time-response curve, as illustrated in Fig. 2.11.

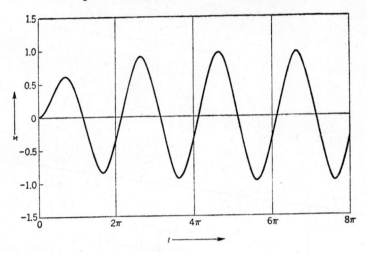

FIGURE 2.11   Time-response curve converted from the phase-plane
trajectory of Fig. 2.10.

(c) Double-delta Method

Let us consider second-order differential equations of the form

$$g\left(\frac{dx}{dt}, x, t\right)\frac{d^2x}{dt^2} + f\left(\frac{dx}{dt}, x, t\right) = 0 \tag{2.76}$$

where $g(dx/dt, x, t)$ and $f(dx/dt, x, t)$ are continuous and single-valued functions of dx/dt, x, and t. Dividing throughout this equation by $g(dx/dt, x, t)$ gives an equation of the form of Eq. (2.65); hence the delta method is applicable. However, the graphical construction becomes impractical, owing to the presence of the complicated term $f(dx/dt, x, t)/g(dx/dt, x, t)$.

We consider a somewhat different method of the graphical solution which is more appropriate for solving such equations. Through addition and subtraction of the terms d^2x/dt^2 and $\omega_0^2 x$, Eq. (2.76) is rewritten as

$$\left[1 + g\left(\frac{dx}{dt}, x, t\right) - 1\right]\frac{d^2x}{dt^2} + \omega_0^2 x + f\left(\frac{dx}{dt}, x, t\right) - \omega_0^2 x = 0$$

Introducing new variables τ and v as defined by Eqs. (2.67) yields

$$\frac{dv}{dx} = -\frac{x + \delta_1}{v + \delta_2} \tag{2.77}$$

where $\quad \delta_1 = \frac{1}{\omega_0^2} f\left(\omega_0 v, x, \frac{\tau}{\omega_0}\right) - x \qquad \delta_2 = \left[g\left(\omega_0 v, x, \frac{\tau}{\omega_0}\right) - 1\right] v$

If these δ functions, δ_1 and δ_2, are assumed to be constant, Eqs. (2.77)

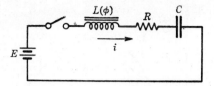

FIGURE 2.12   *LRC* series circuit with step input voltage.

may be integrated to give

$$(x + \delta_1)^2 + (v + \delta_2)^2 = \text{const} = r^2 \tag{2.78}$$

This is the equation of a circle of radius r with its center at the point $(x = -\delta_1, v = -\delta_2)$. There is no longer the constraint that the center of the circular arc has to be located on the x axis. However, it should be noticed that the simple relation between $\Delta\theta$ and $\Delta\tau$ given by Eq. (2.72) does not apply in the double-delta method.

NUMERICAL EXAMPLE

We consider the response of an *LRC* series circuit to a step input voltage (Fig. 2.12). Following the notations in the figure, the circuit equation may be written as

$$n \frac{d\phi}{dt} + Ri + \frac{q}{C} = E \qquad i = \frac{dq}{dt} \tag{2.79}$$

where ϕ is the magnetic flux in the core, n is the number of turns of the coil around the core, and q is the charge on the capacitor. The saturation curve of the core is assumed to be

$$\phi = c_1 ni + c_2 \tanh (ni) \tag{2.80}$$

where c_1 and c_2 are constants dependent on the core. By taking the numerical values of the parameters

$$n = 1 \qquad R = 0.2 \qquad C = 2.5$$
$$c_1 = 0.08 \qquad c_2 = 0.4$$

we obtain, from Eqs. (2.79) and (2.80),

$$\left(1.2 - \tanh^2 \frac{dq}{dt}\right) \frac{d^2q}{dt^2} + 0.5 \frac{dq}{dt} + q = 2.5E \tag{2.81}$$

Equation (2.81) may be rewritten in the form of Eqs. (2.77), that is,

$$\frac{di}{dq} = -\frac{q + \delta_1}{i + \delta_2} \tag{2.82}$$

where $\qquad \delta_1 = 0.5i - 2.5E \qquad \delta_2 = 0.2i - i \tanh^2 i$

In the derivation of these equations, ω_0 in Eqs. (2.67) has been chosen

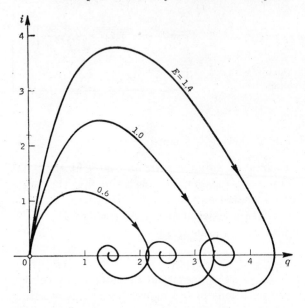

FIGURE 2.13   Phase-plane diagram for Eqs. (2.82).

equal to unity so that $\tau = t$. The phase-plane trajectories starting from the origin ($q = 0$, $i = 0$) are shown in Fig. 2.13 for different values of E.

2.7 Slope-line Method

This section describes the slope-line method of graphical construction for solving certain types of nonlinear differential equations, including van der Pol's equation and Duffing's equation.   The basic notions have been in use for some time by several investigators [97, 99].   A good summary of this method with its application to the hydraulic transient studies has been reported by Paynter [79].[1]   A modification of the basic method permits its application to the solutions of nonautonomous equations.

(a) Development of the Method

As a preliminary example, let it be desired to determine the solution of the first-order equation

$$\frac{dx}{d\tau} = f(\tau) \tag{2.83}$$

under the initial condition that $x = x_0$ at $\tau = \tau_0$.   The incremental rela-

[1] The author is indebted to H. M. Paynter for the discussion of this method.

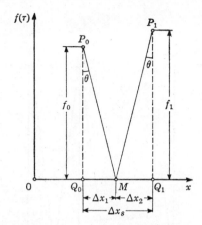

FIGURE 2.14   Graphical construction for obtaining Δx_s of Eq. (2.86).

tion of the variables may be written as

$$\Delta x = [f(\tau)]_{avg} \cdot \Delta \tau$$

where
$$[f(\tau)]_{avg} = \frac{1}{\Delta \tau} \int_{\tau_0}^{\tau_1} f(\tau) \, d\tau \qquad (2.84)$$

$\Delta \tau = \tau_1 - \tau_0$ small change in τ
$\Delta x = x_1 - x_0$ small change in x during the increment $\Delta \tau$

The basic assumption of the slope-line method lies in the use of the arithmetic mean for $[f(\tau)]_{avg}$, that is,

$$[f(\tau)]_{avg} = \tfrac{1}{2}(f_0 + f_1)$$
where
$$f_0 = f(\tau_0) \quad \text{and} \quad f_1 = f(\tau_1) \qquad (2.85)$$

Then an approximation Δx_s for Δx is given by

$$\Delta x_s = [f(\tau_0) + f(\tau_0 + \Delta \tau)] \frac{\Delta \tau}{2} \qquad (2.86)$$

This approximation implies that the trapezoidal method of approximation has been used.

The approximate increment Δx_s may be graphically obtained on the $x f(\tau)$ plane of Fig. 2.14.   An initial point $P_0(x_0, f_0)$ is first located.   Starting from P_0, draw the straight line, i.e., the slope line P_0M, to intersect the x axis at the point M.   The angle θ is so chosen that

$$\tan \theta = \frac{\Delta \tau}{2} \qquad (2.87)$$

for a predetermined value of $\Delta \tau$.   From M draw another slope line, making the same angle θ with the vertical line, to the point P_1 whose

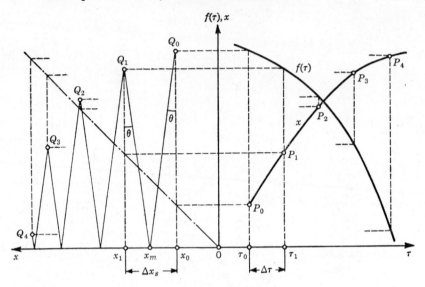

FIGURE 2.15   Graphical process for solving Eq. (2.83).

ordinate is $f(\tau_1)$.   Then

$$Q_0Q_1 = Q_0M + MQ_1 = f(\tau_0) \tan \theta + f(\tau_1) \tan \theta$$
$$= (f_0 + f_1) \frac{\Delta\tau}{2}$$

This gives the increment Δx_s of Eq. (2.86).

A practical procedure of graphical construction is illustrated in Fig. 2.15.   The process is as follows:

1. The function $f(\tau)$ is first plotted on the right half plane, the coordinates being τ and $f(\tau)$.   An initial point P_0 with coordinates τ_0 and x_0 is also plotted on the same plane.

2. Starting from P_0 one may locate Q_0 with coordinates x_0 and f_0 on the left half plane.   The 45° line (chain line) indicated merely serves to permit the graphical transfer of the x values between the horizontal and the vertical scale.

3. The slope line is then drawn starting from Q_0, going to the x axis, then back to Q_1, the ordinate of which is given by $f_1 = f(\tau_0 + \Delta\tau)$.

4. From Q_1 one may readily locate P_1, which, as illustrated in the figure, is on the solution curve $x(\tau)$ at $\tau_1 = \tau_0 + \Delta\tau$.

5. Successive points $P_2, P_3, \ldots$ on the solution curve are to be found by repeating steps 1 to 4.

The accuracy of this method corresponds to the precision of the trapezoidal approximation. The errors may be not so small if the curvature of $f(\tau)$ is large and the increment $\Delta\tau$ is inappropriately chosen. The local error produced by this method in the increment of x is estimated to be

$$\epsilon_s = \Delta x_s - \Delta x = \tfrac{1}{12}\ddot{f}_0(\Delta\tau)^3 + 0_4(\Delta\tau) \tag{2.88}$$

where the dot denotes differentiation with respect to τ. Equation (2.88) gives a measure of the appropriate increment in the independent variable τ. Thus, the interval $\Delta\tau$ may preferably be chosen as

$$\Delta\tau < \left(\frac{12\epsilon_a}{\ddot{f}_0}\right)^{\frac{1}{3}}$$

where ϵ_a is the allowable error.

(b) Second-order Equations of the Autonomous Type

We shall obtain the graphical solution of a set of simultaneous equations of the form

$$\frac{dx}{d\tau} + g(x) - y = 0 \qquad \frac{dy}{d\tau} + h(y) + x = 0 \tag{2.89}$$

Equations (2.89) may be transformed to the second-order equation

$$\frac{d^2x}{d\tau^2} + \frac{dg}{dx}\frac{dx}{d\tau} + h\left(\frac{dx}{d\tau} + g\right) + x = 0 \tag{2.90}$$

This equation represents some of the well-known types of differential equations, namely,

1. If $g(x) = c$ and $h(y) = ky$ (c, k = constants), we obtain a linear equation

$$\frac{d^2x}{d\tau^2} + k\frac{dx}{d\tau} + x + kc = 0 \tag{2.91}$$

2. If $g(x) = -\mu x + \tfrac{1}{3}\mu x^3$ (μ = constant) and $h(y) = 0$, we obtain van der Pol's equation

$$\frac{d^2x}{d\tau^2} - \mu(1 - x^2)\frac{dx}{d\tau} + x = 0 \tag{2.92}$$

3. If $g(x) = 0$ and $h(y) = -\alpha y + \beta y^3$ (α, β = constants), we obtain Rayleigh's equation

$$\frac{d^2x}{d\tau^2} - \left[\alpha - \beta\left(\frac{dx}{d\tau}\right)^2\right]\frac{dx}{d\tau} + x = 0 \tag{2.93}$$

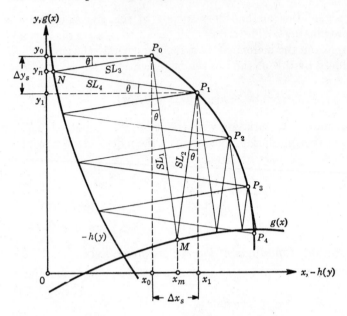

FIGURE 2.16   Graphical process for solving Eqs. (2.89).

The method of graphical construction of the solution curves of Eqs. (2.89) is shown in Fig. 2.16.   The procedure is as follows:

1. The functions $g(x)$ and $-h(y)$ are first plotted, and an initial point P_0 with coordinates x_0 and y_0 for $\tau = \tau_0$ is also prescribed.

2. Starting from P_0 the slope line SL_1 is drawn to intersect the curve $g(x)$ at the point M, where the angle θ is equal to $\tan^{-1}(\Delta\tau/2)$.   From M, the slope line SL_2 is drawn.

3. Starting again from P_0 the slope line SL_3 is drawn to intersect the curve $-h(y)$ at the point N.   From N, the slope line SL_4 is drawn.   The intersection $P_1(x_1, y_1)$ of SL_4 with SL_2 gives the point P_1 on the solution curve at $\tau_1 = \tau_0 + \Delta\tau$.

4. Successive points $P_2, P_3, \ldots$ on the solution curve are obtained by the iteration of steps 1 to 3.

It is clear, from Fig. 2.16, that

$$\Delta x_s = x_1 - x_0 = (x_1 - x_m) + (x_m - x_0)$$
$$= [y_0 - g(x_m)]\frac{\Delta\tau}{2} + [y_1 - g(x_m)]\frac{\Delta\tau}{2} \qquad (2.94)$$

and
$$\Delta y_s = -[x_0 + h(y_n)]\frac{\Delta\tau}{2} - [x_1 + h(y_n)]\frac{\Delta\tau}{2}$$

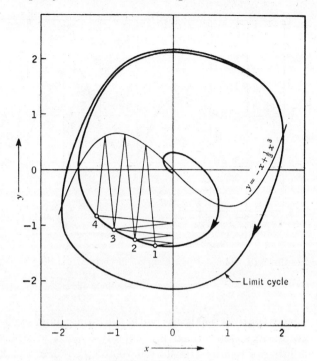

FIGURE 2.17   Phase-plane diagram for van der Pol's equations (2.96).

These values give a good approximation for the increments, Δx and Δy, since $2g(x_m) \cong g(x_0) + g(x_1)$ and $2h(y_n) \cong h(y_0) + h(y_1)$.

NUMERICAL EXAMPLE

As a typical example, let us consider van der Pol's equation. Taking the parameter $\mu = 1.0$ in Eq. (2.92), we have

$$\frac{d^2x}{d\tau^2} - (1 - x^2)\frac{dx}{d\tau} + x = 0 \qquad (2.95)$$

or
$$\frac{dx}{d\tau} = x - \tfrac{1}{3}x^3 + y \qquad \frac{dy}{d\tau} = -x \qquad (2.96)$$

The curve $g(x) = -x + x^3/3$ is plotted in Fig. 2.17. An initial point is prescribed at $x = 0$, $y = 0.05$. Construction then proceeds from this point with $\theta = \tan^{-1}(\Delta\tau/2) = \tan^{-1}(0.2/2)$. Some of the slope lines, from the point 1 to 4, are shown by fine lines in the figure. The integral curve, because of the negative damping for small values of x, spirals outward and finally moves onto the limit-cycle trajectory. Similarly, an initial point outside the limit cycle leads to a curve spiraling

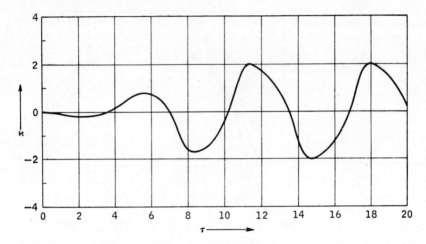

FIGURE 2.18   Time-response curve converted from the phase-plane trajectory of
Fig. 2.17.

inward until it coalesces with the same limit cycle. Since the points
graphically determined are equally spaced in time τ, data from these
points are readily transferred to the axes of τ and x of Fig. 2.18. The
time required for the representative point to complete one revolution
along the limit cycle is 6.64 $\cdots$, and the amplitude of x is 2.01 $\cdots$.
These values agree well with the known exact values of 6.687 and 2.009,
which were correctly calculated to three decimal places by M. Urabe [108].

(c) Second-order Equations of the Nonautonomous Type

A modification of the above method for autonomous systems permits
its extended application to the graphical solution of nonautonomous systems such as

$$\frac{dx}{d\tau} + g_1(x) - y = 0 \qquad \frac{dy}{d\tau} + h(y) + g_2(x) = f(\tau) \qquad (2.97)$$

or
$$\frac{d^2x}{d\tau^2} + \frac{dg_1}{dx}\frac{dx}{d\tau} + h\left[\frac{dx}{d\tau} + g_1(x)\right] + g_2(x) = f(\tau) \qquad (2.98)$$

Among equations of this type, there are the following examples.

1. If $g_3(x) = dg_1/dx$ and $h(y) = 0$, we obtain the equation with nonlinear damping

$$\frac{d^2x}{d\tau^2} + g_3(x)\frac{dx}{d\tau} + g_2(x) = f(\tau) \qquad (2.99)$$

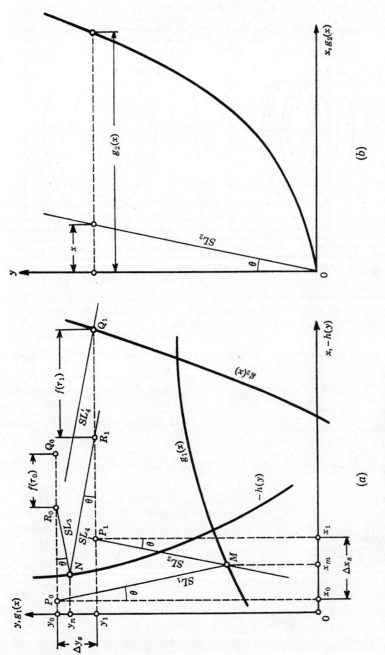

FIGURE 2.19 Graphical process for solving Eqs. (2.97).

65

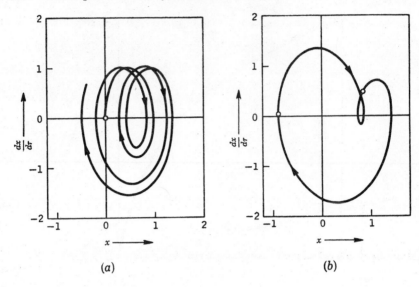

FIGURE 2.20 Phase-plane trajectories of the ½-harmonic oscillation. (a) Transient state. (b) Steady state.

2. If $g_1(x) = kx$ (k = constant) and $h(y) = 0$, we obtain Duffing's equation

$$\frac{d^2x}{d\tau^2} + k\frac{dx}{d\tau} + g_2(x) = f(\tau) \qquad (2.100)$$

Figure 2.19 shows the graphical construction of the solution curve of Eqs. (2.97). The procedure is as follows:

1. The functions $g_1(x)$ and $-h(y)$ are first plotted and an initial point P_0 is prescribed.

2. Starting from P_0 the slope line SL_1 is drawn to intersect the curve $g_1(x)$ at M. From the intersection the slope line SL_2 is drawn.

3. The value of $g_2(x)$ for the abscissa x of each point on the slope line SL_2 is calculated. The curve $g_2(x)$, on which the abscissa of each point is $g_2(x)$ calculated above, is plotted.

4. The point $Q_0(g_2(x_0),y_0)$ is located and shifted to the left by $f(\tau_0)$ to locate the point R_0.

5. Starting from R_0, the line SL_3 is drawn to intersect the curve $-h(y)$ at N. From N, the line SL_4 is drawn.

6. The line SL_4 is shifted to the right by $f(\tau_1)$ to obtain the line SL'_4. It intersects the curve $g_2(x)$ at Q_1.*

* The graphical work can be simplified by reproducing the slope line SL_2 and the curve $g_2(x)$ on a sheet of transparent paper as illustrated in Fig. 2.19b. By putting the y axis and the slope line SL_2 of (b) upon the corresponding lines of (a), one may readily locate the intersection Q_1 of the line SL'_4 with the curve $g_2(x)$.

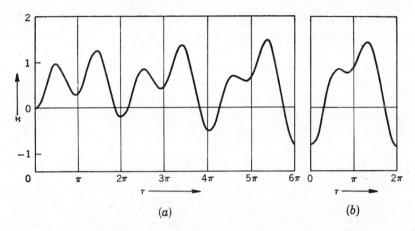

FIGURE 2.21   Waveforms of the ½-harmonic oscillation converted from the phase-plane trajectories of Fig. 2.20. (*a*) Transient state. (*b*) Steady state.

7. Passing through Q_1, the horizontal line as shown dashed is drawn. Its intersection with SL_2 locates the point P_1 on the solution curve at $\tau_1 = \tau_0 + \Delta\tau$.

8. Successive points on the solution curve are obtained by the iteration of steps 1 to 7.

The construction yields the approximate increments in x and y as given by

$$\Delta x_s = [y_0 - g_1(x_m)]\frac{\Delta\tau}{2} + [y_1 - g_1(x_m)]\frac{\Delta\tau}{2}$$

$$\Delta y_s = [f(\tau_0) - g_2(x_0) - h(y_n)]\frac{\Delta\tau}{2} + [f(\tau_1) - g_2(x_1) - h(y_n)]\frac{\Delta\tau}{2}$$

(2.101)

for the change $\Delta\tau$ in τ. The local errors are of order higher than the second in $\Delta\tau$.

NUMERICAL EXAMPLE

We deal with Duffing's equation

$$\frac{d^2x}{d\tau^2} + k\frac{dx}{d\tau} + |x|x = B\cos 2\tau + B_0$$

(2.102)

with $k = 0.20$ $B = 1.50$ $B_0 = 0.50$

or, in the equivalent simultaneous form of equations,

$$\frac{dx}{d\tau} = y \qquad \frac{dy}{d\tau} = -ky - |x|x + B\cos 2\tau + B_0$$

(2.103)

Figure 2.20a shows an integral curve which has started from the initial condition $x = 0$, $dx/d\tau\ (= y) = 0$ at $\tau = 0$. The time interval $\Delta\tau$ is $\pi/12$. After a sufficiently long period of time, the integral curve ultimately leads to a closed curve shown in Fig. 2.20b. Since the time required for the representative point to complete one revolution along the closed curve is 2π, or equal to twice the period of the external force, a subharmonic oscillation of order $\frac{1}{2}$ results. The time-response curves are readily obtained and shown in Fig. 2.21.

Chapter 3

Stability of Nonlinear Systems

3.1 Introduction

The question of stability is concerned with what happens if a system is disturbed slightly near an equilibrium condition. In general terms, any disturbance near an unstable equilibrium condition leads to a larger and larger departure from this condition. Near a stable equilibrium condition, the opposite is the case. An equilibrium condition may be either stationary or oscillatory. When it is stationary, the variables of the system remain constant; when it is oscillatory, the variables are undergoing continuous periodic change.

The question of stability is relatively simple in a linear system in which a single equilibrium condition exists. If the system is nonlinear, more than a single equilibrium condition may appear. A stable equilibrium condition exists actually, whereas an unstable one cannot be maintained. Hence the question of stability is particularly important in the study of nonlinear systems. A periodic solution obtained analytically or graphically in the preceding chapters merely represents a state of equilibrium; its actual existence must be confirmed by stability investigation. As mentioned in Chap. 2, an equilibrium condition may be correlated with a singular point or a limit cycle in the state plane (or state space). Hence the stability of an equilibrium condition depends upon the topological configuration of integral curves near the singularity or the limit cycle.

It is usually not difficult to define exactly what is meant by stability in a linear system. Because of new types of phenomena which may arise in a nonlinear system, it is not possible to use a single definition for stability which is meaningful in every case. For this reason, stability will be defined in a number of ways in the following section. The main purpose of this chapter is to describe some of the criteria for stability which will be used in later chapters. The present chapter also deals with the variational equations of the second order—Mathieu's equation and Hill's equation—which are derived from the original differential equation by considering a small deviation from the periodic solution. Solutions for these variational equations determine the stability of the original system.

3.2 Definition of Stability (Liapunov)

We consider a physical system described by n differential equations of the first order:

$$\frac{dx_i}{d\tau} = X_i(x_1, x_2, \ldots, x_n, \tau) \qquad i = 1, 2, \ldots, n \qquad (3.1)$$

where τ is the independent variable, here considered to be nondimensional time; $x_1, x_2, \ldots, x_n$ are the dependent variables; and the functions $X_1, X_2, \ldots, X_n$ are generally nonlinear functions of those variables. If the time τ does not appear explicitly in the functions $X_1, X_2, \ldots, X_n$, the system is autonomous. Then we may write

$$\frac{dx_i}{d\tau} = X_i(x_1, x_2, \ldots, x_n) \qquad i = 1, 2, \ldots, n \qquad (3.2)$$

First, we shall be concerned with the stability of solution $x_i(\tau)$, $i = 1, 2, \ldots, n$, that satisfies Eqs. (3.1) or (3.2).

1. We shall say that $x_i(\tau)$ is stable if, given $\epsilon > 0$ and τ_0, there is $\eta = \eta(\epsilon, \tau_0)$ such that any solution $x_i'(\tau)$ for which $|x_i(\tau_0) - x_i'(\tau_0)| < \eta$ satisfies $|x_i(\tau) - x_i'(\tau)| < \epsilon$ for $\tau \geqq \tau_0$. If no such η exists, $x_i(\tau)$ is unstable.

2. If $x_i(\tau)$ is stable, and in addition $|x_i(\tau) - x_i'(\tau)| \to 0$ as τ tends to infinity, we say that it is asymptotically stable.

In plain words, the above definition is equivalent to saying that a solution is stable if all solutions coming near it remain in its neighborhood; it is asymptotically stable if the solutions approach it asymptotically.

As discussed in the preceding chapter, a solution of Eqs. (3.2) may be sought not as an explicit function of the time τ but, after eliminating τ among Eqs. (3.2), as a trajectory in the $x_1, \ldots, x_n$ space. In this case the time is associated with the velocity of the representative point along the trajectory. Thus, secondly, the following definition is concerned with trajectories or orbits in the phase space.

3. Let C be a trajectory of Eqs. (3.2). We say that C is orbitally stable if, given $\epsilon > 0$, there is $\eta > 0$ such that, if R' is a representative point (on another trajectory C') which is within a distance η of C at time τ_0, then R' remains within a distance ϵ of C for $\tau \geqq \tau_0$. If no such η exists, C is orbitally unstable.

4. If C is orbitally stable and, in addition, the distance between R' and C tends to zero as $\tau \to \infty$, it is said to be asymptotically orbitally stable.

Summing up, orbital stability requires that the trajectories C and C' remain near each other, whereas stability of the solution $x_i(\tau)$ requires that, in addition, the representative points R and R' (on C and C', respectively) should remain close to each other if they were close to each other initially. An analogous distinction holds for asymptotic stability and asymptotic orbital stability. It is worth noting that, in an autonomous system, if $x_i(\tau)$ is a periodic solution, $x_i(\tau + \delta)$ is another such solution for every value of δ. Hence it is impossible for a periodic solution of an autonomous system to be asymptotically stable. If a solution of an autonomous system is represented by a stable limit cycle, it is asymptotically orbitally stable.

The definition of stability, thus far, is concerned with the behavior of the variables x_i of Eqs. (3.1) or (3.2) in the neighborhood of an equilibrium condition. In contrast with this, the method of investigation known as Liapunov's second method may discuss the question of stability in the large or in a finite region of the state space. An elementary theory of this method will be introduced in Sec. 3.4.

3.3 Routh-Hurwitz Criterion for Nonlinear Systems

We consider a physical system described by a set of simultaneous differential equations

$$\frac{dx_1}{d\tau} = X_1(x_1, x_2, \ldots, x_n)$$

$$\frac{dx_2}{d\tau} = X_2(x_1, x_2, \ldots, x_n)$$

$$\cdots\cdots\cdots\cdots\cdots\cdots\cdots$$

$$\frac{dx_n}{d\tau} = X_n(x_1, x_2, \ldots, x_n)$$

(3.3)

where τ is the nondimensional time and the functions $X_1, X_2, \ldots$ are generally nonlinear functions of the dependent variables $x_1, x_2, \ldots, x_n$. Thus the system under consideration is of the autonomous type.[1]

A state of equilibrium may be represented by a singular point or a limit cycle of Eqs. (3.3). The Routh-Hurwitz criterion is applicable only

[1] By making use of the averaging method, a second-order differential equation of the nonautonomous type may be transformed to the form of Eqs. (3.3) with two dependent variables x_1 and x_2. However, the stability condition obtained by the Routh-Hurwitz criterion for this autonomous system may not be sufficient for the original nonautonomous system. For further details, see Appendix IV.

to the former case, i.e., to an equilibrium point where all the derivatives of $x_1, x_2, \ldots, x_n$ with respect to τ are simultaneously zero.[1]   Under this condition we obtain

$$
\begin{aligned}
X_1(x_1, x_2, \ldots, x_n) &= 0 \\
X_2(x_1, x_2, \ldots, x_n) &= 0 \\
&\cdots\cdots\cdots \\
X_n(x_1, x_2, \ldots, x_n) &= 0
\end{aligned} \tag{3.4}
$$

If the system is linear, a single set of values for the variables x satisfying Eqs. (3.4) is obtained.   Hence the state of equilibrium is uniquely fixed. However, since we are concerned with a nonlinear system, Eqs. (3.4) are nonlinear algebraic equations.   These equations may be satisfied for more than a single set of values for the variables x.   Nonlinear systems, therefore, may have a number of equilibrium states.

In order to investigate the stability of a system near a chosen equilibrium point, we apply a sufficiently small disturbance to the system by changing the x's from their equilibrium values.   If, as the time τ increases indefinitely, all the x's return to their original equilibrium values, the system is asymptotically stable at this equilibrium point.[2]   On the other hand, if all or some of the x's depart further from their original equilibrium values with increasing τ, the system is unstable.   Let us denote a set of equilibrium values for the x's by $x_{10}, x_{20}, \ldots, x_{n0}$ and consider small variations ξ defined by

$$
\begin{aligned}
x_1 &= x_{10} + \xi_1 \\
x_2 &= x_{20} + \xi_2 \\
&\cdots\cdots\cdots \\
x_n &= x_{n0} + \xi_n
\end{aligned} \tag{3.5}
$$

Substituting Eqs. (3.5) into (3.3) and discarding terms of order higher than the first in the ξ's gives

$$
\begin{aligned}
\frac{d\xi_1}{d\tau} &= a_{11}\xi_1 + a_{12}\xi_2 + \cdots + a_{1n}\xi_n \\
\frac{d\xi_2}{d\tau} &= a_{21}\xi_1 + a_{22}\xi_2 + \cdots + a_{2n}\xi_n \\
&\cdots\cdots\cdots\cdots\cdots\cdots\cdots \\
\frac{d\xi_n}{d\tau} &= a_{n1}\xi_1 + a_{n2}\xi_2 + \cdots + a_{nn}\xi_n
\end{aligned} \tag{3.6}
$$

[1] When an equilibrium state is represented by a stable limit cycle, it has asymptotic orbital stability.   There is, however, no straightforward method of finding a limit cycle for Eqs. (3.3).   The method of contact curve due to Poincaré (Sec. 2.2e) and the method of Liapunov given in the following section are sometimes useful in proving the existence of a limit cycle.

[2] In what follows we shall be concerned with this type of stability unless specified otherwise.

where a_{ij} stands for $\partial X_i / \partial x_j$ at the equilibrium state $x_1 = x_{10}$, $x_2 = x_{20}$, . . . , $x_n = x_{n0}$. It has been shown by Liapunov [61; 75, p. 52] that, if the real parts of the roots of the characteristic equation of the system (3.6) are negative, the corresponding equilibrium state is stable; if at least one root has a positive real part, the equilibrium is unstable.[1] The characteristic equation may be written as

$$\begin{vmatrix} a_{11} - \lambda & a_{12} & \cdots & a_{1n} \\ a_{21} & a_{22} - \lambda & \cdots & a_{2n} \\ \cdots\cdots\cdots\cdots\cdots\cdots \\ a_{n1} & a_{n2} & \cdots & a_{nn} - \lambda \end{vmatrix} = 0 \qquad (3.7)$$

When expanded, this nth-order determinant leads to an equation of the form

$$a_0 \lambda^n + a_1 \lambda^{n-1} + \cdots + a_{n-1} \lambda + a_n = 0 \qquad (3.8)$$

The determination of signs of the real parts of the roots λ may be carried out by making use of the Routh-Hurwitz criterion [45, 95]. In applying this criterion, we first construct a set of n determinants set up from the coefficients of the nth-degree characteristic equation (3.8). These determinants are formed as follows:

$$\begin{aligned} \Delta_1 &= |a_1| \\[4pt] \Delta_2 &= \begin{vmatrix} a_1 & a_0 \\ a_3 & a_2 \end{vmatrix} \\[4pt] \Delta_3 &= \begin{vmatrix} a_1 & a_0 & 0 \\ a_3 & a_2 & a_1 \\ a_5 & a_4 & a_3 \end{vmatrix} \\[4pt] & \cdots\cdots\cdots\cdots\cdots \\[4pt] \Delta_n &= \begin{vmatrix} a_1 & a_0 & 0 & 0 & \cdots & \cdots \\ a_3 & a_2 & a_1 & a_0 & \cdots & \cdots \\ a_5 & a_4 & a_3 & a_2 & \cdots & \cdots \\ \cdots\cdots\cdots\cdots\cdots\cdots \\ 0 & 0 & 0 & 0 & \cdots & a_n \end{vmatrix} \end{aligned} \qquad (3.9)$$

The Routh-Hurwitz criterion states that the real parts of the roots λ are negative provided that all the coefficients a_0, a_1, . . . , a_n are positive and that all the determinants Δ_1, Δ_2, . . . , Δ_n are positive also. Since the bottom row of the determinant Δ_n is composed entirely of zeros, except for the last element a_n, it follows that $\Delta_n = a_n \Delta_{n-1}$. Thus, for

---

[1] We consider the case in which the real parts of the roots are different from zero. If the roots are imaginary or zero, Eqs. (3.6) of the first approximation cease to be applicable. The singular points correlated with such equilibrium states are discussed in Sec. 2.2d.

stability it is required that both $a_n > 0$ and $\Delta_{n-1} > 0$, and Δ_n need not actually be evaluated.[1]

As a simplest application of this method we consider a set of differential equations

$$\frac{dx}{d\tau} = X(x,y) \qquad \frac{dy}{d\tau} = Y(x,y) \tag{3.10}$$

An equilibrium point at which x and y are constant is obtained by solving $X(x,y) = 0$ and $Y(x,y) = 0$ simultaneously. In order to investigate the stability of this equilibrium, we write the characteristic equation

$$\lambda^2 - (a_{11} + a_{22})\lambda + a_{11}a_{22} - a_{12}a_{21} = 0$$

$$\text{with } a_{11} = \left(\frac{\partial X}{\partial x}\right)_0, \quad a_{12} = \left(\frac{\partial X}{\partial y}\right)_0, \quad a_{21} = \left(\frac{\partial Y}{\partial x}\right)_0, \quad a_{22} = \left(\frac{\partial Y}{\partial y}\right)_0 \tag{3.11}$$

where the suffix 0 denotes the evaluation of the partial derivatives at the equilibrium point. Application of the Routh-Hurwitz criterion gives the stability conditions

$$-a_{11} - a_{22} > 0 \qquad a_{11}a_{22} - a_{12}a_{21} > 0 \tag{3.12}$$

3.4 Liapunov's Criterion for Stability[2]

The basic concept of this method is first explained by a topological analysis of the following simple example. We consider a system described by a set of differential equations

$$\frac{dx}{d\tau} = y - x(x^2 + y^2) \qquad \frac{dy}{d\tau} = -x - y(x^2 + y^2) \tag{3.13}$$

and investigate the behavior of trajectories in the state plane. Clearly, the origin, $x = 0$, $y = 0$, is a singular point which is correlated with a state of equilibrium of the system. We shall be interested in the stability of this singularity. Assume a solution curve, i.e., a trajectory of Eqs. (3.13) as shown by a curve OPQ of Fig. 3.1. If the equilibrium state is asymptotically stable, the representative point $(x(\tau),y(\tau))$ tends to the origin with the lapse of the time τ. Also plotted in the figure is a circle

---

[1] As the degree n of the characteristic equation increases, the evaluation of the determinants becomes very laborious. Under such circumstances, the use of the Markov determinant [50], an alternative of the Hurwitz determinant, would facilitate the computation to some extent.

[2] This criterion, known as the second (or direct) method of Liapunov, has been developed in the Soviet Union [60, 61, 67, 71]. The book by Hahn [29] presents a good summary of the Liapunov theory. See also Refs. 23, 24, 28, 52, 57, and 98.

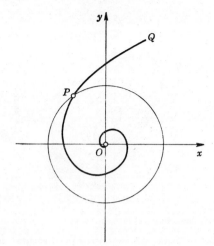

FIGURE 3.1   Trajectory OPQ and a circle
centered at the origin.

defined by

$$V(x,y) = x^2 + y^2 = r^2 \qquad (3.14)$$

where r is the distance between the point of intersection P and the origin
O.   We now consider the behavior of the intersection P as the time τ
increases.   Since both x and y are functions of the time τ, V is also a
function of τ.   The origin is an asymptotically stable singularity if $V(\tau)$
is a decreasing function of τ and reduces to zero for $\tau \to \infty$.   From Eq.
(3.14) we obtain

$$\frac{dV}{d\tau} = 2x\frac{dx}{d\tau} + 2y\frac{dy}{d\tau}$$

Substituting Eqs. (3.13) into this gives

$$\frac{dV}{d\tau} = 2x[y - x(x^2 + y^2)] + 2y[-x - y(x^2 + y^2)]$$
$$= -2(x^2 + y^2)^2 \qquad (3.15)$$

Hence $dV/d\tau < 0$ for all values of τ and the coordinates (x,y) except at
the origin.   We can thus conclude that the representative point, which
has started at any initial point $(x(0),y(0))$ on the xy plane, tends to the
origin with increasing τ.   The system (3.13) therefore has only one stable
equilibrium at the origin.

The form of Eq. (3.14) is generally so chosen that the trajectory of
the system and $V =$ constant have a single point of intersection.   In
Fig. 3.2, the equation $V =$ constant represents a family of concentric
ellipses.   Further, when a system is governed by a set of simultaneous
equations of n variables, $V =$ constant represents a family of closed
surfaces in the n-dimensional space.

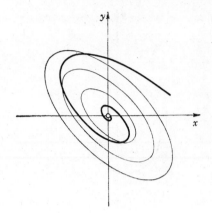

FIGURE 3.2   A trajectory with concentric ellipses.

Now we explain this method in a general way. Let us consider a system of differential equations

$$\frac{dx_i}{d\tau} = X_i(x_1, x_2, \ldots, x_n) \qquad i = 1, 2, \ldots, n \qquad (3.16)$$

and assume that the origin is a singular point of this system.

We introduce a function $V(x_1, \ldots, x_n)$ said to be positive definite with the following properties:

1. V is continuous together with its first partial derivatives in a certain domain D about the origin.

2. Outside the origin V is positive; it vanishes only at the origin.

Then the time derivative of V along the trajectory of the system (3.16) is given by[1]

$$\frac{dV}{d\tau} = \sum_{i=1}^{n} \frac{\partial V}{\partial x_i} \frac{dx_i}{d\tau} = \sum_{i=1}^{n} \frac{\partial V}{\partial x_i} X_i = W(x_1, \ldots, x_n) \qquad (3.17)$$

The (positive definite) function $V(x_1, \ldots, x_n)$ is called a Liapunov function provided that $dV/d\tau = W(x_1, \ldots, x_n) \leqq 0$ in the domain D.

With the aid of these definitions Liapunov's stability theorem[2] may be stated as follows:

If there exists in a certain domain D about the origin a Liapunov

---

[1] It is noted that the derivative $dV/d\tau$ is a function of x_i vanishing at the origin.

[2] It is intuitively clear that if near an equilibrium state of a physical system the energy of the system is always decreasing, then the equilibrium is stable. Liapunov's theorem is a generalization of this idea. A Liapunov function is simply an extension of the energy concept. For the mathematical justification of Liapunov's theorem see Ref. 57, p. 37, or Ref. 76, p. 137.

function $V(x_1, \ldots, x_n)$, then the origin is stable. Furthermore, if $-dV/d\tau = -W(x_1, \ldots, x_n)$ is likewise positive definite in the domain D, then the stability is asymptotic.

When we apply this theorem to a concrete physical system, a difficulty may occur in the determination of Liapunov functions. It is observed that the method does not give any means for determining such functions, but merely states that if such a function exists, the stability condition is fulfilled. We shall show, in what follows, some examples of the Liapunov function which may be useful in view of practical applications of the method.

First, let us consider a second-order equation of the form

$$\frac{d^2x}{d\tau^2} + f(x)\frac{dx}{d\tau} + g(x) = 0 \tag{3.18}$$

To simplify matters, we shall assume that $f(x)$ and $g(x)$ are polynomials, $f(x)$ even and $g(x)$ odd. Moreover, we shall suppose that $g(x)$ is monotone increasing with x. Even under such restrictions, Eq. (3.18) may represent a number of interesting examples such as van der Pol's equation, Rayleigh's equation, and the LRC equation of electric circuits (see Secs. 2.6 and 2.7).

It is convenient to introduce the integrals

$$F(x) = \int_0^x f(x)\ dx \qquad G(x) = \int_0^x g(x)\ dx$$

Then $F(x)$ is odd and $G(x)$ even and $F(0) = G(0) = 0$. Instead of Eqs. (3.18) we shall discuss the equivalent system

$$\frac{dx}{d\tau} = y - F(x) \qquad \frac{dy}{d\tau} = -g(x) \tag{3.19}$$

Clearly, the origin is a singular point of this system. In this case we conveniently assume a Liapunov function of the form

$$V(x,y) = \tfrac{1}{2}y^2 + G(x) \tag{3.20}$$

It is in fact the total energy when $f(x) = 0$, that is, when there is no dissipation in the system. We find at once

$$\frac{dV}{d\tau} = -g(x)F(x) \tag{3.21}$$

Hence we may conclude that, if $-g(x)F(x) \leqq 0$ in a certain domain D about the origin, the origin is stable. If, in particular, $-g(x)F(x) < 0$ for nonvanishing x in the domain D, the stability is asymptotic.

Second, we consider a system of equations of the first order:

$$\frac{dx}{d\tau} = a_1 x + a_2 y + f(x,y) \qquad \frac{dy}{d\tau} = b_1 x + b_2 y + g(x,y) \qquad (3.22)$$

where we assume that $f(x,y)$ and $g(x,y)$ are polynomials containing terms of order higher than the first in x and y. Moreover we assume that the linear system

$$\frac{dx}{d\tau} = a_1 x + a_2 y \qquad \frac{dy}{d\tau} = b_1 x + b_2 y \qquad (3.23)$$

has an asymptotically stable singularity at the origin. This is equivalent to saying that the characteristic equation

$$(a_1 - \lambda)(b_2 - \lambda) - a_2 b_1 = 0$$

has two roots with negative real parts, that is,

$$a_1 + b_2 < 0 \qquad \text{and} \qquad a_1 b_2 - a_2 b_1 > 0 \qquad (3.24)$$

We try to determine a Liapunov function for the linear system (3.23) and assume a quadratic form

$$V(x,y) = \tfrac{1}{2}(\alpha x^2 + 2\beta xy + \gamma y^2) \qquad (3.25)$$

The unknown coefficients α, β, and γ will be determined in such a way that $dV/d\tau$ is a negative definite function of the form

$$\frac{dV}{d\tau} = -(x^2 + y^2) \qquad (3.26)$$

From Eqs. (3.23) and (3.25) we obtain

$$\frac{dV}{d\tau} = (\alpha x + \beta y)(a_1 x + a_2 y) + (\beta x + \gamma y)(b_1 x + b_2 y) \qquad (3.27)$$

Comparison of Eq. (3.26) with (3.27) yields

$$a_1 \alpha + b_1 \beta = -1 \qquad a_2 \alpha + (a_1 + b_2)\beta + b_1 \gamma = 0 \qquad a_2 \beta + b_2 \gamma = -1$$

Solving these equations for α, β, and γ gives

$$\alpha = -\frac{(a_1 b_2 - a_2 b_1) + b_1{}^2 + b_2{}^2}{(a_1 + b_2)(a_1 b_2 - a_2 b_1)}$$

$$\beta = \frac{a_1 b_1 + a_2 b_2}{(a_1 + b_2)(a_1 b_2 - a_2 b_1)} \qquad (3.28)$$

$$\gamma = -\frac{(a_1 b_2 - a_2 b_1) + a_1{}^2 + a_2{}^2}{(a_1 + b_2)(a_1 b_2 - a_2 b_1)}$$

Thus the unknown coefficients in the Liapunov function (3.25) are fixed. One readily sees, from the conditions (3.24), that $\alpha\gamma - \beta^2 > 0$ and $\gamma > 0$. Therefore $V(x,y)$ is positive definite for any values of x and y.

Now we return to the original nonlinear system (3.22). By using the same Liapunov function $V(x,y)$, we calculate $dV/d\tau$ and obtain

$$\frac{dV}{d\tau} = -(x^2 + y^2) + (\alpha x + \beta y)f(x,y) + (\beta x + \gamma y)g(x,y) \quad (3.29)$$

Since the polynomials $f(x,y)$ and $g(x,y)$ contain terms of order higher than the first, $dV/d\tau$ is negative for sufficiently small values of x and y. Hence the origin is asymptotically stable. If we can determine a domain D where the right side of Eq. (3.29) is negative, we may also conclude that any trajectories inside the domain D cross the curve $V = $ constant from outside to inside as τ increases. Determination of such a domain D is useful for the study of trajectories which are not in the neighborhood of the origin.[1]

Thus far the system of two first-order equations is considered. Extension to a larger number of equations may be carried out in much the same way. Let a system of n variables be given by

$$\begin{aligned}
\dot{x}_1 &= a_{11}x_1 + a_{12}x_2 + \cdots + a_{1n}x_n + f_1(x_1, x_2, \ldots, x_n) \\
\dot{x}_2 &= a_{21}x_1 + a_{22}x_2 + \cdots + a_{2n}x_n + f_2(x_1, x_2, \ldots, x_n) \\
&\cdots\cdots\cdots\cdots\cdots\cdots\cdots\cdots\cdots\cdots\cdots \\
\dot{x}_n &= a_{n1}x_1 + a_{n2}x_2 + \cdots + a_{nn}x_n + f_n(x_1, x_2, \ldots, x_n)
\end{aligned} \quad (3.30)$$

where the dots over $x_1, \ldots, x_n$ refer to differentiation with respect to τ and $f_1, \ldots, f_n$ are polynomials containing terms of order higher than the first in those variables. The system may be expressed in a matrix form

$$\begin{bmatrix} \dot{x}_1 \\ \dot{x}_2 \\ \cdot \\ \dot{x}_n \end{bmatrix} = \begin{bmatrix} a_{11} & a_{12} & \cdots & a_{1n} \\ a_{21} & a_{22} & \cdots & a_{2n} \\ \multicolumn{4}{c}{\cdots\cdots\cdots\cdots} \\ a_{n1} & a_{n2} & \cdots & a_{nn} \end{bmatrix} \begin{bmatrix} x_1 \\ x_2 \\ \cdot \\ x_n \end{bmatrix} + \begin{bmatrix} f_1 \\ f_2 \\ \cdot \\ f_n \end{bmatrix}$$

or
$$\dot{\mathbf{x}} = A\mathbf{x} + \mathbf{f} \quad (3.31)$$

We assume that the linear system

$$\dot{\mathbf{x}} = A\mathbf{x} \quad (3.32)$$

has an asymptotically stable singularity at the origin. A Liapunov func-

---

[1] Following this method of analysis, limit cycles correlated with almost periodic oscillations in a self-oscillatory system will be studied in Sec. 13.6a.

tion for this linear system is assumed to take a quadratic form

$$V = [x_1 \ \ x_2 \ \ \cdots \ \ x_n] \begin{bmatrix} b_{11} & b_{12} & \cdots & b_{1n} \\ b_{21} & b_{22} & \cdots & b_{2n} \\ \cdot & \cdot & \cdot & \cdot \\ b_{n1} & b_{n2} & \cdots & b_{nn} \end{bmatrix} \begin{bmatrix} x_1 \\ x_2 \\ \cdot \\ x_n \end{bmatrix}$$

$$= \mathbf{x}'B\mathbf{x} \tag{3.33}$$

where $\mathbf{x}'$ is the transposed matrix of $\mathbf{x}$ and B is a symmetric matrix (that is, $b_{ij} = b_{ji}$). The time derivative of V is

$$\dot{V} = \dot{\mathbf{x}}'B\mathbf{x} + \mathbf{x}'B\dot{\mathbf{x}} = \mathbf{x}'(A'B + BA)\mathbf{x} \tag{3.34}$$

Proceeding analogously as before [see Eq. (3.26)], we put

$$\dot{V} = -(x_1{}^2 + \cdots + x_n{}^2) = -\mathbf{x}'E\mathbf{x} \tag{3.35}$$

where E is a unit matrix. It follows from Eqs. (3.34) and (3.35) that

$$A'B + BA = -E \tag{3.36}$$

The unknown $n(n + 1)/2$ elements of the matrix B are determined by the simultaneous linear equations (3.36) of the same number. It has been shown by Alimov [2] that the function V determined in this way is positive definite[1] and may be used as a Liapunov function.

By making use of this Liapunov function, we treat the original system (3.30). The time derivative of V becomes

$$\begin{aligned} \dot{V} &= \dot{\mathbf{x}}'B\mathbf{x} + \mathbf{x}'B\dot{\mathbf{x}} \\ &= (A\mathbf{x} + \mathbf{f})'B\mathbf{x} + \mathbf{x}'B(A\mathbf{x} + \mathbf{f}) \\ &= -\mathbf{x}'E\mathbf{x} + 2\mathbf{f}'B\mathbf{x} \end{aligned} \tag{3.37}$$

Hence, as before, we may conclude that the origin is asymptotically stable, and that, inside the domain where $-\mathbf{x}'E\mathbf{x} + 2\mathbf{f}'B\mathbf{x}$ is negative, any trajectories cross the surface V = constant from outside to inside as τ increases.

The principal advantage of Liapunov's criterion may be summarized as follows. First, the difficult (and often impossible) problem of integration of a system of variational equations is obviated and replaced by a much simpler problem of an algebraic character. Second, the criterion gives directly stability in a finite or infinite region[2] of the state space

---

[1] The necessary and sufficient conditions for V to be positive definite are that the successive principal minors of the symmetric determinant B are positive.

[2] In order to discuss the stability in the large, $V = c$ (c = constant) must represent a closed surface which may contain any point of the state space if c is large enough. The condition that $V = c$ is a closed surface is guaranteed if V ceases to be bounded when $\sum\limits_{i=1}^{n} x_i{}^2 \to \infty$.

instead of stability in the neighborhood of an equilibrium point. Consequently, the criterion may also furnish topological information concerning the behavior of trajectories in certain domains of the state space.

As mentioned before, the difficulty of this criterion lies in the construction of a Liapunov function.[1] The stability condition obtained in this way is naturally a sufficient condition. Therefore, if the construction is not appropriate, the criterion for stability will result in too severe constraints upon the system parameters.

Example 1

Consider van der Pol's equation

$$\frac{d^2x}{d\tau^2} - \epsilon(1 - x^2)\frac{dx}{d\tau} + x = 0 \qquad \epsilon > 0 \qquad (3.38)$$

or its equivalent

$$\frac{dx}{d\tau} = y + \epsilon\left(x - \frac{x^3}{3}\right) \qquad \frac{dy}{d\tau} = -x \qquad (3.39)$$

The characteristic roots of the linear system obtained by discarding the nonlinear term in Eqs. (3.39) are

$$\lambda_{1,2} = \frac{\epsilon \pm \sqrt{\epsilon^2 - 4}}{2}$$

Since ϵ is positive, the origin is an unstable singularity.

To investigate the behavior of trajectories about the origin, we apply the Liapunov criterion to the system (3.39). By making use of Eq. (3.20), we take the Liapunov function

$$V = \tfrac{1}{2}y^2 + \int_0^x x\,dx = \tfrac{1}{2}(x^2 + y^2)$$

Then

$$\dot{V} = x\dot{x} + y\dot{y} = \epsilon x^2\left(1 - \frac{x^2}{3}\right)$$

Therefore, $\dot{V} > 0$ for $x^2 < 3$. We may conclude that any trajectory of the system (3.39) points, whatever ϵ (>0), toward the exterior of the circle $x^2 + y^2 =$ constant if the radius is less than $\sqrt{3}$.*

---

[1] A systematic method of constructing a Liapunov function for a broad class of systems was given by Schultz and Gibson [98].

* The existence of a limit cycle for the system (3.39) will be proved by the following procedure. Since a limit cycle, if it exists, is exterior to the circle of radius $\sqrt{3}$, we construct a ring-shaped domain with inner boundary $x^2 + y^2 = 3$. The outer boundary will be so chosen that the trajectory of the system (3.39) points toward the interior of the ring on its boundary. It is not difficult, in fact, to find the outer boundary with the desired property; see, for instance, Ref. 100, pp. 242–245. Since there is no singularity in the ring domain—the origin being the only singularity—the existence of a limit cycle in this domain may be concluded.

EXAMPLE 2

Consider the following system of differential equations:

$$\frac{dx}{d\tau} = -x - y + (x^2 + y^2)x \qquad \frac{dy}{d\tau} = x - y + (x^2 + y^2)y \quad (3.40)$$

We first ignore the nonlinear terms and write

$$\frac{dx}{d\tau} = -x - y \qquad \frac{dy}{d\tau} = x - y \qquad\qquad (3.41)$$

Since the characteristic roots of this system are $-1 \pm i$, the origin is an asymptotically stable focus. We construct a Liapunov function of the form

$$V(x,y) = \tfrac{1}{2}(\alpha x^2 + 2\beta xy + \gamma y^2) \qquad\qquad (3.42)$$

and assume that

$$\dot{V} = -(x^2 + y^2)$$

Then the unknown coefficients in Eq. (3.42) are determined from Eqs. (3.28). Thus we obtain

$$V(x,y) = \tfrac{1}{2}(x^2 + y^2)$$

We now apply this Liapunov function to the original system (3.40). Then

$$\dot{V} = x[-x - y + (x^2 + y^2)x] + y[x - y + (x^2 + y^2)y]$$
$$= r^2(r^2 - 1)$$

where $r^2 = x^2 + y^2$. One sees, therefore, that if $r < 1$ all the trajectories of Eqs. (3.40) spiral into the origin as the time τ increases. If $r > 1$, on the other hand, the trajectories unwind from the circle, $r = 1$, outward. Hence the conclusion that the system (3.40) has a single stable focus at the origin and a single unstable limit cycle at $r = 1$.

3.5 *Stability of Periodic Oscillations* [34]

As mentioned in Sec. 3.3, the variational equations form a linear system with constant coefficients when they are related to the behavior of a solution in the neighborhood of an equilibrium state. In this section we consider a case in which the solution of a system is periodic with period T. The system may not necessarily be autonomous, but we assume that the time appears only in the form of periodic functions of period T. The variational equations based on such a solution take the form

$$\dot{\xi}_1 = a_{11}(\tau)\xi_1 + a_{12}(\tau)\xi_2 + \cdots + a_{1n}(\tau)\xi_n$$
$$\dot{\xi}_2 = a_{21}(\tau)\xi_1 + a_{22}(\tau)\xi_2 + \cdots + a_{2n}(\tau)\xi_n$$
$$\cdots \cdots \cdots \cdots \cdots \cdots \cdots \cdots \cdots \cdots \cdots \quad (3.43)$$
$$\dot{\xi}_n = a_{n1}(\tau)\xi_1 + a_{n2}(\tau)\xi_2 + \cdots + a_{nn}(\tau)\xi_n$$

where $\qquad a_{ij}(\tau + T) = a_{ij}(\tau) \qquad i, j = 1, 2, \ldots, n$

A theory, known as Floquet's theory [22], gives certain basic information concerning solutions for these equations, although determination of solutions may generally be a difficult process. A brief discussion of this theory is given in what follows [76].

Using the vector notation as mentioned in Sec. 1.1, let

$$\mathbf{f}^i(\tau) = [f_1{}^i(\tau), \ldots, f_n{}^i(\tau)] \qquad i = 1, \ldots, n$$

be a fundamental set of solutions for Eqs. (3.43). In this notation, subscripts are reserved for components of a vector, different vectors being distinguished by superscripts. Since Eqs. (3.43) are unchanged if we replace τ by $\tau + T$, the vectors $\mathbf{f}^i(\tau + T)$ form another set of solutions. Thus we may write

$$\mathbf{f}^i(\tau + T) = \sum_{j=1}^{n} b_j{}^i \mathbf{f}^j(\tau) \qquad (3.44)$$

where the vectors $\mathbf{b}^i$ are linearly independent.

Now we try to find a solution $\mathbf{F}(\tau)$ with the property that

$$\mathbf{F}(\tau + T) = m\mathbf{F}(\tau) \qquad (3.45)$$

Writing

$$\mathbf{F}(\tau) = \sum_{j=1}^{n} c_j \mathbf{f}^j(\tau)$$

and using Eq. (3.44) leads to

$$\mathbf{F}(\tau + T) = \sum_{j=1}^{n} c_j \mathbf{f}^j(\tau + T) = \sum_{j=1}^{n} c_j \sum_{i=1}^{n} b_i{}^j \mathbf{f}^i(\tau) = \sum_{j=1}^{n} \sum_{i=1}^{n} c_i b_j{}^i \mathbf{f}^j(\tau)$$

We also obtain, from Eq. (3.45),

$$\mathbf{F}(\tau + T) = m \sum_{j=1}^{n} c_j \mathbf{f}^j(\tau)$$

Comparison of the last two equations gives

$$\sum_{j=1}^{n} \left(\sum_{i=1}^{n} c_i b_j{}^i - mc_j \right) \mathbf{f}^j(\tau) = 0 \qquad (3.46)$$

Since the solutions $\mathbf{f}^j(\tau)$ are linearly independent, the coefficients in Eq. (3.46) must vanish, giving rise to the system

$$
\begin{aligned}
(b_1{}^1 - m)c_1 + b_1{}^2 c_2 + \cdots + b_1{}^n c_n &= 0 \\
b_2{}^1 c_1 + (b_2{}^2 - m)c_2 + \cdots + b_2{}^n c_n &= 0 \\
\cdots\cdots\cdots\cdots\cdots\cdots\cdots\cdots\cdots\cdots \\
b_n{}^1 c_1 + b_n{}^2 c_2 + \cdots + (b_n{}^n - m)c_n &= 0
\end{aligned}
\tag{3.47}
$$

The condition for a nonvanishing solution c is

$$
\begin{vmatrix}
b_1{}^1 - m & b_1{}^2 & b_1{}^3 & \cdots & b_1{}^n \\
b_2{}^1 & b_2{}^2 - m & b_2{}^3 & \cdots & b_2{}^n \\
\cdots & \cdots & \cdots & \cdots & \cdots \\
b_n{}^1 & b_n{}^2 & b_n{}^3 & \cdots & b_n{}^n - m
\end{vmatrix} = 0
\tag{3.48}
$$

The solutions $m_1, \ldots, m_n$ of this characteristic equation are called the characteristic multipliers of the system (3.43). It can be shown that they are independent of the solutions $\mathbf{f}^i(\tau)$. Since the column vectors $\mathbf{b}^i$ are linearly independent, $m = 0$ cannot be a solution of Eq. (3.48). Substituting a solution of (3.48) into (3.47) and solving for c yields a desired solution $\mathbf{F}(\tau)$ of Eqs. (3.43). We assume that the roots m_i of Eq. (3.48) are distinct, and thus we obtain a fundamental set of solutions $\mathbf{F}^i(\tau)$ with the property that $\mathbf{F}^i(\tau + T) = m_i \mathbf{F}^i(\tau)$.

Defining $\boldsymbol{\phi}(\tau)$ by

$$
\boldsymbol{\phi}(\tau) = \left[\exp\left(-\frac{\log m}{T}\tau \right) \right] \mathbf{F}(\tau)
$$

we have

$$
\boldsymbol{\phi}(\tau + T) = \left[\exp\left(-\frac{\log m}{T}\tau \right) \right] \frac{1}{m} m\mathbf{F}(\tau) = \boldsymbol{\phi}(\tau)
$$

so that $\boldsymbol{\phi}(\tau)$ is periodic of period T. Hence we may write

$$
\mathbf{F}^i(\tau) = e^{\mu_i \tau} \boldsymbol{\phi}^i(\tau)
$$

where
$$
\mu_i = \frac{\log m_i}{T} \qquad i = 1, 2, \ldots, n
\tag{3.49}
$$

This result is known as Floquet's theorem, and the numbers μ_i are called the characteristic exponents of the system (3.43).

We now speak of the stability of periodic oscillations. If all the characteristic exponents of (3.43) have negative real parts, the variations $\xi_1, \ldots, \xi_n$ tend to zero with increasing τ; whereas if at least one characteristic exponent has a positive real part, the variations diverge as τ increases. Therefore we may in general state: If all the characteristic exponents of the variational equation based on a periodic solution have negative real parts, the solution is asymptotically stable. If at least one characteristic exponent has a positive real part, the solution is unstable.

The principal difficulty in this analysis is that the fundamental set of solutions, that is, $f^i(\tau)$, with which we began and from which we derived the characteristic exponents, is in general unknown. At present, practical calculation of the analysis is mostly confined to second-order systems.

As a typical but relatively general example, we consider the following equation:

$$\frac{d^2v}{d\tau^2} + f\left(v, \frac{dv}{d\tau}\right) = e(\tau) \tag{3.50}$$

with

$$e(\tau + \tau_0) = e(\tau)$$

Equations of this type, including Duffing's equation, are often encountered in physical systems.

It might be of interest to refer to an investigation due to Trefftz [107] concerning the property of periodic solutions for Eq. (3.50). His conclusion is given here without proof: Under the condition that it is bounded and asymptotically stable, the solution must ultimately lead to a periodic solution of which the least period is equal to either the period τ_0 of the external force or an integral multiple (different from unity) of τ_0. Corresponding to these two cases, the terms *harmonic oscillation* and *subharmonic oscillation* are respectively applied.

By use of Floquet's theory, we investigate the stability of a periodic solution for Eq. (3.50). Let the solution be expressed by

$$v(\tau) = v_0(\tau) \tag{3.51}$$

where $v_0(\tau + n\tau_0) = v_0(\tau)$, n being a positive integer. A small variation from this periodic state is denoted by ξ; then substitution of $v_0(\tau) + \xi$ for $v(\tau)$ in Eq. (3.50) leads to the variational equation

$$\ddot{\xi} + \left(\frac{\partial f}{\partial \dot{v}}\right)_0 \dot{\xi} + \left(\frac{\partial f}{\partial v}\right)_0 \xi = 0 \qquad \cdot \equiv \frac{d}{d\tau} \tag{3.52}$$

in which the symbol $(\ )_0$ denotes the insertion of $v_0(\tau)$ and $\dot{v}_0(\tau)$ after the differentiation. Since the coefficients of ξ and $\dot{\xi}$ are periodic functions of τ, Eq. (3.52) may be rewritten as

$$\ddot{\xi} + F(\tau)\dot{\xi} + G(\tau)\xi = 0 \tag{3.53}$$

By introducing a new variable η defined by

$$\xi = \{\exp[-\tfrac{1}{2}\textstyle\int F(\tau)\,d\tau]\}\eta \tag{3.54}$$

and thus eliminating the first-derivative term in Eq. (3.53), we obtain

$$\ddot{\eta} + \left\{G(\tau) - \frac{1}{2}\frac{dF}{d\tau} - \frac{1}{4}[F(\tau)]^2\right\}\eta = 0 \tag{3.55}$$

This is a linear equation in which the coefficient of η is a periodic function of τ and may be developed into a Fourier series.

By Floquet's theory we may assume a particular solution $h(\tau)$ such that

$$h(\tau + T) = mh(\tau) \tag{3.56}$$

where T is the fundamental period of the Fourier series and m is the characteristic multiplier as defined before. Let $\eta_1(\tau)$ and $\eta_2(\tau)$ be linearly independent solutions with the initial conditions

$$\eta_1(0) = 1 \qquad \dot{\eta}_1(0) = 0$$
$$\eta_2(0) = 0 \qquad \dot{\eta}_2(0) = 1$$

Then we have

$$m^2 - [\eta_1(T) + \dot{\eta}_2(T)]m + 1 = 0 \tag{3.57}$$

By definition, m is related to the characteristic exponent μ by $m = e^{\mu T}$, so that

$$\cosh \mu T = \tfrac{1}{2}[\eta_1(T) + \dot{\eta}_2(T)] \tag{3.58}$$

Since we are concerned with the real functions $\eta_1(\tau)$ and $\eta_2(\tau)$, $\cosh \mu T$ is also real. Hence μT is real or imaginary or complex but with imaginary part $jn\pi$, n being an integer. Therefore the general solution of Eq. (3.55) may be written as

$$\eta(\tau) = c_1 e^{\mu\tau}\phi(\tau) + c_2 e^{-\mu\tau}\psi(\tau) \tag{3.59}$$

where, without loss of generality, we may consider that μ is real or imaginary (but not complex), and $\phi(\tau)$, $\psi(\tau)$ are periodic in τ of period T or $2T$. It is noted that, since the product of the two roots of Eq. (3.57) is equal to unity, the two characteristic exponents in the solution (3.59) have opposite signs.

As one readily sees from Eqs. (3.54) and (3.59), the variation ξ tends to zero with increasing τ if both the real parts of $-\delta \pm \mu$ are negative, where 2δ is the constant term in the series of $F(\tau)$. Therefore, the corresponding periodic solution is asymptotically stable. On the contrary, if at least one of the real parts of $-\delta \pm \mu$ is positive, the variation ξ diverges without limit as τ increases, and the corresponding periodic solution is unstable. Hence, for establishing the stability criterion, it is necessary to evaluate the characteristic exponent in Eq. (3.59). Some representative equations which take the form of (3.55) will be discussed in the following sections.

3.6 Mathieu's Equation [51, 69, 73, 74]

As an example of Eq. (3.55) we commence with Mathieu's equation, since it will be a suitable introduction to our later consideration. Following Whittaker [113, p. 405], the equation takes the form

$$\frac{d^2x}{d\tau^2} + (a + 16q \cos 2\tau)x = 0 \tag{3.60}$$

where the parameters a, q will be limited to real numbers. The equation is a particular case of a linear type of the second order with periodic coefficients. Thus Floquet's theory is applicable; a particular solution of Eq. (3.60) is given by

$$x = e^{\mu\tau}\phi(\tau) \tag{3.61}$$

where μ is the characteristic exponent dependent upon the parameters a and q and $\phi(\tau)$ is a periodic function of τ with period π or 2π. Since Eq. (3.60) is unchanged if $-\tau$ be written for τ, $e^{-\mu\tau}\phi(-\tau)$ is another independent solution. Hence, the general solution of Eq. (3.60) may be written as

$$x = c_1 e^{\mu\tau}\phi(\tau) + c_2 e^{-\mu\tau}\phi(-\tau) \tag{3.62}$$

where c_1 and c_2 are arbitrary constants.

We now discuss the stability of solutions. The following cases will be considered with the increase of τ:

1. A solution is defined to be unstable if it grows unboundedly as τ tends to infinity.

2. A solution is defined to be stable if it remains bounded as τ tends to infinity.

3. A solution with period π or 2π is said to be neutral; it is regarded as a special case of a stable solution.

Since $\phi(\tau)$ and $\phi(-\tau)$ in Eq. (3.62) are periodic in τ, the stability depends upon $e^{\mu\tau}$, or upon μ. Assuming the form of solution (3.62), where $\phi(\tau)$ has the period π or 2π, the characteristic exponent μ may be considered to be real or imaginary, but not complex. Hence the solution (3.62) is unstable if μ is real and stable if μ is imaginary.

(a) *Mathieu Functions and Correlated Characteristic Numbers of the Parameters* [26, 69, 74, 113]

We shall consider case 3, that is, the periodic solutions of Eq. (3.60) with period π or 2π. These solutions are, by definition, called the Mathieu functions. In order that such solutions may exist, a must have one of an infinite sequence of values for each value of q; that is, a must be one of an infinite sequence of functions of q. When q is zero, the solutions required are 1, $\cos \tau$, $\sin \tau$, $\cos 2\tau$, $\sin 2\tau$, and so on; the corresponding values of a are the squares of the integers. For other values of q, the Mathieu functions are denoted by $ce_0(\tau,q)$, $ce_1(\tau,q)$, $se_1(\tau,q)$, $ce_2(\tau,q)$,

$se_2(\tau,q)$, etc. $ce_n(\tau,q)$ and $se_n(\tau,q)$ are those Mathieu functions that reduce respectively to $\cos n\tau$ and $\sin n\tau$ when $q \to 0$.

The Fourier series for the Mathieu functions are preferably written in the form

$$ce_{2n}(\tau,q) = \sum_{r=0}^{\infty} A_{2r}(q) \cos 2r\tau$$

$$ce_{2n+1}(\tau,q) = \sum_{r=0}^{\infty} A_{2r+1}(q) \cos (2r+1)\tau$$

$$se_{2n}(\tau,q) = \sum_{r=0}^{\infty} B_{2r}(q) \sin 2r\tau$$

$$se_{2n+1}(\tau,q) = \sum_{r=0}^{\infty} B_{2r+1}(q) \sin (2r+1)\tau$$

(3.63)

In these series the Fourier coefficients A and B are functions of q. If $|q|$ is sufficiently small, these coefficients and accordingly the Mathieu functions are developed in power series of q.* For a given q, the value of a is definite for each Mathieu function, hence is called the characteristic number of the corresponding Mathieu function [47]. Denoting, in general, the characteristic numbers by a_{cn} and a_{sn} corresponding to $ce_n(\tau,q)$ and $se_n(\tau,q)$ respectively, we have the following expansions:[1]

$$a_{c1} = 1 - 8q - 8q^2 + 8q^3 - \tfrac{8}{3}q^4 + \cdots$$
$$a_{s1} = 1 + 8q - 8q^2 - 8q^3 - \tfrac{8}{3}q^4 + \cdots$$
$$a_{c2} = 4 + \tfrac{80}{3}q^2 - \tfrac{6104}{27}q^4 + \cdots$$
$$a_{s2} = 4 - \tfrac{16}{3}q^2 + \tfrac{400}{27}q^4 + \cdots$$
$$a_{c3} = 9 + 4q^2 - 8q^3 + \tfrac{13}{5}q^4 + \cdots$$
$$a_{s3} = 9 + 4q^2 + 8q^3 + \tfrac{13}{5}q^4 + \cdots$$

(3.64)

. . . .

Tables of the Mathieu functions $ce_0(\tau,q)$, . . . , $ce_5(\tau,q)$; $se_1(\tau,q)$, . . . , $se_6(\tau,q)$ and the corresponding characteristic numbers are given by Ince and others [47, 48, 64]. The characteristic curves showing the relation between q and a_{cn}, a_{sn} ($n = 1, 2, 3$) are plotted in Fig. 3.3, where the broad curves are obtained from the tables and the fine curves are calculated by Eqs. (3.64). These characteristic curves divide the plane into regions of stability and instability; that is, if a point (a,q) lies on the unstable region interposed between a_{cn} and a_{sn}, the original equation (3.60) has an unstable solution; and if a point (a,q) lies on the remaining region, a stable solution results.

* Very little is known on the convergence of the series. Only for $ce_0(\tau,q)$ Watson [110] showed, by constructing a majorant series, that the power series converge when $32|q|^2 < 1$.

[1] Expansions of the Mathieu functions are shown in Appendix I.

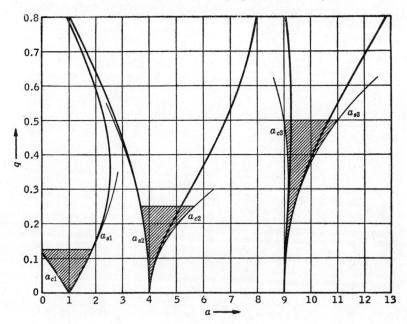

FIGURE 3.3   Stability chart for Mathieu's equation.

(b) Solutions in the Unstable Regions

As we have mentioned in Sec. 3.5, the stability of equilibrium in nonlinear systems may be investigated by computing the characteristic exponent of an unstable solution for the variational equation (3.55). In order to solve this kind of equation, the method of Hill's infinite determinant [44] and many other methods are proposed; among them the one introduced by Whittaker [112, 113] is useful in finding the unstable solution when $|q|$ is small.

Following Whittaker, we shall first seek the unstable solution associated with the first unstable region which lies between a_{c1} and a_{s1} of Fig. 3.3. As one sees in Eqs. (3.64), if the parameters a and q satisfy the relation

$$a = a_{c1} = 1 - 8q - 8q^2 + 8q^3 + \cdots$$

Eq. (3.60) has a solution (see Appendix I)

$$x = ce_1(\tau,q) = \cos \tau + q \cos 3\tau + q^2(-\cos 3\tau + \tfrac{1}{3} \cos 5\tau)$$
$$+ q^3(\tfrac{1}{3} \cos 3\tau - \tfrac{4}{9} \cos 5\tau + \tfrac{1}{18} \cos 7\tau) + \cdots$$

and if

$$a = a_{s1} = 1 + 8q - 8q^2 - 8q^3 + \cdots$$

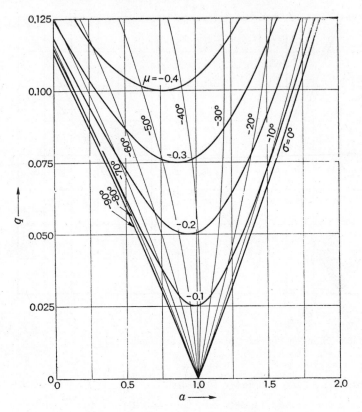

FIGURE 3.4   Iso-μ and iso-σ curves in the first unstable region.

then

$$x = se_1(\tau,q) = \sin \tau + q \sin 3\tau + q^2(\sin 3\tau + \tfrac{1}{3} \sin 5\tau)$$
$$+ q^3(\tfrac{1}{3} \sin 3\tau + \tfrac{4}{9} \sin 5\tau + \tfrac{1}{18} \sin 7\tau) + \cdots$$

Whittaker has pointed out that they are degenerate cases of a quasi-periodic solution of Mathieu's equation having the form

$$x = e^{\mu\tau}\phi(\tau)$$

with

(3.65)

$$\phi(\tau) = \sin (\tau - \sigma) + a_3 \cos (3\tau - \sigma) + b_3 \sin (3\tau - \sigma)$$
$$+ a_5 \cos (5\tau - \sigma) + b_5 \sin (5\tau - \sigma) + \cdots$$

where σ is a new parameter taking a value between 0 and $-\pi/2$ for the unstable solution.   Hence, the Mathieu functions $ce_1(\tau,q)$ and $se_1(\tau,q)$ are simply particular cases corresponding to $\sigma = -\pi/2$ and 0, respectively. The characteristic exponent μ and the unknown coefficients a_3, b_3, . . . in Eqs. (3.65) may be determined as follows.   Since a, q and μ, σ are inter-

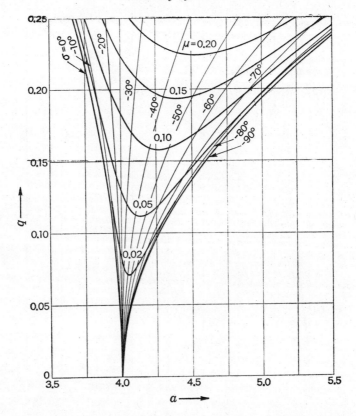

FIGURE 3.5 Iso-μ and iso-σ curves in the second unstable region.

related, we may assume, for small values of $|q|$, that

$$\mu = q\kappa(\sigma) + q^2\lambda(\sigma) + q^3\mu(\sigma) + \cdots$$
$$a = 1 + q\alpha(\sigma) + q^2\beta(\sigma) + q^3\gamma(\sigma) + \cdots \qquad (3.66)$$

Substituting Eqs. (3.65) and (3.66) into (3.60) and equating coefficients of the same powers of q to zero ultimately leads to

$$a_3 = 3q^2 \sin 2\sigma + 3q^3 \sin 4\sigma + \cdots$$
$$b_3 = q + q^2 \cos 2\sigma + q^3(-1\tfrac{1}{3} + 5\cos 4\sigma) + \cdots$$
$$a_5 = 1\tfrac{4}{9}q^3 \sin 2\sigma + \cdots \qquad (3.67)$$
$$b_5 = \tfrac{1}{3}q^2 + \tfrac{4}{9}q^3 \cos 2\sigma + \cdots$$

$$\cdots \cdots \cdots \cdots \cdots \cdots \cdots$$

$$\mu = 4q \sin 2\sigma - 12q^3 \sin 2\sigma + \cdots$$
$$a = 1 + 8q \cos 2\sigma + q^2(-16 + 8\cos 4\sigma) - 8q^3 \cos 2\sigma + \cdots \qquad (3.68)$$

As far as the stability is considered, it is necessary to evaluate the characteristic exponent μ only. To this end, we calculate σ by using the

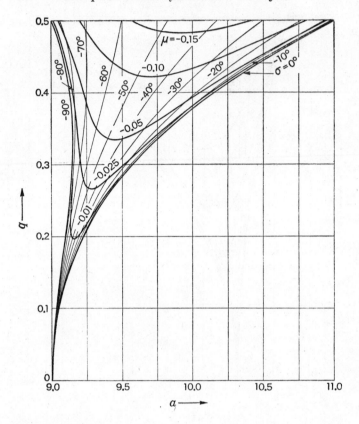

FIGURE 3.6   Iso-μ and iso-σ curves in the third unstable region.

second equation of (3.68); then substituting this into the first, we obtain the value of μ.

By proceeding likewise for the unstable solution associated with the second unstable region which lies between a_{c2} and a_{s2} [114], we obtain

$$\phi(\tau) = \sin (2\tau - \sigma) + q[2 \sin \sigma + \tfrac{2}{3} \sin (4\tau - \sigma)] + \tfrac{1}{6}q^2 \sin (6\tau - \sigma)$$
$$+ q^3[\tfrac{8}{3} \sin \sigma - 16 \sin^3 \sigma - \tfrac{16}{9} \sin 2\sigma \cos (4\tau - \sigma)$$
$$+ (-\tfrac{5}{27} + \tfrac{16}{9} \sin^2 \sigma) \sin (4\tau - \sigma) + \tfrac{1}{45} \sin (8\tau - \sigma)] + \cdots \quad (3.69)$$

$$\mu = -4q^2 \sin 2\sigma + \cdots \qquad a = 4 - q^2(\tfrac{16}{3} - 32 \sin^2 \sigma) + \cdots \quad (3.70)$$

and for the third unstable region which lies between a_{c3} and a_{s3}, we obtain

$$\phi(\tau) = \sin (3\tau - \sigma) + q[- \sin (\tau - \sigma) + \tfrac{1}{2} \sin (5\tau - \sigma)]$$
$$+ q^2[- \sin (\tau + \sigma) + \tfrac{1}{10} \sin (7\tau - \sigma)]$$
$$+ q^3[-\tfrac{1}{2} \sin (\tau - \sigma) + \tfrac{7}{40} \sin (5\tau - \sigma)$$
$$+ \tfrac{1}{90} \sin (9\tau - \sigma)] + \cdots \quad (3.71)$$

$$\mu = \tfrac{4}{3}q^3 \sin 2\sigma + \cdots \qquad a = 9 + 4q^2 + 8q^3 \cos 2\sigma + \cdots \quad (3.72)$$

From these results the characteristic exponent μ is readily calculated when the new parameter σ is known. Practically, however, when Mathieu's equation is to be solved, the parameters a and q are first known, and it is rather difficult to find σ from given values of a and q. To avoid this situation, a and μ are first calculated from Eqs. (3.68), (3.70), and (3.72) by varying q and σ, and then the iso-μ and iso-σ curves are plotted in the aq plane, as illustrated in Figs. 3.4 to 3.6. These figures respectively correspond to the first, the second, and the third unstable regions, which are shown shaded in Fig. 3.3. Thus, μ and σ may immediately be evaluated when a and q are given.

3.7 Hill's Equation [12, 44, 69, 101, 113]

(a) Stability Problem for Hill's Equation

Hill's equation takes the form

$$\frac{d^2x}{d\tau^2} + \left(\theta_0 + 2 \sum_{\nu=1}^{\infty} \theta_\nu \cos 2\nu\tau \right) x = 0 \qquad (3.73)$$

where θ_0, θ_1, θ_2, ... are assigned parameters and $\sum_{\nu=1}^{\infty} |\theta_\nu|$ converges. The theory in the preceding section applies to Hill's equation also; hence the solution of Eq. (3.73) may be either stable or unstable.

We shall now briefly discuss the unstable solution by making use of Whittaker's method of change of parameter. By Floquet's theory, a particular solution of Eq. (3.73) is given by

$$x = e^{\mu\tau}\phi(\tau) \qquad (3.74)$$

Substitution of Eq. (3.74) into (3.73) gives

$$\frac{d^2\phi}{d\tau^2} + 2\mu \frac{d\phi}{d\tau} + \left(\theta_0 + \mu^2 + 2 \sum_{\nu=1}^{\infty} \theta_\nu \cos 2\nu\tau \right) \phi = 0 \qquad (3.75)$$

Following Whittaker, the periodic function $\phi(\tau)$ in the nth unstable region may be assumed, to a first approximation, in the form[1]

$$\phi(\tau) = \sin (n\tau - \sigma) \qquad n = 1, 2, 3, \ldots \qquad (3.76)$$

in which σ is a new parameter to be determined presently. Substituting Eq. (3.76) into (3.75) and equating the coefficients of $\sin n\tau$ and $\cos n\tau$

---

[1] It is assumed that the parameters θ_ν are small.

separately to zero gives us

$$2\mu n \sin \sigma + (\theta_0 + \mu^2 - n^2) \cos \sigma - \theta_n \cos \sigma = 0$$
$$2\mu n \cos \sigma - (\theta_0 + \mu^2 - n^2) \sin \sigma - \theta_n \sin \sigma = 0 \qquad (3.77)$$

Hence the characteristic exponent μ and the parameter σ are given by

$$\mu = \frac{\theta_n}{2n} \sin 2\sigma \qquad \text{and} \qquad \theta_0 = n^2 + \theta_n \cos 2\sigma - \left(\frac{\theta_n}{2n}\right)^2 \sin^2 2\sigma \quad (3.78)$$

from which, upon eliminating σ, we obtain

$$\mu^2 = -(\theta_0 + n^2) + \sqrt{4n^2\theta_0 + \theta_n{}^2} \qquad (3.79)$$

From Eqs. (3.78), μ and σ are obtained for given θ_0 and θ_n, so that a particular solution (3.74) with (3.76) is determined. Furthermore, if we write $-\sigma$ for σ in Eqs. (3.78), the value of θ_0 is unchanged, but this change alters the sign of μ. Hence we see that the second independent solution may take the form

$$x = e^{-\mu\tau} \sin (n\tau + \sigma)$$

and the complete solution, with two arbitrary constants, is

$$x = c_1 e^{\mu\tau} \sin (n\tau - \sigma) + c_2 e^{-\mu\tau} \sin (n\tau + \sigma) \qquad (3.80)$$

Since the characteristic exponent μ may be taken to be imaginary or real according as the solution is stable or unstable, we have $\mu^2 > 0$ for the unstable solution. By virtue of Eq. (3.79) this condition is transformed to

$$(\theta_0 - n^2 + \theta_n)(\theta_0 - n^2 - \theta_n) < 0 \qquad \text{or} \qquad |\theta_n| > |\theta_0 - n^2| \quad (3.81)$$

Since $\mu = 0$ on the boundary between the stable and unstable regions, the boundary lines of the nth unstable region are given by

$$\theta_0 = n^2 \pm \theta_n \qquad (3.82)$$

which may also be derived directly by putting $\sigma = -\pi/2$ and $\sigma = 0$ in the second equation of (3.78).[1]

(b) Extended Form of Hill's Equation

Thus far we have dealt with Mathieu's equation and Hill's equation as the representative forms of Eq. (3.55) in Sec. 3.5. More generally,

---

[1] It is seen in these equations that the values of μ, σ and consequently the boundary of the nth unstable region are determined only by θ_0 and θ_n, and are not affected by other parameters. This is because we have confined our calculation to a first approximation. By closer approximation, however, the remaining parameters are related to them, as will be shown in Appendix II, where the parameters are taken into account up to the third power (see also Ref. 46).

however, the variational equation takes the form

$$\frac{d^2x}{d\tau^2} + \left(\theta_0 + 2\sum_{\nu=1}^{\infty}\theta_{\nu s}\sin 2\nu\tau + 2\sum_{\nu=1}^{\infty}\theta_{\nu c}\cos 2\nu\tau\right)x = 0 \quad (3.83)$$

which may be considered as an extended form of Hill's equation. Whittaker's method of solution may likewise be applicable in this case. Thus, assuming for x the following first approximation,

$$x = e^{\mu\tau}\sin(n\tau - \sigma) \qquad n = 1, 2, 3, \ldots \quad (3.84)$$

we obtain

$$2\mu n = \theta_{nc}\sin 2\sigma - \theta_{ns}\cos 2\sigma$$
$$\theta_0 = n^2 - \mu^2 + \theta_{ns}\sin 2\sigma + \theta_{nc}\cos 2\sigma \quad (3.85)$$

Elimination of σ in Eqs. (3.85) gives

$$\mu^2 = -(\theta_0 + n^2) + \sqrt{4n^2\theta_0 + \theta_n^2} \qquad \theta_n^2 = \theta_{ns}^2 + \theta_{nc}^2 \quad (3.86)$$

This is identical with Eq. (3.79); but we have two different values in magnitude of σ which satisfy Eqs. (3.85). Hence, denoting them by σ_1 and σ_2 respectively, we obtain the general solution[1]

$$x = c_1 e^{\mu\tau}\sin(n\tau - \sigma_1) + c_2 e^{-\mu\tau}\sin(n\tau - \sigma_2) \quad (3.87)$$

3.8 Improved Approximation of the Characteristic Exponent for Hill's Equation

In the preceding section the characteristic exponent was evaluated to a first approximation by assuming the unstable solution of the form $x = e^{\mu\tau}\phi(\tau)$, where $\phi(\tau) = \sin(n\tau - \sigma)$. As one sees in Sec. 3.6b and in Appendix III, however, the periodic function $\phi(\tau)$ generally contains terms of frequency other than n. Hence one may expect a closer evaluation of the characteristic exponent if some of these additional terms are taken into consideration.

In this section we shall be concerned particularly with the characteristic exponents associated with solutions in the first, second, and third unstable regions, since in most applications such regions of lower order become of practical importance. For convenience' sake, we begin with the second unstable region and then treat the first and third unstable regions.

[1] The above consideration is confined to the first approximation. In Appendix III a closer approximation is carried out in which terms are taken into account up to the second power of the θ's.

(a) Characteristic Exponent Associated with the Second Unstable Region

Since the periodic function $\phi(\tau)$ in unstable regions of even order generally contains a zero-frequency term, it may be preferable to add a constant term in $\phi(\tau)$. Thus we write

$$x = e^{\mu\tau}\phi(\tau) = e^{\mu\tau}[c + \sin(2\tau - \sigma)] \qquad (3.88)$$

Substituting this into Eq. (3.83) and equating the constant term and the coefficients of $\sin 2\tau$ and $\cos 2\tau$ separately to zero gives us

$$
\begin{aligned}
c(\theta_0 + \mu^2) - \theta_{1c}\sin\sigma + \theta_{1s}\cos\sigma &= 0 \\
2c\theta_{1s} - (\theta_{2s} - 4\mu)\sin\sigma + (\theta_0 + \mu^2 - 4 - \theta_{2c})\cos\sigma &= 0 \\
2c\theta_{1c} - (\theta_0 + \mu^2 - 4 + \theta_{2c})\sin\sigma + (\theta_{2s} + 4\mu)\cos\sigma &= 0
\end{aligned} \qquad (3.89)
$$

Upon elimination of c and σ in Eqs. (3.89), we have

$$
\Delta(\mu) \equiv \begin{vmatrix} \theta_0 + \mu^2 & \theta_{1s} & \theta_{1c} \\ 2\theta_{1s} & \theta_0 + \mu^2 - 4 - \theta_{2c} & \theta_{2s} - 4\mu \\ 2\theta_{1c} & \theta_{2s} + 4\mu & \theta_0 + \mu^2 - 4 + \theta_{2c} \end{vmatrix} = 0 \qquad (3.90)
$$

As mentioned before, the condition $\mu^2 > 0$ holds for the second unstable region, and μ vanishes at the boundary of this region. Hence, since we are concerned with the domain in the neighborhood of the boundary, μ^2 is so small that we may neglect higher powers of μ than the second in computing the condition $\mu^2 > 0$. Furthermore, we assume that the parameters θ_1, θ_2 are sufficiently small. Then, by virtue of Eq. (3.90), this condition leads to

$$
16\mu^2 = -(\theta_0 - 4)^2 + \theta_2{}^2 + 2\frac{\theta_0 - 4}{\theta_0}\theta_1{}^2
$$
$$
- \frac{2}{\theta_0}[(\theta_{1c}{}^2 - \theta_{1s}{}^2)\theta_{2c} + 2\theta_{1c}\theta_{1s}\theta_{2s}] > 0 \qquad (3.91)
$$

or

$$
\Delta(0) \equiv \begin{vmatrix} \theta_0 & \theta_{1s} & \theta_{1c} \\ 2\theta_{1s} & \theta_0 - 4 - \theta_{2c} & \theta_{2s} \\ 2\theta_{1c} & \theta_{2s} & \theta_0 - 4 + \theta_{2c} \end{vmatrix} < 0 \qquad (3.92)
$$

This is the condition that the characteristic exponent μ is real, so that the solution of Eq. (3.83) is in the second unstable region. It follows directly from Eq. (3.90) that the determinant $\Delta(0)$ vanishes when the parameters θ lie on the boundary of the second unstable region.

(b) Characteristic Exponents Associated with the First and Third Unstable Regions

Since the periodic function $\phi(\tau)$ in unstable regions of odd order contains odd-frequency terms only (see Appendix III), we assume a solution of the form

$$x = e^{\mu\tau}\phi(\tau) = e^{\mu\tau}[a_1 \sin (\tau - \sigma_1) + a_3 \sin (3\tau - \sigma_3)] \qquad (3.93)$$

Substituting this into Eq. (3.83) and equating the coefficients of $\sin \tau$, $\cos \tau$, $\sin 3\tau$, and $\cos 3\tau$ separately to zero gives us

$$(\theta_{1s} - 2\mu)a_1 \sin \sigma_1 - (\theta_0 + \mu^2 - 1 - \theta_{1c})a_1 \cos \sigma_1$$
$$- (\theta_{1s} - \theta_{2s})a_3 \sin \sigma_3 - (\theta_{1c} - \theta_{2c})a_3 \cos \sigma_3 = 0$$
$$(\theta_0 + \mu^2 - 1 + \theta_{1c})a_1 \sin \sigma_1 - (\theta_{1s} + 2\mu)a_1 \cos \sigma_1$$
$$+ (\theta_{1c} + \theta_{2c})a_3 \sin \sigma_3 - (\theta_{1s} + \theta_{2s})a_3 \cos \sigma_3 = 0$$
$$(\theta_{1s} + \theta_{2s})a_1 \sin \sigma_1 - (\theta_{1c} - \theta_{2c})a_1 \cos \sigma_1 \qquad (3.94)$$
$$+ (\theta_{3s} - 6\mu)a_3 \sin \sigma_3 - (\theta_0 + \mu^2 - 9 - \theta_{3c})a_3 \cos \sigma_3 = 0$$
$$(\theta_{1c} + \theta_{2c})a_1 \sin \sigma_1 + (\theta_{1s} - \theta_{2s})a_1 \cos \sigma_1$$
$$+ (\theta_0 + \mu^2 - 9 + \theta_{3c})a_3 \sin \sigma_3 - (\theta_{3s} + 6\mu)a_3 \cos \sigma_3 = 0$$

Upon elimination of a_1, a_3, σ_1, and σ_3 in Eqs. (3.94), we have

$$\Delta(\mu) \equiv \begin{vmatrix} \theta_0 + \mu^2 - 1 - \theta_{1c} & \theta_{1s} - 2\mu & \theta_{1c} - \theta_{2c} & -\theta_{1s} + \theta_{2s} \\ \theta_{1s} + 2\mu & \theta_0 + \mu^2 - 1 + \theta_{1c} & \theta_{1s} + \theta_{2s} & \theta_{1c} + \theta_{2c} \\ \theta_{1c} - \theta_{2c} & \theta_{1s} + \theta_{2s} & \theta_0 + \mu^2 - 9 - \theta_{3c} & \theta_{3s} - 6\mu \\ -\theta_{1s} + \theta_{2s} & \theta_{1c} + \theta_{2c} & \theta_{3s} + 6\mu & \theta_0 + \mu^2 - 9 + \theta_{3c} \end{vmatrix} = 0 \quad (3.95)$$

The condition $\mu^2 > 0$ holds for the first and third unstable regions, and μ vanishes at the boundaries of these regions. Proceeding in the same manner as in part (a), we may conclude that the condition $\mu^2 > 0$ leads to

$$\Delta(0) = \begin{vmatrix} \theta_0 - 1 - \theta_{1c} & \theta_{1s} & \theta_{1c} - \theta_{2c} & -\theta_{1s} + \theta_{2s} \\ \theta_{1s} & \theta_0 - 1 + \theta_{1c} & \theta_{1s} + \theta_{2s} & \theta_{1c} + \theta_{2c} \\ \theta_{1c} - \theta_{2c} & \theta_{1s} + \theta_{2s} & \theta_0 - 9 - \theta_{3c} & \theta_{3s} \\ -\theta_{1s} + \theta_{2s} & \theta_{1c} + \theta_{2c} & \theta_{3s} & \theta_0 - 9 + \theta_{3c} \end{vmatrix} < 0$$
$$(3.96)$$

This is the condition that the characteristic exponent μ is real, so that the solution of Eq. (3.83) is in the first or the third unstable region. It follows directly from Eq. (3.95) that $\Delta(0)$ vanishes at the boundaries of the first and third unstable regions.

Part II

Forced Oscillations in Steady States

Chapter 4

Stability of Periodic Oscillations in Second-order Systems

4.1 Introduction

When a periodic force is applied to a linear system, the resulting motion is obtained by a superposition of the transient- and the steady-state components of the oscillation. The former is due to the free oscillation of the system, and the latter is related to the forced oscillation which arises from the action of the external force. When the system is (asymptotically) stable, the free oscillation is damped out after a sufficiently long period of time. Hence only the forced oscillation having the same frequency as the external force would be observed. Thus, as far as linear systems are concerned, the forced oscillation is uniquely determined once the system and the external force are given, and it is affected in no way by the initial condition with which the oscillation was started. In nonlinear systems, however, the circumstances may be quite different in this respect. The principle of superposition will no longer be applicable. A nonlinear system may possess a wide variety of periodic oscillations in addition to those which have the same period as the external force.

In this chapter we confine our attention to the stability problem of periodic oscillations. We again consider the second-order differential equation [Eq. (3.50)]

$$\frac{d^2v}{d\tau^2} + f\left(v, \frac{dv}{d\tau}\right) = e(\tau) \tag{4.1}$$

in which the function $f(v, dv/d\tau)$ is in general nonlinear and $e(\tau)$ represents a periodic external force.[1]

It is known from the theory of differential equations that Eq. (4.1) possesses a solution $v(\tau)$ that is uniquely determined once the values of $v(0)$ and $(dv/d\tau)_{\tau=0}$, that is, the initial conditions, are prescribed. It is,

[1] As mentioned in Sec. 3.5, a stable solution of Eq. (4.1) is either harmonic or subharmonic even if higher harmonics may predominate.

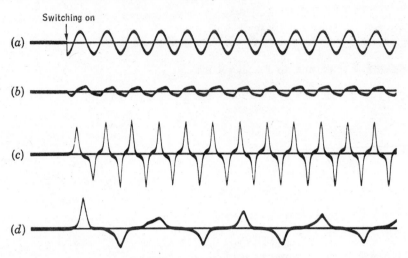

FIGURE 4.1   Different types of oscillation in a nonlinear system.   (*a*) Applied
voltage.   (*b*) Nonresonant oscillation.   (*c*) Resonant oscillation.
(*d*) Subharmonic oscillation.

however, a distinctive feature of the nonlinear system that various types
of periodic oscillations may exist for the same system, depending on
different values of the initial conditions.   Experimental results displayed
in Fig. 4.1 show the waveforms of an applied voltage (60 cps) and the
resulting currents in an electric circuit containing a saturable inductor
and capacitor in series.   As will be shown in Sec. 5.1, the system is
described by an equation included in the general form of (4.1).   Upon
application of the same sinusoidal voltage (*a*), the waveforms of (*b*) and (*c*)
in the steady state have the same frequency as the applied voltage, and
hence these are harmonic oscillations.   However, the response (*d*) is a subharmonic oscillation of order $\frac{1}{3}$, since the steady-state frequency is
one-third the driving frequency.   These waveforms are obtained by
starting the oscillation with different values of the initial charge on the
capacitor.[1]

Contrary to the case of linear differential equations, it is hardly
possible to obtain the general solution of Eq. (4.1) for a given initial condition.   Moreover, since explicit solutions in terms of elementary functions may not be expected, Eq. (4.1) is treated by various approximate
methods.   In so far as we deal with periodic solutions, our conventional
method of solution is to assume for $v(\tau)$ a Fourier series development with
undetermined coefficients and then to fix them by the method of harmonic

---

[1] The relationship between the initial conditions and the resulting oscillations
will be studied in Part III.

balance.   However, it should be noticed that by this method of solution we merely find the periodic states of equilibrium, which are not always realized, but are able to exist only so long as they are stable.   The circumstances under which this condition obtains are determined by the stability investigation of periodic solutions.

4.2 Condition for Stability of Periodic Oscillations [34]

It was mentioned in Sec. 3.5 that the stability condition for the periodic solution of Eq. (4.1) is given by

$$\text{Real parts of } (-\delta \pm \mu) < 0 \tag{4.2}$$

where 2δ is the constant term in the series $F(\tau)$ in Eq. (3.53) and μ is the characteristic exponent of the solution (3.59).   If μ is imaginary, this condition is simply reduced to $\delta > 0$.   If μ is real, the condition (4.2) is equivalent to the following two conditions:

$$\delta > 0 \quad \text{and} \quad \delta^2 > \mu^2 \tag{4.3}$$

We now discuss these conditions by making use of the characteristic exponent evaluated in Secs. 3.6 to 3.8.   For convenience, we consider the following two cases.

CASE 1.   WHEN THE VARIATIONAL EQUATION (3.55) TAKES THE FORM

$$\frac{d^2\eta}{d\tau^2} + \left[\theta_0 + 2 \sum_{\nu=1}^{\infty} \theta_\nu \cos (2\nu\tau - \epsilon_\nu) \right] \eta = 0 \tag{4.4}$$

Substituting Eq. (3.86) into (4.2) gives

$$\delta > 0 \tag{4.5}$$
$$\text{and} \quad (\theta_0 - n^2)^2 + 2(\theta_0 + n^2)\delta^2 + \delta^4 > \theta_n^2 \quad n = 1, 2, 3, \ldots \tag{4.6}$$

The latter is the stability condition (of the first approximation[1]) for the nth unstable region.   In order that the periodic state of equilibrium be stable, the condition (4.6) must be satisfied for all values of n simultaneously.

CASE 2.   WHEN THE VARIATIONAL EQUATION (3.55) TAKES THE FORM

$$\frac{d^2\eta}{d\tau^2} + \left[\theta_0 + 2 \sum_{\nu=1}^{\infty} \theta_\nu \cos (\nu\tau - \epsilon_\nu) \right] \eta = 0 \tag{4.7}$$

The stability conditions may be derived in the same manner as above;

---

[1] In the case when the stability condition of a higher-order approximation is desirable, refer to Appendixes II and III for a closer evaluation of μ.

thus we obtain

$$\delta > 0 \tag{4.8}$$

and

$$\left[\theta_0 - \left(\frac{n}{2}\right)^2\right]^2 + 2\left[\theta_0 + \left(\frac{n}{2}\right)^2\right]\delta^2 + \delta^4 > \theta_n{}^2 \quad n = 1, 2, 3, \ldots \tag{4.9}$$

It is noted that the above conditions apply to the system described by

$$\frac{d^2v}{d\tau^2} + 2\delta \frac{dv}{d\tau} + f(v) = e(\tau) \tag{4.10}$$

where $f(v)$ is a nonlinear function of v.   As will be shown in later chapters, case 1 is generally applied when $f(v)$ is odd and case 2 when $f(v)$ is nonodd.

4.3 Improved Stability Conditions

Following the consideration in Sec. 3.8, we shall improve the stability condition (4.6) or (4.9) for the unstable regions of lower order.

(a) Stability Condition for $n = 2$

CASE 1

When the variational equation leads to (4.4), η may be assumed to take the form

$$\eta = e^{\mu\tau}[c + \sin(2\tau - \sigma)] \tag{4.11}$$

The characteristic exponent μ is given by Eq. (3.90), that is, $\Delta(\mu) = 0$. Since the stability condition is given by $\delta^2 - \mu^2 > 0$,* we have $\Delta(\delta) = 0$ at the boundary of the second unstable region. We expand Eq. (3.90) in powers of $\delta^2 - \mu^2$ and neglect higher powers of $\delta^2 - \mu^2$ than the first.   Then, as before (see Sec. 3.8a), we obtain

$$16(\delta^2 - \mu^2) = (\theta_0 + \delta^2 - 4)^2 - \theta_2{}^2 - 2\frac{\theta_0 + \delta^2 - 4}{\theta_0 + \delta^2}\theta_1{}^2$$

$$+ \frac{2}{\theta_0 + \delta^2}[(\theta_{1c}{}^2 - \theta_{1s}{}^2)\theta_{2c} + 2\theta_{1c}\theta_{1s}\theta_{2s}] + 16\delta^2 > 0 \tag{4.12}$$

or

$$\Delta(\delta) = \begin{vmatrix} \theta_0 + \delta^2 & \theta_{1s} & \theta_{1c} \\ 2\theta_{1s} & \theta_0 + \delta^2 - 4 - \theta_{2c} & \theta_{2s} - 4\delta \\ 2\theta_{1c} & \theta_{2s} + 4\delta & \theta_0 + \delta^2 - 4 + \theta_{2c} \end{vmatrix} > 0 \tag{4.13}$$

This is the stability condition for the second unstable region.

* We assume that $\delta > 0$.

CASE 2

When the variational equation leads to Eq. (4.7), η takes the form

$$\eta = e^{\mu\tau}[c + \sin(\tau - \sigma)] \tag{4.14}$$

When the above procedure is followed, the improved stability condition for the second unstable region becomes

$$4(\delta^2 - \mu^2) = (\theta_0 + \delta^2 - 1)^2 - \theta_2{}^2 - 2\frac{\theta_0 + \delta^2 - 1}{\theta_0 + \delta^2}\theta_1{}^2$$

$$+ \frac{2}{\theta_0 + \delta^2}[(\theta_{1c}{}^2 - \theta_{1s}{}^2)\theta_{2c} + 2\theta_{1c}\theta_{1s}\theta_{2s}] + 4\delta^2 > 0 \tag{4.15}$$

or

$$\begin{vmatrix} \theta_0 + \delta^2 & \theta_{1s} & \theta_{1c} \\ 2\theta_{1s} & \theta_0 + \delta^2 - 1 - \theta_{2c} & \theta_{2s} - 2\delta \\ 2\theta_{1c} & \theta_{2s} + 2\delta & \theta_0 + \delta^2 - 1 + \theta_{2c} \end{vmatrix} > 0 \tag{4.16}$$

(b) Stability Condition for n = 1 and 3

When the variational equation leads to (4.4), η takes the form

$$\eta = e^{\mu\tau}[a_1 \sin(\tau - \sigma_1) + a_3 \sin(3\tau - \sigma_3)] \tag{4.17}$$

Assuming that $\delta > 0$, an improved stability condition is obtained by substitution of Eq. (3.95) into $\delta^2 - \mu^2 > 0$. The result is

$$\Delta(\delta) = \begin{vmatrix} \theta_0 + \delta^2 - 1 - \theta_{1c} & \theta_{1s} - 2\delta & \theta_{1c} - \theta_{2c} & -\theta_{1s} + \theta_{2s} \\ \theta_{1s} + 2\delta & \theta_0 + \delta^2 - 1 + \theta_{1c} & \theta_{1s} + \theta_{2s} & \theta_{1c} + \theta_{2c} \\ \theta_{1c} - \theta_{2c} & \theta_{1s} + \theta_{2s} & \theta_0 + \delta^2 - 9 - \theta_{3c} & \theta_{3s} - 6\delta \\ -\theta_{1s} + \theta_{2s} & \theta_{1c} + \theta_{2c} & \theta_{3s} + 6\delta & \theta_0 + \delta^2 - 9 + \theta_{3c} \end{vmatrix} > 0 \tag{4.18}$$

4.4 Complementing Remarks on the Stability Conditions

(a) Stability Condition (4.6) for n = 1

We shall first consider the condition (4.6) for the first unstable region. Since, in this case, the solution of the variational equation (4.4) takes the form $e^{\mu\tau}[\sin(\tau - \sigma) + \text{higher-order terms in } \theta_1, \theta_2, \ldots]$, the condition (4.6) ascertains the stability against buildup of an unstable oscillation in which the fundamental-frequency component predominates over higher harmonics.

In order to discuss the stability condition concretely, let us take an example:

$$\frac{d^2v}{d\tau^2} + 2\delta\frac{dv}{d\tau} + f(v) = B\cos\tau \tag{4.19}$$

where $f(v)$ is odd in v, and consider the case in which a periodic solution is approximated by the harmonic oscillation

$$v = x \sin \tau + y \cos \tau \qquad (4.20)$$

The amplitudes x and y may readily be found by the method of harmonic balance. Namely, substituting Eq. (4.20) into (4.19) and equating the coefficients of $\sin \tau$ and $\cos \tau$ on both sides respectively gives the following simultaneous equations to determine x and y, that is,

$$X(x,y) = -x - 2\,\delta y + \frac{1}{\pi} \int_0^{2\pi} f(v) \sin \tau \, d\tau = 0$$

$$Y(x,y) = 2\,\delta x - y + \frac{1}{\pi} \int_0^{2\pi} f(v) \cos \tau \, d\tau = B \qquad (4.21)$$

The variational equation, in this case, takes the form (4.4), and in fact we obtain

$$\frac{d^2\eta}{d\tau^2} + \left(\theta_0 + 2\sum_{\nu=1}^{\infty} \theta_{\nu s} \sin 2\nu\tau + 2\sum_{\nu=1}^{\infty} \theta_{\nu c} \cos 2\nu\tau\right)\eta = 0 \qquad (4.22)$$

where

$$\theta_0 = \frac{1}{2\pi} \int_0^{2\pi} \frac{df}{dv} \, d\tau - \delta^2$$

$$\theta_{1s} = \frac{1}{2\pi} \int_0^{2\pi} \frac{df}{dv} \sin 2\tau \, d\tau = \frac{1}{\pi} \int_0^{2\pi} \frac{df}{dv} \cos \tau \sin \tau \, d\tau$$

$$= \frac{1}{\pi} \int_0^{2\pi} \frac{df}{dv} \frac{\partial v}{\partial y} \sin \tau \, d\tau = \frac{\partial X}{\partial y} + 2\delta$$

$$= \frac{1}{\pi} \int_0^{2\pi} \frac{df}{dv} \frac{\partial v}{\partial x} \cos \tau \, d\tau = \frac{\partial Y}{\partial x} - 2\delta \qquad (4.22a)$$

$$\theta_{1c} = \frac{1}{2\pi} \int_0^{2\pi} \frac{df}{dv} \cos 2\tau \, d\tau = \frac{1}{\pi} \int_0^{2\pi} \frac{df}{dv} \cos^2 \tau \, d\tau$$

$$\qquad\qquad - \frac{1}{2\pi} \int_0^{2\pi} \frac{df}{dv} \, d\tau = \frac{\partial Y}{\partial y} + 1 - \theta_0 - \delta^2$$

$$= \frac{1}{2\pi} \int_0^{2\pi} \frac{df}{dv} \, d\tau - \frac{1}{\pi} \int_0^{2\pi} \frac{df}{dv} \sin^2 \tau \, d\tau = \theta_0 + \delta^2 - 1 - \frac{\partial X}{\partial x}$$

. . . .

Now, the stability condition (4.6) may be written in an alternative form:

$$\begin{vmatrix} \theta_0 + \delta^2 - n^2 - \theta_{nc} & \theta_{ns} - 2n\delta \\ \theta_{ns} + 2n\delta & \theta_0 + \delta^2 - n^2 + \theta_{nc} \end{vmatrix} > 0 \qquad (4.23)$$

Putting $n = 1$ and substituting the parameters θ_{1s} and θ_{1c} as given by

Eqs. (4.22a), the stability condition (4.6) ultimately leads to

$$
\begin{vmatrix}
\dfrac{\partial X}{\partial x} & \dfrac{\partial X}{\partial y} \\[2mm]
\dfrac{\partial Y}{\partial x} & \dfrac{\partial Y}{\partial y}
\end{vmatrix} > 0
\tag{4.24}
$$

The last form of the stability condition may also be derived from the Routh-Hurwitz criterion (Sec. 3.3). To verify this, let the variations of the amplitudes x and y be $\xi(\tau)$ and $\eta(\tau)$, respectively, and write

$$
v(\tau) = x(\tau) \sin \tau + y(\tau) \cos \tau
\tag{4.25}
$$

with $\quad x(\tau) = x + \xi(\tau) \qquad y(\tau) = y + \eta(\tau)$

Then, if $\xi(\tau)$ and $\eta(\tau)$ tend to zero with the lapse of time τ, the corresponding periodic solution (4.20) is stable. Substituting Eqs. (4.25) into (4.19) and equating the coefficients of $\sin \tau$ and $\cos \tau$ on both sides respectively gives

$$
-2 \frac{d\eta}{d\tau} + \frac{\partial X}{\partial x} \xi + \frac{\partial X}{\partial y} \eta = 0 \qquad 2 \frac{d\xi}{d\tau} + \frac{\partial Y}{\partial x} \xi + \frac{\partial Y}{\partial y} \eta = 0 \tag{4.26}
$$

under the assumptions that $\xi(\tau)$ and $\eta(\tau)$ are slowly varying functions of τ, so that $d^2\xi/d\tau^2$ and $d^2\eta/d\tau^2$ may be neglected, and that δ is a sufficiently small quantity and, hence, $\delta \, d\xi/d\tau$ and $\delta \, d\eta/d\tau$ are also discarded. The solutions of these simultaneous equations have the form $e^{\lambda\tau}$, where λ is determined by the characteristic equation

$$
\begin{vmatrix}
\dfrac{\partial X}{\partial x} & \dfrac{\partial X}{\partial y} - 2\lambda \\[2mm]
\dfrac{\partial Y}{\partial x} + 2\lambda & \dfrac{\partial Y}{\partial y}
\end{vmatrix} = 0
\tag{4.27}
$$

Hence the periodic solution (4.20) is stable provided that the real part of λ is negative. This stability condition is given by[1]

$$
\frac{\partial Y}{\partial x} - \frac{\partial X}{\partial y} > 0 \qquad \frac{\partial X}{\partial x} \frac{\partial Y}{\partial y} - \frac{\partial X}{\partial y} \frac{\partial Y}{\partial x} > 0 \tag{4.28}
$$

The first condition is always fulfilled, because it is reduced to $\delta > 0$ by virtue of Eqs. (4.22a). The second condition is the one we have already derived by making use of Hill's equation as the stability criterion.

We shall further verify that the characteristic curve (which shows the amplitude of the periodic oscillation against B) has a vertical tangent at the stability limit of the first unstable region.

[1] See the stability conditions (3.12) in Chap. 3.

To prove this, we put $r^2 = x^2 + y^2$. Then the characteristic curve, i.e., the Br relation, is readily obtained by solving Eqs. (4.21). Further, differentiating Eqs. (4.21) with respect to B gives us

$$\frac{\partial X}{\partial x}\frac{dx}{dB} + \frac{\partial X}{\partial y}\frac{dy}{dB} = 0 \qquad \frac{\partial Y}{\partial x}\frac{dx}{dB} + \frac{\partial Y}{\partial y}\frac{dy}{dB} = 1 \qquad (4.29)$$

Solving these simultaneous equations gives us

$$\frac{dx}{dB} = -\frac{\partial X/\partial y}{\Delta} \qquad \frac{dy}{dB} = \frac{\partial X/\partial x}{\Delta}$$

where
$$\Delta = \frac{\partial X}{\partial x}\frac{\partial Y}{\partial y} - \frac{\partial X}{\partial y}\frac{\partial Y}{\partial x} \qquad (4.30)$$

Consequently we have

$$\frac{dr}{dB} = \frac{\partial r}{\partial x}\frac{dx}{dB} + \frac{\partial r}{\partial y}\frac{dy}{dB} = \frac{\frac{1}{r}\left(y\frac{\partial X}{\partial x} - x\frac{\partial X}{\partial y}\right)}{\Delta} \qquad (4.31)$$

Hence, the vertical tangency ($dr/dB \to \infty$) results at the stability limit $\Delta = 0$ of the first unstable region.

Thus far we have dealt with a particular case as specified by the periodic solution (4.20).[1] However, it is mentioned that the same conclusion as obtained in this part may also be valid for more general cases which give rise to different types of oscillations.[2]

(b) Stability Condition (4.6) for $n \geq 2$

We lay stress on this condition. Although it has not been explicitly discussed hitherto, it has an important role in the investigation of the stability of nonlinear periodic oscillations. Stability investigations in nonlinear systems are to be found in many physical and technical journals. In Appendix IV, for example, the one reported by Mandelstam and Papalexi [72] is cited and compared with our stability condition. The analysis reveals that their criterion corresponds to our condition for $n = 1$ and offers no information for $n \geq 2$. In our investigation the generalized stability condition (4.6) for the nth unstable region will furnish the criterion to distinguish the stability for the nth harmonic of the fundamental oscillation, because the solution of the variational equation in the nth unstable region has the form $e^{\mu\tau}[\sin(n\tau - \sigma) + $ higher-order terms in $\theta_1, \theta_2, \ldots]$.

---

[1] A more detailed investigation of the harmonic oscillation will appear in Chaps. 5 and 8.

[2] See Appendix IV for the discussion of subharmonic oscillations.

Our generalized condition will be effective when we investigate the harmonic oscillation in which higher harmonics are predominant, as we shall see in Sec. 6.1. Our condition will also be useful when we investigate the higher-harmonic oscillations (see Sec. 6.3) and the subharmonic oscillations (Secs. 7.4 and 7.6). In studying the subharmonic oscillation of order $\frac{1}{3}$, for example, we frequently encounter the self-excitation of the second harmonic of the subharmonic oscillation, i.e., the oscillation of order $\frac{2}{3}$, which will no longer permit the continuation of the original subharmonic oscillation. Thus, the stability condition for the first unstable region is not sufficient in this case; the instability mentioned above may be detected by putting $n = 2$ in the stability condition (4.6).

(c) Stability Condition (4.9)

The discussion in parts (a) and (b) also applies if necessary alteration for n is made. Since the solution of the variational equation (4.7) has the form $e^{\mu\tau} [\sin (\frac{1}{2}n\tau - \sigma) + \cdots]$ in the nth unstable region, the $n/2$th harmonic of the fundamental oscillation will be excited in this region.

Thus, if we consider the differential equation (4.19), the stability condition (4.9) for $n = 1$ ascertains the stability against buildup of the unstable oscillation having the frequency one-half the driving frequency, and the result obtained in part (a) will be derived by putting $n = 2$ in (4.9). However, the variational equation takes the form (4.7) in such a case that the nonlinear function $f(v)$ in Eq. (4.19) is not odd in v, or that the external force contains a unidirectional component. In such cases the periodic solution usually contains a nonoscillatory term, and consequently it is preferable to resort to the improved stability condition investigated in Sec. 4.3a.

(d) Improved Stability Condition (4.15)

We shall particularly be concerned with this condition, because it is useful in investigating the nonlinear oscillations with unsymmetrical characteristic (Secs. 5.2 and 7.6).

By a procedure analogous to that used in part (a), we consider the differential equation

$$\frac{d^2v}{d\tau^2} + 2\delta \frac{dv}{d\tau} + f(v) = B \cos \tau + B_0 \tag{4.32}$$

where the asymmetry is given by a nonodd function $f(v)$ or by the presence of a unidirectional force B_0. Assuming a periodic solution of the form

$$v = z + x \sin \tau + y \cos \tau \tag{4.33}$$

we obtain the following relations to determine x, y, and z, namely,

$$X(x,y,z) = -x - 2 \, \delta y + \frac{1}{\pi} \int_0^{2\pi} f(v) \sin \tau \, d\tau = 0$$

$$Y(x,y,z) = 2 \, \delta x - y + \frac{1}{\pi} \int_0^{2\pi} f(v) \cos \tau \, d\tau = B \qquad (4.34)$$

$$Z(x,y,z) = \frac{1}{2\pi} \int_0^{2\pi} f(v) \, d\tau = B_0$$

Under such circumstances the variational equation takes the form (4.7) and, in fact, we obtain

$$\frac{d^2\eta}{d\tau^2} + \left(\theta_0 + 2 \sum_{\nu=1}^{\infty} \theta_{\nu s} \sin \nu\tau + 2 \sum_{\nu=1}^{\infty} \theta_{\nu c} \cos \nu\tau \right) \eta = 0 \qquad (4.35)$$

where

$$\theta_0 = \frac{1}{2\pi} \int_0^{2\pi} \frac{df}{dv} \, d\tau - \delta^2 = \frac{\partial Z}{\partial z} - \delta^2$$

$$\theta_{1s} = \frac{1}{2\pi} \int_0^{2\pi} \frac{df}{dv} \sin \tau \, d\tau = \frac{1}{2} \frac{\partial X}{\partial z} = \frac{\partial Z}{\partial x}$$

$$\theta_{1c} = \frac{1}{2\pi} \int_0^{2\pi} \frac{df}{dv} \cos \tau \, d\tau = \frac{1}{2} \frac{\partial Y}{\partial z} = \frac{\partial Z}{\partial y} \qquad (4.35a)$$

$$\theta_{2s} = \frac{1}{2\pi} \int_0^{2\pi} \frac{df}{dv} \sin 2\tau \, d\tau = \frac{\partial X}{\partial y} + 2\delta = \frac{\partial Y}{\partial x} - 2\delta$$

$$\theta_{2c} = \frac{1}{2\pi} \int_0^{2\pi} \frac{df}{dv} \cos 2\tau \, d\tau = \frac{\partial Y}{\partial y} + 1 - \theta_0 - \delta^2 = -\frac{\partial X}{\partial x} - 1 + \theta_0 + \delta^2$$

. . . .

By virtue of these relations, the stability condition (4.15) leads to

$$\Delta \equiv \begin{vmatrix} \dfrac{\partial X}{\partial x} & \dfrac{\partial X}{\partial y} & \dfrac{\partial X}{\partial z} \\[1.2em] \dfrac{\partial Y}{\partial x} & \dfrac{\partial Y}{\partial y} & \dfrac{\partial Y}{\partial z} \\[1.2em] \dfrac{\partial Z}{\partial x} & \dfrac{\partial Z}{\partial y} & \dfrac{\partial Z}{\partial z} \end{vmatrix} > 0 \qquad (4.36)$$

This form of stability condition may be derived from the Routh-Hurwitz criterion as it was in part (a).

We now discuss the vertical tangency of the characteristic curve. Putting $r^2 = x^2 + y^2$, the relationship between B, B_0, and r is readily obtained by solving Eqs. (4.34). By differentiating Eqs. (4.34) with respect to B while holding B_0 constant, we obtain

$$\frac{\partial X}{\partial x}\frac{dx}{dB} + \frac{\partial X}{\partial y}\frac{dy}{dB} + \frac{\partial X}{\partial z}\frac{dz}{dB} = 0$$

$$\frac{\partial Y}{\partial x}\frac{dx}{dB} + \frac{\partial Y}{\partial y}\frac{dy}{dB} + \frac{\partial Y}{\partial z}\frac{dz}{dB} = 1 \qquad (4.37)$$

$$\frac{\partial Z}{\partial x}\frac{dx}{dB} + \frac{\partial Z}{\partial y}\frac{dy}{dB} + \frac{\partial Z}{\partial z}\frac{dz}{dB} = 0$$

Solving these simultaneous equations gives us

$$\frac{dx}{dB} = \frac{\dfrac{\partial X}{\partial z}\dfrac{\partial Z}{\partial y} - \dfrac{\partial X}{\partial y}\dfrac{\partial Z}{\partial z}}{\Delta} \qquad \frac{dy}{dB} = \frac{\dfrac{\partial X}{\partial x}\dfrac{\partial Z}{\partial z} - \dfrac{\partial X}{\partial z}\dfrac{\partial Z}{\partial x}}{\Delta} \qquad (4.38)$$

and consequently,

$$\frac{dr}{dB} = \frac{\partial r}{\partial x}\frac{dx}{dB} + \frac{\partial r}{\partial y}\frac{dy}{dB} = \frac{x}{r}\frac{dx}{dB} + \frac{y}{r}\frac{dy}{dB}$$

$$= \frac{1}{r\Delta}\left[x\left(\frac{\partial X}{\partial z}\frac{\partial Z}{\partial y} - \frac{\partial X}{\partial y}\frac{\partial Z}{\partial z} \right) + y\left(\frac{\partial X}{\partial x}\frac{\partial Z}{\partial z} - \frac{\partial X}{\partial z}\frac{\partial Z}{\partial x} \right) \right] \qquad (4.39)$$

Hence the characteristic curve (Br relation) has a vertical tangent at the stability limit $\Delta = 0$ of the second unstable region.[1]

(e) Improved Stability Condition (4.18)

We once more deal with Eq. (4.19), but assume a periodic solution of the form

$$v = x_1 \sin \tau + y_1 \cos \tau + x_3 \sin 3\tau + y_3 \cos 3\tau \qquad (4.40)$$

Thus the third harmonic is also taken into consideration.   By proceeding as before, we obtain the following relations to determine the unknown coefficients in Eq. (4.40):

$$X_1(x_1,y_1,x_3,y_3) = -x_1 - 2\,\delta y_1 + \frac{1}{\pi}\int_0^{2\pi} f(v)\sin\tau\,d\tau = 0$$

$$Y_1(x_1,y_1,x_3,y_3) = 2\,\delta x_1 - y_1 + \frac{1}{\pi}\int_0^{2\pi} f(v)\cos\tau\,d\tau = B$$

$$\qquad\qquad\qquad\qquad\qquad\qquad\qquad\qquad\qquad\qquad (4.41)$$

$$X_3(x_1,y_1,x_3,y_3) = -9x_3 - 6\,\delta y_3 + \frac{1}{\pi}\int_0^{2\pi} f(v)\sin 3\tau\,d\tau = 0$$

$$Y_3(x_1,y_1,x_3,y_3) = 6\,\delta x_3 - 9y_3 + \frac{1}{\pi}\int_0^{2\pi} f(v)\cos 3\tau\,d\tau = 0$$

The variational equation is given by

$$\frac{d^2\eta}{d\tau^2} + \left(\theta_0 + 2\sum_{\nu=1}^{\infty} \theta_{\nu s}\sin 2\nu\tau + 2\sum_{\nu=1}^{\infty} \theta_{\nu c}\cos 2\nu\tau \right)\eta = 0 \qquad (4.42)$$

[1] Similar conclusion may be obtained in the case where B is constant and B_0 varies.

where

$$\theta_0 = \frac{1}{2\pi} \int_0^{2\pi} \frac{df}{dv} \, d\tau - \delta^2$$

$$\theta_{1s} = \frac{1}{2\pi} \int_0^{2\pi} \frac{df}{dv} \sin 2\tau \, d\tau = \frac{\partial X_1}{\partial y_1} + 2\delta = \frac{\partial Y_1}{\partial x_1} - 2\delta$$

$$= -\frac{1}{2}\left(\frac{\partial X_1}{\partial y_3} - \frac{\partial Y_1}{\partial x_3}\right) = \frac{1}{2}\left(\frac{\partial X_3}{\partial y_1} - \frac{\partial Y_3}{\partial x_1}\right)$$

$$\theta_{1c} = \frac{1}{2\pi} \int_0^{2\pi} \frac{df}{dv} \cos 2\tau \, d\tau = \frac{\partial Y_1}{\partial y_1} + 1 - \theta_0 - \delta^2 = -\frac{\partial X_1}{\partial x_1} - 1$$

$$+ \theta_0 + \delta^2 = \frac{1}{2}\left(\frac{\partial X_1}{\partial x_3} + \frac{\partial Y_1}{\partial y_3}\right) = \frac{1}{2}\left(\frac{\partial X_3}{\partial x_1} + \frac{\partial Y_3}{\partial y_1}\right)$$

$$\theta_{2s} = \frac{1}{2\pi} \int_0^{2\pi} \frac{df}{dv} \sin 4\tau \, d\tau = \frac{1}{2}\left(\frac{\partial X_1}{\partial y_3} + \frac{\partial Y_1}{\partial x_3}\right) = \frac{1}{2}\left(\frac{\partial X_3}{\partial y_1} + \frac{\partial Y_3}{\partial x_1}\right) \qquad (4.42a)$$

$$\theta_{2c} = \frac{1}{2\pi} \int_0^{2\pi} \frac{df}{dv} \cos 4\tau \, d\tau = -\frac{1}{2}\left(\frac{\partial X_1}{\partial x_3} - \frac{\partial Y_1}{\partial y_3}\right)$$

$$= -\frac{1}{2}\left(\frac{\partial X_3}{\partial x_1} - \frac{\partial Y_3}{\partial y_1}\right)$$

$$\theta_{3s} = \frac{1}{2\pi} \int_0^{2\pi} \frac{df}{dv} \sin 6\tau \, d\tau = \frac{\partial X_3}{\partial y_3} + 6\delta = \frac{\partial Y_3}{\partial x_3} - 6\delta$$

$$\theta_{3c} = \frac{1}{2\pi} \int_0^{2\pi} \frac{df}{dv} \cos 6\tau \, d\tau = \frac{\partial Y_3}{\partial y_3} + 9 - \theta_0 - \delta^2$$

$$= -\frac{\partial X_3}{\partial x_3} - 9 + \theta_0 + \delta^2$$

. . . .

We now discuss the vertical tangency of the characteristic curve. Since the approximation of the periodic solution is improved by adding the third harmonic, it will be desirable to apply the improved stability condition (4.18) instead of (4.6). By virtue of Eqs. (4.42a), the stability condition (4.18) leads to

$$\Delta \equiv \begin{vmatrix} \dfrac{\partial X_1}{\partial x_1} & \dfrac{\partial X_1}{\partial y_1} & \dfrac{\partial X_1}{\partial x_3} & \dfrac{\partial X_1}{\partial y_3} \\[2mm] \dfrac{\partial Y_1}{\partial x_1} & \dfrac{\partial Y_1}{\partial y_1} & \dfrac{\partial Y_1}{\partial x_3} & \dfrac{\partial Y_1}{\partial y_3} \\[2mm] \dfrac{\partial X_3}{\partial x_1} & \dfrac{\partial X_3}{\partial y_1} & \dfrac{\partial X_3}{\partial x_3} & \dfrac{\partial X_3}{\partial y_3} \\[2mm] \dfrac{\partial Y_3}{\partial x_1} & \dfrac{\partial Y_3}{\partial y_1} & \dfrac{\partial Y_3}{\partial x_3} & \dfrac{\partial Y_3}{\partial y_3} \end{vmatrix} > 0 \qquad (4.43)$$

Differentiating Eqs. (4.41) with respect to B yields

$$\frac{\partial X_1}{\partial x_1}\frac{dx_1}{dB} + \frac{\partial X_1}{\partial y_1}\frac{dy_1}{dB} + \frac{\partial X_1}{\partial x_3}\frac{dx_3}{dB} + \frac{\partial X_1}{\partial y_3}\frac{dy_3}{dB} = 0$$

$$\frac{\partial Y_1}{\partial x_1}\frac{dx_1}{dB} + \frac{\partial Y_1}{\partial y_1}\frac{dy_1}{dB} + \frac{\partial Y_1}{\partial x_3}\frac{dx_3}{dB} + \frac{\partial Y_1}{\partial y_3}\frac{dy_3}{dB} = 1$$

$$\frac{\partial X_3}{\partial x_1}\frac{dx_1}{dB} + \frac{\partial X_3}{\partial y_1}\frac{dy_1}{dB} + \frac{\partial X_3}{\partial x_3}\frac{dx_3}{dB} + \frac{\partial X_3}{\partial y_3}\frac{dy_3}{dB} = 0$$

$$\frac{\partial Y_3}{\partial x_1}\frac{dx_1}{dB} + \frac{\partial Y_3}{\partial y_1}\frac{dy_1}{dB} + \frac{\partial Y_3}{\partial x_3}\frac{dx_3}{dB} + \frac{\partial Y_3}{\partial y_3}\frac{dy_3}{dB} = 0$$

(4.44)

Solving these simultaneous equations gives us

$$\frac{dx_1}{dB} = \frac{\Delta_{21}}{\Delta} \qquad \frac{dy_1}{dB} = \frac{\Delta_{22}}{\Delta} \qquad \frac{dx_3}{dB} = \frac{\Delta_{23}}{\Delta} \qquad \frac{dy_3}{dB} = \frac{\Delta_{24}}{\Delta} \qquad (4.45)$$

where Δ_{2i} ($i = 1 \sim 4$) is the cofactor of row 2 and column i of the determinant Δ. Consequently,

$$\frac{dr_1}{dB} = \frac{1}{r_1\Delta}(x_1\Delta_{21} + y_1\Delta_{22}) \qquad \frac{dr_3}{dB} = \frac{1}{r_3\Delta}(x_3\Delta_{23} + y_3\Delta_{24}) \qquad (4.46)$$

where $\qquad\qquad r_1{}^2 = x_1{}^2 + y_1{}^2 \qquad r_3{}^2 = x_3{}^2 + y_3{}^2$

Hence the vertical tangency of the characteristic curves (Br_1 and Br_3 relations) occurs at the stability limit $\Delta = 0$ of the first and the third unstable regions.

Chapter 5

Harmonic Oscillations

5.1 Harmonic Oscillations with Symmetrical Nonlinear Characteristic

In this chapter we deal with the harmonic oscillations [19–21, 94] in which the fundamental component having a period the same as that of the external force is predominant, so that the higher harmonics may be neglected. In order to get the physical idea concretely, we consider an electric oscillatory circuit and derive the nonlinear differential equation of the form (4.10) [31, 33].

(a) Fundamental Equation

The schematic diagram in Fig. 5.1 shows an electric circuit in which the nonlinear oscillation takes place owing to the presence of a saturable-core coil of inductance L under the impression of an alternating voltage $E \sin \omega t$. As shown in the figure, the resistor R is paralleled with the capacitor C, so that the circuit is dissipative. With the notation of Fig. 5.1, we have

$$n \frac{d\phi}{dt} + R i_R = E \sin \omega t \qquad R i_R = \frac{1}{C} \int i_C \, dt \qquad i = i_R + i_C \quad (5.1)$$

where n is the number of turns of the inductor coil and ϕ is the magnetic flux in the core. Let the saturation curve of the core be expressed by

$$ni = a_1 \phi + a_3 \phi^3 + a_5 \phi^5 + \cdots \qquad (5.2)$$

where $a_1, a_3, a_5, \ldots$ are constants characterizing the core. It is noted that the effect of hysteresis is neglected in Eq. (5.2). In order to take

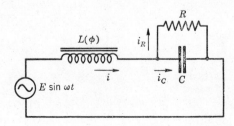

FIGURE 5.1   Oscillatory circuit with nonlinear inductor.

care of the analysis when the frequency ω varies, we put $\omega = \nu\,\omega_0$, where ω_0 is a fixed frequency and ν is a variable parameter. Furthermore, we introduce nondimensional variables u and v, defined by

$$i = I_n u \qquad \text{and} \qquad \phi = \Phi_n v \tag{5.3}$$

where I_n and Φ_n are appropriate base quantities of the current and the flux, respectively. Then Eq. (5.2) may be rewritten as

$$
\begin{aligned}
u &= \frac{1}{nI_n}\,(a_1\Phi_n v + a_3\Phi_n{}^3 v^3 + a_5\Phi_n{}^5 v^5 + \cdots) \\
&= c_1 v + c_3 v^3 + c_5 v^5 + \cdots
\end{aligned}
\tag{5.4}
$$

Although the base quantities I_n and Φ_n can be chosen quite arbitrarily, it is preferable, for the brevity of calculation, to fix them by the relations

$$n\omega_0{}^2 C\Phi_n = I_n \qquad c_1 + c_3 + c_5 + \cdots = 1 \tag{5.5}$$

Then, after retaining i and ϕ and eliminating i_R and i_C in Eqs. (5.1) and further using Eqs. (5.3) to (5.5), we obtain

$$\frac{d^2v}{d\tau^2} + k\frac{dv}{d\tau} + c_1 v + c_3 v^3 + c_5 v^5 + \cdots = B \cos \nu\tau$$

where
$$\tau = \omega_0 t - \frac{1}{\nu}\tan^{-1}\frac{k}{\nu} \qquad k = \frac{1}{\omega_0 C R} \tag{5.6}$$

$$B = \frac{E}{n\omega_0\Phi_n}\sqrt{\nu^2 + k^2}$$

This takes the form of Eq. (4.10), in which the nonlinear function $f(v)$ is symmetrical, i.e., an odd function of v. A similar equation may also be obtained for a mechanical system in which the restoring force is nonlinear [21, p. 78].

(b) Periodic States of Equilibrium

In order to facilitate the computation, we consider the following simple but typical case of Eq. (5.4), that is,

$$c_5 = c_7 = \cdots = 0$$
so that
$$u = c_1 v + c_3 v^3 * \tag{5.7}$$

In the case of harmonic oscillations in which the fundamental component having the period $2\pi/\nu$ predominates over the higher harmonics, the

---

* The saturation curve, defined by Eq. (5.4), of an iron core usually has terms $c_n v^n$, in which n extends to 9 or even higher. However, as far as the harmonic oscillation is concerned, neglect of such higher powers of v results in no significant change in the following analysis. It is possible to realize the cubic characteristic by imposing an appropriate form of air gap in the core or by making use of a composite core (Sec. 7.5).

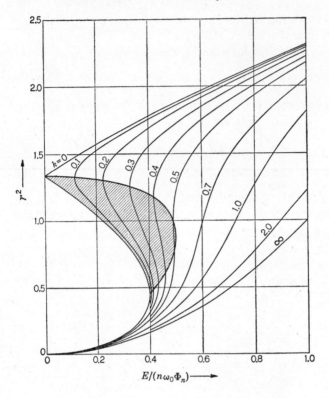

FIGURE 5.2   Amplitude characteristic of the harmonic
oscillation.

periodic solution of Eq. (5.6) may be assumed by the form[1]

$$v_0 = x \sin \nu\tau + y \cos \nu\tau \tag{5.8}$$

Substituting Eq. (5.8) into Eq. (5.6), in which the nonlinearity is
given by Eq. (5.7), and equating the coefficients of the terms containing
$\sin \nu\tau$ and $\cos \nu\tau$ separately to zero yields

$$Ax + \nu ky = 0 \qquad \nu kx - Ay = B$$

where $\qquad A = \nu^2 - c_1 - \tfrac{3}{4}c_3 r^2 \qquad r^2 = x^2 + y^2$ $\qquad$ (5.9)

or eliminating x and y in Eqs. (5.9) gives us

$$(A^2 + \nu^2 k^2)r^2 = B^2 \tag{5.10}$$

[1] When B is large in Eqs. (5.6), we have to take into account terms pertaining to
higher harmonics, and we shall do so in the following chapter.

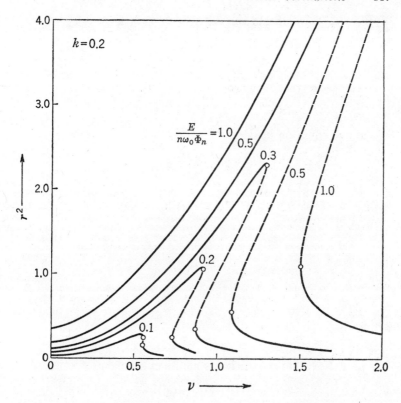

FIGURE 5.3   Frequency response curves of the harmonic oscillation.

This relation is rather simple, yet it shows a good agreement with the experimental result [31, vol. 29, p. 601; 94, p. 728]. Figure 5.2 shows an amplitude characteristic of the harmonic oscillation, i.e., the relationship between $E/(n\omega_0 \Phi_n)$ and r^2, in the case where $c_1 = 0.1$, $c_3 = 0.9$, and $\nu = 1$. Also plotted in Fig. 5.3 is a frequency response of the harmonic oscillation, i.e., the relationship between ν and r^2 in the case where $c_1 = 0.1$, $c_3 = 0.9$, and $k = 0.2$.

(c) Stability Condition of the Periodic Solutions

The periodic states of equilibrium determined by Eqs. (5.9) or (5.10) are not always realized, but are actually able to exist only so long as they are stable. Following the analysis of the preceding chapter, we shall investigate the stability of the equilibrium states and find the periodic solutions which are sustained in the stable state. To do this, we consider the small variation ξ from the equilibrium states and substitute $v_0 + \xi$ in place of v in Eqs. (5.6). Then, remembering the nonlinearity as given by Eq. (5.7), we obtain the variational equation

$$\frac{d^2\xi}{d\tau^2} + k\frac{d\xi}{d\tau} + c_1\xi + 3c_3v_0^2\xi = 0$$

By use of the transformation

$$\xi = e^{-\frac{1}{2}k\tau}\eta$$

the above equation leads to

$$\frac{d^2\eta}{d\tau^2} + \left(c_1 - \frac{1}{4}k^2 + 3c_3v_0^2\right)\eta = 0$$

Now, by substituting the periodic solution (5.8), we ultimately obtain Mathieu's equation

$$\frac{d^2\eta}{d\tau^2} + [\theta_0 + 2\theta_1 \cos(2\nu\tau - \epsilon)]\eta = 0$$

where   (5.11)

$$\theta_0 = c_1 - \frac{1}{4}k^2 + \frac{3}{2}c_3r^2 \qquad \theta_1 = \frac{3}{4}c_3r^2 \qquad \epsilon = \tan^{-1}\frac{2xy}{y^2 - x^2}$$

The stability condition for the first unstable region is given by inequality (4.6) with $n = 1$. Since the periodic coefficient of η in Eqs. (5.11) has period π/ν, this condition is modified to

$$(\theta_0 - \nu^2)^2 + \frac{1}{2}(\theta_0 + \nu^2)k^2 + \frac{1}{16}k^4 > \theta_1^2$$

Substituting the values of θ_0 and θ_1 of Eqs. (5.11) gives us

$$A^2 - \frac{3}{2}c_3Ar^2 + \nu^2k^2 > 0 \tag{5.12}$$

We shall consider what this condition means physically. First we assume that the amplitude E varies while the frequency ν is held constant. Then, by virtue of Eq. (5.10), we obtain

$$\frac{dB^2}{dr^2} = A^2 - \frac{3}{2}c_3Ar^2 + \nu^2k^2$$

Therefore, the stability condition may be rewritten as

$$\frac{dB^2}{dr^2} > 0 \qquad \text{or} \qquad \frac{dE^2}{dr^2} > 0 \tag{5.13}$$

This shows that the periodic solution is stable under such conditions that the amplitude r increases with the increase of the amplitude B. From the physical point of view, this is a plausible conclusion. In Fig. 5.2 the region of unstable solutions is represented by hatching where dB^2/dr^2 is negative. It is also noted that the boundary curve between the stable and the unstable regions is given by the condition

$$\frac{dB^2}{dr^2} = 0$$

which shows the vertical tangency of the characteristic curves at the stability limit (Sec. 4.4a).

Thus we see that there are three kinds of equilibrium state under certain values of k and E. Two of them are stable, and the remaining one is unstable. As mentioned in Sec. 4.4a, this unstable state is, with increasing time, transferred to one of the two stable states, owing to the buildup of the variation having the form $e^{\lambda\tau}[\sin(\nu\tau - \sigma) + \cdots]$, where the real part of λ is positive. In order to distinguish between the two stable states, we shall refer to the one with larger amplitude as the resonant state and to the other with smaller amplitude as the nonresonant state. Now, starting from the origin of Fig. 5.2, the amplitude r increases slowly with the increase of E (k being held constant); thus the oscillation is in the nonresonant state. When the equilibrium state comes to the boundary of the hatched region, a slight increase in E will cause a discontinuous jump to the resonant state with accompanying increase in the amplitude r, after which r increases rather slowly with E. Upon a reversal of the process, i.e., with decreasing E, the oscillation jumps down from the resonant to the nonresonant state.[1] This transition takes place at a lower value of E than before, thus exhibiting a hysteresis phenomenon. If we choose E between these two critical values and apply a voltage of this amplitude to the circuit, either the resonant or the nonresonant oscillation results. Which of the two does result depends upon the prescribed initial condition.[2]

Second, we consider the case in which the frequency ν varies while the amplitude E is held constant. Then we obtain

$$\frac{d\nu^2}{dr^2} = \frac{A^2 - \frac{3}{2}c_3 A r^2 + \nu^2 k^2}{-2Ar^2 - k^2 r^2 + (E/n\omega_0 \Phi_n)^2} \tag{5.14}$$

Hence, as before, the vertical tangency of the response curves results at the stability limit. In Fig. 5.3 the dashed parts of the response curves are unstable, and hysteresis occurs here also as ν varies.

Thus far we have discussed the stability condition for $n = 1$. Generally speaking, the conditions for $n \geq 2$ must be considered also. But, as will be discussed in Sec. 6.1, this comes into question only when θ_0 is very large. Under this condition, r^2 and B also become large, so that higher

---

[1] Since this is a resonance phenomenon associated with a saturable iron core, the jump phenomenon is sometimes given the name *ferroresonance* [90, 94].

[2] This problem will be discussed in Part III.

harmonics must be added to the solution of Eq. (5.8).   A detailed investigation of the higher-harmonic oscillation is deferred to Chap. 6.   Here we mention that the condition for $n = 1$ alone will be sufficient to ascertain the stability in the case when B is not so large.

5.2 *Harmonic Oscillations with Unsymmetrical Nonlinear Characteristic*

(a) *Fundamental Equation*

When terms of v^2, v^4, . . . are present in the saturation curve (5.4) of the iron core, the nonlinear characteristic becomes unsymmetrical, and the resulting oscillation presents some aspects different from the symmetrical case.   We consider the following case:

$$u = c_1 v + c_2 v^2 + c_3 v^3 \tag{5.15}$$

Then Eq. (5.6) is modified to[1]

$$\frac{d^2 v}{d\tau^2} + k \frac{dv}{d\tau} + c_1 v + c_2 v^2 + c_3 v^3 = B \cos \tau \tag{5.16}$$

This equation is readily transformed to the alternative form

$$\frac{d^2 v'}{d\tau^2} + k \frac{dv'}{d\tau} + c_1' v' + c_3' v'^3 = B \cos \tau + B_0 \tag{5.17}$$

where $v' = v + c_2/(3c_3)$ and c_1', c_3', B_0 are constants determined from c_1, c_2, and c_3.   Thus we see in Eq. (5.17) that the nonlinear characteristic is symmetrical, but the external force is unsymmetrical, since it contains the unidirectional component B_0.

We can actually realize an electrical system the circuit equation of which takes the form (5.16) or (5.17).   The schematic diagrams of such systems are illustrated in Fig. 5.4a and b, respectively.   The saturable reactor in Fig. 5.4a has a secondary coil through which the biasing direct current i_0 flows, so that the saturation curve takes the form (5.15).   In Fig. 5.4b the d-c voltage E_0 is superposed on the alternating voltage $E \sin \omega t$, and so Eq. (5.17) results.

We shall, in what follows, consider the system of Fig. 5.4b and assume the fundamental equation[2]

$$\frac{d^2 v}{d\tau^2} + k \frac{dv}{d\tau} + v^3 = B \cos \tau + B_0 \tag{5.18}$$

---

[1] We consider only the case in which $\nu = 1$.

[2] When the iron core has a large air gap, a linear term in v must be taken into consideration.   This additional term, however, results in no significant change in the analysis to follow.

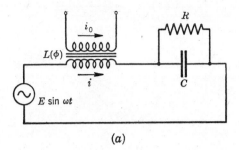

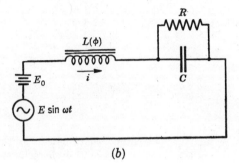

FIGURE 5.4   Oscillatory circuits with
unsymmetrical charac-
teristic.

(b)

(b) Periodic States of Equilibrium

When the system is unsymmetrical, the flux in the iron core may generally have a unidirectional component; therefore, we assume for v

$$v_0 = z + x \sin \tau + y \cos \tau \tag{5.19}$$

Substituting Eq. (5.19) into (5.18) and equating the constant term and the coefficients of the terms containing $\sin \tau$ and $\cos \tau$ separately to zero yields

$$\tfrac{3}{2}r^2 z + z^3 = B_0 \qquad Ax + ky = 0 \qquad kx - Ay = B$$

where
$$A = 1 - \tfrac{3}{4}r^2 - 3z^2 \qquad r^2 = x^2 + y^2 \tag{5.20}$$

or eliminating x and y from the second and the third equations of (5.20) yields

$$(A^2 + k^2)r^2 = B^2 \tag{5.21}$$

The periodic states of equilibrium will be determined by solving Eqs. (5.20). Figure 5.5 shows the amplitude y of the harmonic oscillation for the nondissipative case, i.e., for the case in which $k = 0$ (x being zero in this case). Further, in Fig. 5.6, the relationship between $|B|$ and r^2 is illustrated for $B_0 = 0.36$. We see that there are five kinds of equilibrium state under certain values of $|B|$ and k.

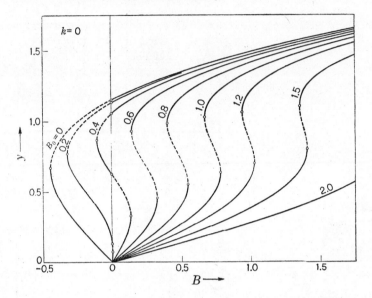

FIGURE 5.5   Amplitude characteristic of the harmonic oscillation
with unsymmetrical characteristic (nondissipative case).

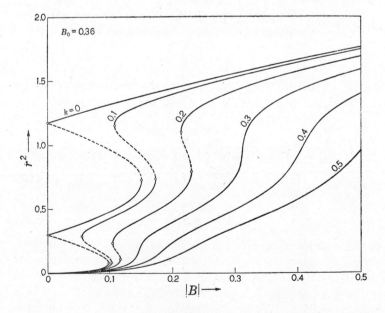

FIGURE 5.6   Amplitude characteristic of the harmonic oscillation with
unsymmetrical characteristic (dissipative case).

(c) Stability Condition of the Periodic Solutions

We shall now determine the stability of the equilibrium states. By a procedure analogous to that for the symmetrical case in Sec. 5.1c, the variation ξ from the equilibrium states satisfies the equation

$$\frac{d^2\xi}{d\tau^2} + k\frac{d\xi}{d\tau} + 3v_0^2\xi = 0$$

By using the transformation

$$\xi = e^{-\frac{1}{2}k\tau}\eta$$

and substituting the periodic solution (5.19), we obtain the variational equation of Hill's type, namely,

$$\frac{d^2\eta}{d\tau^2} + \left[\theta_0 + 2\sum_{n=1}^{2}\theta_n\cos(n\tau - \epsilon_n)\right]\eta = 0$$

where
$$\theta_0 = \tfrac{3}{2}r^2 + 3z^2 - \tfrac{1}{4}k^2$$

$$\theta_n^2 = \theta_{ns}^2 + \theta_{nc}^2 \qquad \epsilon_n = \tan^{-1}\frac{\theta_{ns}}{\theta_{nc}}$$

$$\theta_{1s} = 3xz \qquad\qquad \theta_{1c} = 3yz$$

$$\theta_{2s} = \tfrac{3}{2}xy \qquad\qquad \theta_{2c} = \tfrac{3}{4}(y^2 - x^2)$$

$$(5.22)$$

This takes the form (4.7), and so the stability condition will be given by inequality (4.9). We shall particularly investigate the stability condition for the second unstable region in order to test the stability against buildup of an unstable oscillation having a frequency the same as that of the periodic solution (5.19). We are, however, concerned with the unsymmetrical system and a consideration of the constant term z in Eq. (5.19); therefore, it will be preferable to use the improved stability condition (4.15) instead of (4.9). Thus we write

$$4(\delta^2 - \mu^2) = (\theta_0 + \delta^2 - 1)^2 - \theta_2^2 - 2\frac{\theta_0 + \delta^2 - 1}{\theta_0 + \delta^2}\theta_1^2$$

$$+ \frac{2}{\theta_0 + \delta^2}[(\theta_{1c}^2 - \theta_{1s}^2)\theta_{2c} + 2\theta_{1c}\theta_{1s}\theta_{2s}] + 4\delta^2 > 0 \quad (5.23)$$

where $\delta = k/2$. Substituting the parameters as given by Eqs. (5.22) gives us

$$\left(\frac{3}{4}r^2 - A\right)^2 - \frac{9}{16}r^4 + k^2 + \frac{6Ar^2z^2}{r^2/2 + z^2} > 0 \qquad (5.24)$$

or, by virtue of Eqs. (5.20) and (5.21), this condition leads to[1]

$$\frac{dB^2}{dr^2} > 0 \qquad (5.25)$$

---

[1] If we use the stability condition of (4.9) instead of (4.15), the last term in the left side of (5.24) is dropped and the condition (5.25) of vertical tangency at the stability limit does not result.

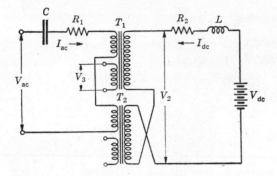

FIGURE 5.7   Oscillatory circuit containing reactors with direct current superposed.

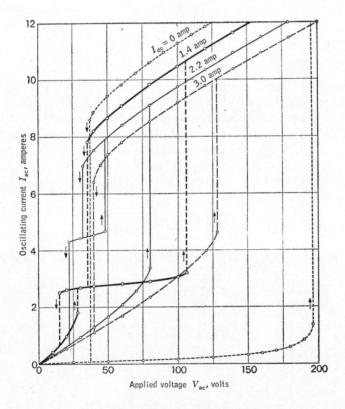

FIGURE 5.8   Amplitude characteristic of the harmonic oscillation with unsymmetrical characteristic.

From these stability conditions one sees that the curves shown dashed in Figs. 5.5 and 5.6 represent the unstable states.

We have also to deal with the stability condition in the other unstable regions. However, as far as we are concerned with the range of B as shown in Figs. 5.5 and 5.6, no significant result is obtained; therefore, further discussion is omitted here. Thus in the end we may conclude that, when the system is unsymmetrical as described by Eq. (5.18), three kinds of periodic oscillation are sustained in a certain range of B.

5.3 *Experimental Investigation*

In this section we show some experimental results in order to compare them with the foregoing analysis. When a secondary winding is provided on the saturable core as illustrated in Fig. 5.4, a pulsating current flows through it owing to induction from the primary oscillating current. To suppress this a-c component in the secondary circuit, we resort to the circuit configuration shown in Fig. 5.7. Two reactors are used; in them the secondary windings are so connected that the a-c voltages induced in the windings counteract each other. However, as the degree of saturation is raised, the second-harmonic current begins to flow in the secondary circuit (as will be shown in the oscillograms of Fig. 5.9); therefore, a coil of high inductance L is inserted to minimize this pulsation.

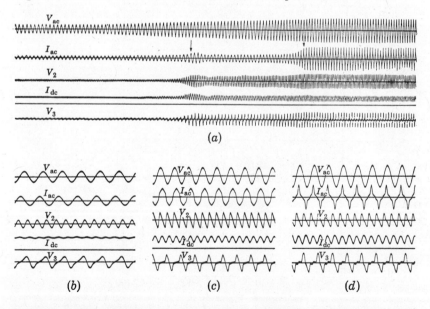

FIGURE 5.9   Waveforms of the voltages and currents in a system with unsymmetrical characteristic. (*a*) Oscillation with increasing V_{ac}. (*b*) Nonresonant state. (*c*) Subresonant state. (*d*) Resonant state.

Referring to the notation in Fig. 5.7, we show, in Fig. 5.8, the relationship between V_{ac} and I_{ac} for several values of I_{dc}. The thick-line curve in the figure is a characteristic for $I_{dc} = 1.4$ amp. In this case, as the voltage V_{ac} is increased, the oscillating current I_{ac} jumps twice with accompanying increase in the magnitude. When the experiment is performed in the reverse direction, i.e., with decreasing V_{ac}, it is found that the same curve is not retraced completely, but each jump from the higher to the lower current takes place at a lower value of V_{ac} than before. Thus we see that between the resonant and the nonresonant states there exists the third stable state which we shall refer to as the subresonant state. This agrees with the result of the preceding analysis.

The waveforms of the voltage and the current in these three states are shown in Fig. 5.9. Oscillogram (*a*) is taken for the case in which the applied voltage V_{ac} is gradually raised. Two jumps of the current I_{ac} are marked by the arrows; the first jump indicates the transition from the

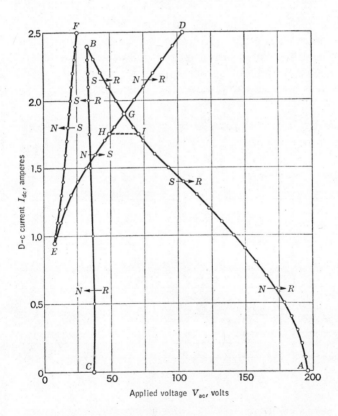

FIGURE 5.10   Regions of different types of oscillation. N = nonresonant, S = subresonant, R = resonant.

nonresonant to the subresonant state and the second from the subresonant to the resonant state. Oscillograms (*b*) to (*d*) show the waveforms in these states more clearly. We see that the current I_{ac} is first leading in the nonresonant state, then lagging in the subresonant state, and again leading in the resonant state. We also see that, as V_{ac} is raised, the second harmonic current increases predominantly in the secondary circuit.

The regions in which these three states obtain are determined by experiment; they are shown in Fig. 5.10. The arrows in the figure show the transition between these states which occurs when V_{ac} is changed in the direction of the arrows. Thus, taking the case of $I_{dc} = 1.5$ amp, for example, the oscillation is transferred from the nonresonant to the sub-resonant state when V_{ac} is raised up to the boundary DE, and it is further transferred from the subresonant to the resonant state when V_{ac} crosses the boundary AB. We mention that when V_{ac} is increased and made to cross the portion GH of the boundary DE, the oscillation jumps directly from the nonresonant to the resonant state. The boundary GI is determined by holding V_{ac} constant and gradually increasing I_{dc} from a lower value for which the oscillation is first in the subresonant state. We see in the end that the subresonant state may take place provided that the direct current I_{dc} is chosen at a value between B and E.

Chapter 6

Higher-harmonic Oscillations

6.1 Higher-harmonic Oscillations in Series-Resonance Circuits

In the preceding chapter the stability of harmonic oscillations was investigated by making use of variational equations of the Hill type. As mentioned in Sec. 5.1c, the stability condition (4.6) for $n \geq 2$ must be considered when the amplitude B of the external force is very large. Oscillations under this condition are worthy of consideration, since anomalous excitation of higher-harmonic components [31, vol. 29, p. 670; 94, p. 778] results if the above stability condition is not satisfied. Since very few investigations have been made in this field, this section will be concerned with such oscillations.

Referring to the electric circuit of Fig. 5.1,[1] we consider the equation

$$\frac{d^2v}{d\tau^2} + k\frac{dv}{d\tau} + v^3 = B \cos \tau \qquad (6.1)$$

Since we are particularly concerned with the case in which B is large, a periodic solution for Eq. (6.1) may preferably be chosen as

$$v_0 = x_1 \sin \tau + y_1 \cos \tau + x_3 \sin 3\tau + y_3 \cos 3\tau \qquad (6.2)$$

Terms of harmonics higher than the third are certain to be present but are ignored to this order of approximation. Substituting Eq. (6.2) into (6.1) and equating the coefficients of the terms containing $\sin \tau$, $\cos \tau$, $\sin 3\tau$, and $\cos 3\tau$ separately to zero yields

$$
\begin{aligned}
A_1x_1 + ky_1 + \tfrac{3}{4}(x_1{}^2 - y_1{}^2)x_3 + \tfrac{3}{2}x_1y_1y_3 &= 0 \\
kx_1 - A_1y_1 - \tfrac{3}{4}(x_1{}^2 - y_1{}^2)y_3 + \tfrac{3}{2}x_1y_1x_3 &= B \\
A_3x_3 + 3ky_3 + \tfrac{1}{4}x_1(x_1{}^2 - 3y_1{}^2) &= 0 \\
3kx_3 - A_3y_3 - \tfrac{1}{4}y_1(3x_1{}^2 - y_1{}^2) &= 0
\end{aligned}
\qquad (6.3)
$$

where $\quad A_1 = 1 - \tfrac{3}{4}(r_1{}^2 + 2r_3{}^2) \qquad A_3 = 9 - \tfrac{3}{4}(2r_1{}^2 + r_3{}^2)$
$\qquad r_1{}^2 = x_1{}^2 + y_1{}^2 \qquad\qquad\qquad r_3{}^2 = x_3{}^2 + y_3{}^2$

---

[1] In contrast with parallel-resonance circuits (which will be treated in Sec. 6.3), we refer to such circuits as series-resonance circuits.

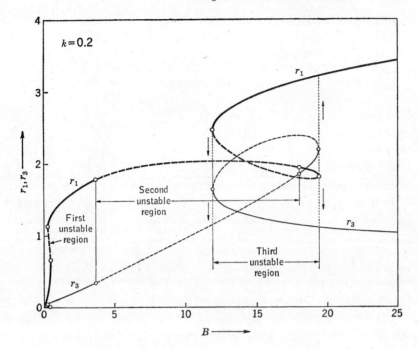

FIGURE 6.1 Amplitude characteristic of the oscillations in a series-resonance circuit.

or eliminating x and y components in the above equations gives us

$$\left[\left(A_1 - \frac{3A_3 r_3^2}{r_1^2}\right)^2 + k^2\left(1 + \frac{9r_3^2}{r_1^2}\right)^2\right]r_1^2 = B^2$$

$$(A_3^2 + 9k^2)r_3^2 = \frac{1}{16}r_1^6 \tag{6.4}$$

Figure 6.1 is obtained by plotting Eqs. (6.4) for $k = 0.2$.

We now determine the stability of the equilibrium states as illustrated in Fig. 6.1. By a procedure like that used before, the variational equation becomes

$$\frac{d^2\xi}{d\tau^2} + k\frac{d\xi}{d\tau} + 3v_0^2\xi = 0$$

By using the transformation

$$\xi = e^{-\frac{1}{2}k\tau}\eta$$

and substituting the periodic solution (6.2), we obtain the variational equation of the Hill type, namely,

$$\frac{d^2\eta}{d\tau^2} + \left[\theta_0 + 2\sum_{n=1}^{3} \theta_n \cos(2n\tau - \epsilon_n)\right]\eta = 0$$

where
$$\theta_0 = \tfrac{3}{2}(r_1^2 + r_3^2) - \tfrac{1}{4}k^2$$

$$\theta_n^2 = \theta_{ns}^2 + \theta_{nc}^2 \qquad\qquad \epsilon_n = \tan^{-1}\frac{\theta_{ns}}{\theta_{nc}} \tag{6.5}$$

$$\theta_{1s} = \tfrac{3}{2}(x_1y_1 - x_1y_3 + x_3y_1) \qquad \theta_{1c} = -\tfrac{3}{4}(x_1^2 - y_1^2) \\ + \tfrac{3}{2}(x_1x_3 + y_1y_3)$$

$$\theta_{2s} = \tfrac{3}{2}(x_1y_3 + x_3y_1) \qquad\qquad \theta_{2c} = \tfrac{3}{2}(-x_1x_3 + y_1y_3)$$

$$\theta_{3s} = \tfrac{3}{2}x_3y_3 \qquad\qquad\qquad \theta_{3c} = -\tfrac{3}{4}(x_3^2 - y_3^2)$$

The stability condition (to a first approximation) is obtained by putting $2\delta = k\ (= 0.2)$ in Eq. (4.6), that is,

$$(\theta_0 - n^2)^2 + 2(\theta_0 + n^2)\delta^2 + \delta^4 > \theta_n^2 \qquad n = 1, 2, 3 \tag{6.6}$$

Consequently, if the parameters θ calculated by Eqs. (6.5) do not satisfy the condition (6.6), the periodic solution (6.2) is unstable and does not exist actually. We turn to a geometrical discussion of the behavior of the θ's on the stability chart for Hill's equation. The boundary curves of the stable and unstable regions are given by

$$(\theta_0 - n^2)^2 + 2(\theta_0 + n^2)\delta^2 + \delta^4 = \theta_n^2 \qquad \text{with } \delta = 0.1 \tag{6.7}$$

These boundary curves are drawn by fine lines in Fig. 6.2. One sees from Eqs. (6.3) and (6.5) that, for a given value of B, a point (θ_0, θ_n) will be

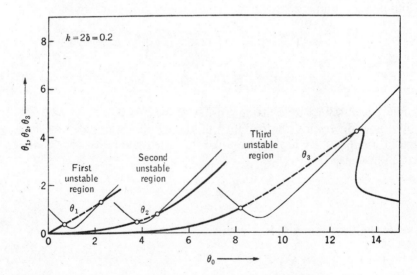

FIGURE 6.2   Loci of θ's in Eqs. (6.5) with varying B.

fixed on the stability chart.   Therefore, as B varies, the point (θ_0, θ_n) will move along the loci of θ_n as illustrated by thick lines in the figure.   Hence, by virtue of the stability condition (6.6), a periodic solution becomes unstable if the point (θ_0, θ_n) lies in the nth unstable region.   The unstable portions are marked by dashed lines in Fig. 6.2.   We shall now consider each unstable region in detail.

CASE 1.   WHEN THE POINT (θ_0, θ_1) LIES IN THE FIRST UNSTABLE REGION

In this case the stability condition (6.6) fails for $n = 1$.   This kind of instability has been discussed in Sec. 5.1c.   In the present section, however, the third harmonic is taken into consideration; hence the periodic solution (6.2) affords a better approximation than that in Sec. 5.1. Therefore it is preferable to make use of a stability condition of the same-order approximation as used for obtaining the periodic solution.   It was mentioned in Sec. 4.4e that, if the improved stability condition (4.18) is used instead of (6.6), the vertical tangency of the characteristic curves results at the stability limit of the first unstable region.   The unstable states determined in this way are indicated by dashed lines in Fig. 6.1.

CASE 2.   WHEN THE POINT (θ_0, θ_2) LIES IN THE SECOND UNSTABLE REGION

The stability condition (6.6) fails for $n = 2$.   We see in Fig. 6.2 that the locus of θ_2 enters into the second unstable region in the neighborhood of $\theta_0 = 4$.   An improved stability condition for the second unstable region was given by (4.12).   By making use of this condition, the unstable portions of the characteristic curves are shown dashed in Fig. 6.1.   As explained in Sec. 4.4b, an unstable oscillation of even harmonics builds up in the second unstable region.[1]

CASE 3.   WHEN THE POINT (θ_0, θ_3) LIES IN THE THIRD UNSTABLE REGION

The stability condition (6.6) is not satisfied for $n = 3$.   Concerning the vertical tangency of the characteristic curves the same remarks as in the first case $(n = 1)$ apply to this case.   Thus, as one sees in Fig. 6.1, a jump phenomenon in the amplitudes r_1 and r_3 occurs as B varies.

To summarize the above consideration, we mention that, when the amplitude B of the external force is very large, the periodic solution (6.2) is featured by the occurrence of even harmonics in the second unstable region and by the anomalous behavior of odd harmonics in the third unstable region.[2]   It may be conjectured that the occurrence of such harmonics will be repeated as B increases further.

[1] In order to find the ultimate amplitudes of even harmonics, one must assume these components in the periodic solution (6.2).

[2] Analog-computer solutions of Eq. (6.1) confirm this conclusion.   See also an experimental result in the following section.

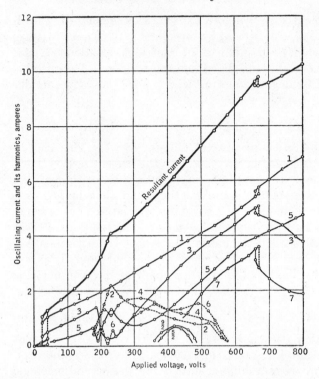

FIGURE 6.3   Oscillating current and its harmonic components in the case when the iron core is highly
saturated.

6.2 Experimental Investigation

In comparison with Fig. 5.2, Fig. 6.1 shows that the anomalous
phenomena occur when B is very large.   Since B is proportional to the
external force, we shall perform an experiment for the series-resonance
circuit (as illustrated in Fig. 5.1) in which the applied voltage is raised
exceedingly.

In Fig. 6.3 the effective value of the oscillating current is plotted (in
thick line) for a wide range of the applied voltage.   By making use of a
heterodyne harmonic analyzer, this current is analyzed into harmonic
components.   These are shown by fine lines, the numbers on which
indicate the order of the harmonics.   The first unstable region ranges
between 24 and 40 volts of the applied voltage, and the jump phenomenon
in this region is what is called the ferroresonance (Sec. 5.1).   The second
unstable region extends from 180 to 580 volts.   As would be expected
from the preceding analysis, the concurrence of even harmonics is a
salient feature of this region.   The third unstable region occurs between

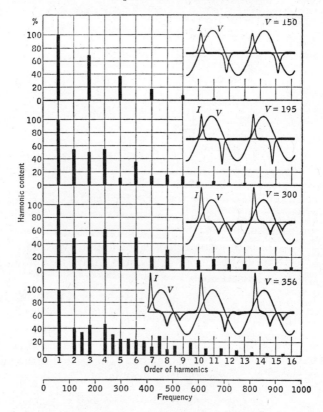

FIGURE 6.4   Typical waveforms of the oscillating currents
and their harmonic analysis.

660 and 670 volts, and the oscillation jumps into another stable state.
Typical waveforms of the oscillation are illustrated in Fig. 6.4, together
with the percentage content of each harmonic.

Since the series capacitor of Fig. 5.1 limits the current which causes
the saturation of the iron-cored coil, the applied voltage must be raised
exceedingly in order to bring the oscillation into the unstable regions of
higher orders.   Therefore, we may expect that a higher-harmonic oscillation is apt to occur in a parallel-resonance circuit; therefore, this will be
studied in the following section.

6.3 Higher-harmonic Oscillations in Parallel-Resonance Circuits
[*31, Vol. 30, p. 418*]

Figure 6.5 shows an electric circuit diagram in which two parallel-
resonance circuits are connected in series.   It is assumed that the circuit

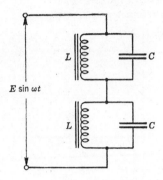

FIGURE 6.5   Parallel-resonance circuit with non-linear inductance.

constants L and C are the same for both circuits.   Under the impression of an alternating voltage, a single oscillation circuit would not present any anomalous phenomenon, because the terminal voltages across L and C are both made equal to the applied voltage.

Denoting the currents in L and C by i_L and i_C, respectively, we have the circuit equations[1]

$$ n\frac{d\phi}{dt} = \frac{1}{C}\int i_C\, dt = \frac{1}{2}E\sin\omega t \qquad (6.8) $$

where n is the number of turns of the inductance coil and ϕ is the magnetic flux in the core.   Proceeding analogously as for the harmonic oscillation in Sec. 5.1a, we put for the currents, etc.,

$$ i_L = I_n u_L \qquad i_C = I_n u_C \qquad \phi = \Phi_n v \qquad \text{and} \qquad C = C_n m \quad (6.9) $$

and assume the saturation curve of the form

$$ u_L = c_1 v + c_3 v^3 \qquad (6.10) $$

and, as before, we set the relations

$$ n\omega^2 C_n \Phi_n = I_n \qquad c_1 + c_3 = 1 \qquad (6.11) $$

Then, by virtue of Eqs. (6.9) and (6.11), Eqs. (6.8) are transformed to

$$ \frac{d^2 v}{d\tau^2} = \frac{1}{m}u_C = B\cos\tau $$

$$ \text{where} \qquad \tau = \omega t \qquad B = \frac{1}{2}\frac{E}{n\omega\Phi_n} \qquad (6.12) $$

Hence the equilibrium state is readily obtained as

$$ v_0 = y\cos\tau = -B\cos\tau \qquad (6.13) $$

---

[1] It is assumed that the two resonance circuits behave in just the same way, so that each circuit is impressed by one-half the applied voltage. We shall shortly investigate the stability of such a state.

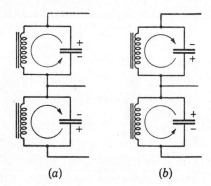

FIGURE 6.6 Direction of the oscillations
caused by perturbation.   (a)   (b)

We now consider the stability of this equilibrium.   Let the variations
of u_L, u_C, and v in one of the resonance circuits be δu_L, δu_C, and ξ, respectively.   Then the corresponding variations $-\delta u_L$, $-\delta u_C$, and $-\xi$ will
result in the other resonance circuit.   Upon applying Kirchhoff's first
law to the junction point of the two resonance circuits, we have

$$\delta u_L + \delta u_C = 0 \tag{6.14}$$

Hence, as indicated in Fig. 6.6, the currents due to this perturbation flow
in both circuits with opposite sense and do not come out to the external
source.

From Eq. (6.10), δu_L and ξ are interrelated by

$$\delta u_L = (c_1 + 3c_3 v_0{}^2)\xi \tag{6.15}$$

Substituting Eqs. (6.14) and (6.15) into (6.12) gives

$$\frac{d^2\xi}{d\tau^2} + \frac{1}{m}(c_1 + 3c_3 v_0{}^2)\xi = 0$$

Further, by substituting the periodic solution (6.13), we obtain the equation of the Mathieu type, that is,

$$\frac{d^2\xi}{d\tau^2} + (\theta_0 + 2\theta_1 \cos 2\tau)\xi = 0$$

$$\tag{6.16}$$

where $$\theta_0 = \frac{1}{m}\left(c_1 + \frac{3}{2}c_3 y^2\right) \qquad \theta_1 = \frac{1}{m}\frac{3}{4}c_3 y^2$$

We see that the equilibrium state (6.13) becomes unstable if Eq.
(6.16) has an unstable solution and that the higher-harmonic oscillation
(mostly the nth harmonic) will be excited if the point (θ_0,θ_1) lies in the
nth unstable region of the stability chart.

As mentioned in Sec. 6.1, a higher-harmonic oscillation is likely to
occur when θ_0 is increased.   In the series-resonance circuit, we have the

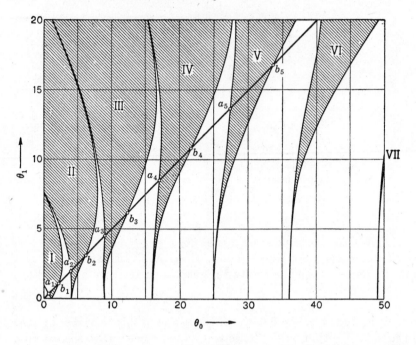

Figure 6.7   Locus of the point (θ_0, θ_1) in Eq. (6.17) with varying B^2/m.

relations given by Eqs. (6.4);[1] therefore, in order to increase θ_0, an excessive value for B would be required. In the parallel-resonance circuit, however, θ_0 in Eqs. (6.16) may readily be increased, since $y = -B$, as one sees in Eq. (6.13). Therefore, a higher-harmonic oscillation may take place under a comparatively low value of the applied voltage in the parallel-resonance circuit. This agrees with the consideration that we noticed at the end of Sec. 6.2.

We are now to determine the regions in which higher-harmonic oscillations are excited. For the brevity of calculation, we take the particular case that

$$c_1 = 0 \qquad \text{and} \qquad c_3 = 1$$

Then we obtain the following relations for the parameters θ_0 and θ_1 in Eqs. (6.16), that is,

$$\theta_0 = 2\theta_1 = \frac{3B^2}{2m} \tag{6.17}$$

This linear relation between θ_0 and θ_1 is plotted on the stability chart for Mathieu's equation, as shown in Fig. 6.7. The unstable regions of orders 1, 2, 3, . . . are indicated by the hatched areas I, II, III, . . . , respec-

[1] See also Fig. 6.1.

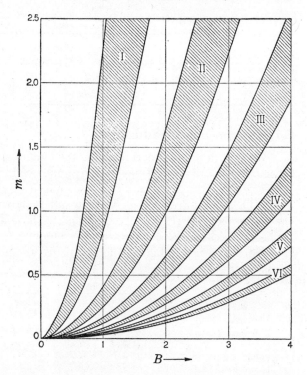

FIGURE 6.8   Regions of self-excitation of the higher-harmonic oscillations (calculated).

tively (cf. Fig. 3.3). Hence we see that, as B^2/m increases, the locus of the point (θ_0, θ_1) passes alternately through the regions of stability and instability. If we denote the intersecting points of the locus with the boundary of the nth unstable region by a_n and b_n (as marked in the figure) and, further, the corresponding values of θ_0 by θ_{an} and θ_{bn}, respectively, we may determine in the Bm plane the boundary curves of the nth region in which the nth harmonic is predominantly excited; they are given by

$$\frac{3}{2}\frac{B^2}{m} = \theta_{an} \quad \text{and} \quad \frac{3}{2}\frac{B^2}{m} = \theta_{bn} \tag{6.18}$$

The particular values of the parameters θ_{an} and θ_{bn} are readily obtained from Fig. 6.7, and they are shown in Table 6.1. By substituting these values into Eqs. (6.18), we obtain the regions of higher-harmonic oscillations as illustrated in Fig. 6.8.[1]

---

[1] The Roman numerals I, II, III, . . . in these regions correspond to those of Fig. 6.7.

Table 6.1   Values of θ_{an} and θ_{bn} in Eqs. (6.18)

n	1	2	3	4	5	6
θ_{an}	0.66	3.71	9.24	17.27	27.76	40.66
θ_{bn}	1.78	6.10	12.88	22.00	33.82	47.80

6.4 Experimental Investigation

We first perform experiments on the circuit as illustrated in Fig. 6.5. Since B and m are proportional to the applied voltage and the capacitance, respectively, we seek, by varying these quantities, the regions in which the higher harmonics are excited. The result is shown in Fig. 6.9, in which the numerals I, II, III, . . . correspond to those in Fig. 6.8. The waveforms of the capacitor current I_C with reference to the applied voltage V are also shown in the figure; their principal frequencies are 1, 2, 3, . . . times the impressed frequency.

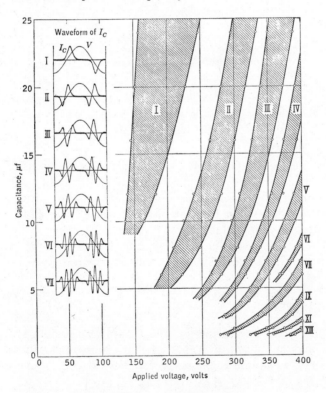

FIGURE 6.9   Regions of self-excitation of the higher-harmonic oscillations (experimental).

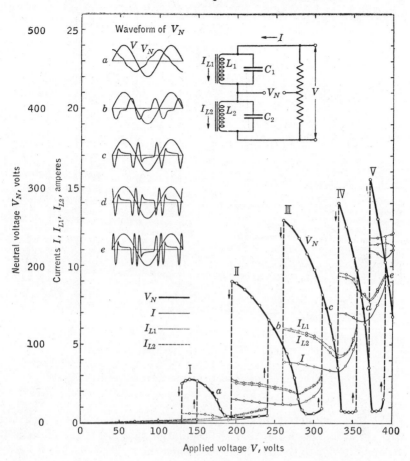

FIGURE 6.10 Neutral instability caused by higher-harmonic excitation.

As a result of the excitation of such a harmonic, the potential of the junction point of the two resonance circuits oscillates with respect to the neutral point of the applied voltage with the frequency of that harmonic. In Fig. 6.10, the anomalous neutral voltage V_n is illustrated against the applied voltage.[1] We see in the figure that, once the self-excitation is started, it may be stopped by decreasing the applied voltage to a value which is lower than before. The preceding analysis gives no information with regard to this hysteresis phenomenon, because, when the self-excitation is built up to some extent, the circuit condition is altered, and the analysis which assumes the equilibrium state (6.13) is no longer applica-

[1] The self-excited oscillation in the first unstable region (marked by I) has the same frequency as that of the applied voltage. This phenomenon is sometimes called the *neutral inversion* in electric transmission lines.

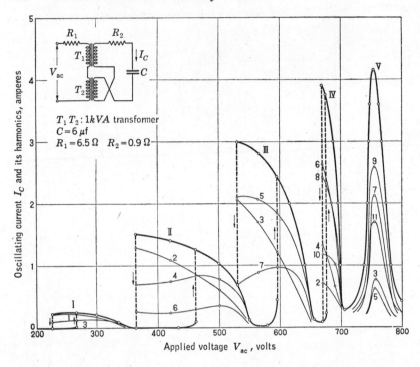

FIGURE 6.11   Higher-harmonic oscillations and their harmonic analysis.

ble.[1]   It is also noted that the regions of Fig. 6.9 are those in which the self-excitation starts and that the oscillation may be sustained in certain regions exterior to the hatched areas when the applied voltage is lowered.

In Fig. 6.11, the effective value of the capacitor current I_c is plotted (in thick lines) against the applied voltage V_{ac}.   This current is analyzed into its harmonic components (drawn by fine lines, the numbers on which show the order of the harmonics).   Hence we see that the odd harmonics are present in the regions of odd orders, that is, in I, III, V, ... , and the even harmonics in the regions of even orders, that is, in II, IV, VI, ...   This is in excellent agreement with the theory of Mathieu's equation, as we may expect from the form of the unstable solutions in Sec. 3.6*b*.

It may be conjectured that, when the alternating voltage is applied to the circuit, the self-excited oscillation will gradually build up with

---

[1] The ultimate amplitude of the self-excited oscillation may be found by assuming this component, as well as the fundamental-frequency component in the periodic solution (6.13), and then by applying the method of harmonic balance to determine the amplitudes of these components.

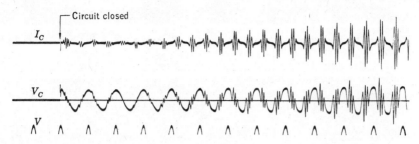

FIGURE 6.12   Buildup of a self-excited oscillation. V = applied voltage, V_C = voltage across capacitor, I_C = current through capacitor.

increasing amplitude, taking the form described by

$$e^{\mu\tau}[\sin{(n\tau - \sigma)} + \cdot \cdot \cdot] \qquad \text{with } \mu > 0$$

and ultimately reach the steady state with constant amplitude which is limited by the nonlinear characteristic of the circuit.   This is confirmed by experiments.   An example of such a transient state is given in Fig. 6.12, which shows the buildup of the seventh harmonic oscillation correlated with the seventh unstable region for Mathieu's equation.

Finally, it is added that, when the self-excitation of those oscillations is undesirable, the following scheme is effective to prevent it.   Namely, we provide the reactors with the secondary coils, as illustrated in Fig. 6.13, and connect them in series but with opposite polarity.   Then, since the anomalous oscillation is excited in the primary circuits with opposite sense (Fig. 6.6), induced voltages in the secondary coils have the same direction and a short circuit results, thus preventing the buildup of the anomalous oscillation.   On the other hand, the forced current due to the applied voltage flows through the primary coils in the same direction and induces no current in the secondary circuit.   Therefore, this secondary circuit is effective only in preventing the anomalous oscillation and exerts no influence upon the primary forced current.

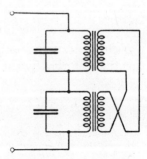

FIGURE 6.13   Suppression of the self-excited oscillation.

Chapter 7

Subharmonic Oscillations

7.1 Introduction

In the preceding two chapters we have investigated the harmonic and the higher-harmonic oscillations whose fundamental frequency is equal to that of the external force. We are now to deal with the subharmonic oscillations whose fundamental frequency is a fraction $1/\nu$ ($\nu = 2, 3, 4, \ldots$) of the driving frequency. As we have briefly noted in Secs. 3.5 and 4.1, these oscillations are of another important type in the field of nonlinear oscillations, and they frequently occur in various branches of engineering and physical sciences [35, 65, 72, 76, 80, 96, 100].

We first take the fundamental equation

$$\frac{d^2v}{d\tau^2} + 2\delta\frac{dv}{d\tau} + f(v) = B\cos\nu\tau \tag{7.1}$$

in which 2δ is a constant damping coefficient and $f(v)$ is a term representing the nonlinear restoring force. It will be noticed that, since the period of the external force is $2\pi/\nu$, the subharmonic oscillation of order $1/\nu$ has a period 2π and may be expressed by a linear combination of $\sin\tau$ and $\cos\tau$.

In the following sections, we shall first investigate the relationship between the nonlinear characteristic expressed by the term $f(v)$ and the order $1/\nu$ of the subharmonic oscillations. Then the subharmonic oscillations of orders $\frac{1}{3}$ and $\frac{1}{2}$ will be investigated in detail, with special attention directed to the stability of the periodic oscillations.[1]

7.2 Relationship between the Nonlinear Characteristics and the Order of Subharmonic Oscillations

In order to investigate this, we consider the following polynomial for the restoring force $f(v)$, that is,

$$f(v) = c_1v + c_2v^2 + c_3v^3 + \cdots \tag{7.2}$$

where $c_1, c_2, c_3, \ldots$ are constants determined by the nonlinear charac-

---

[1] This chapter is confined to the study of periodic oscillations. The transient state of the subharmonic oscillation will be discussed in Chaps. 9 and 10.

teristic. As mentioned in the case of the harmonic oscillations [see Eqs. (5.5)], these constants are subject, without loss of generality, to the condition

$$c_1 + c_2 + c_3 + \cdots = 1 \qquad (7.3)$$

As far as we deal with the steady state of the oscillation, the periodic solution of Eq. (7.1) may be assumed to take the form

$$v_0 = z + x \sin \tau + y \cos \tau + w \cos \nu\tau \qquad (7.4)$$

in which the constant term z, the subharmonic oscillation $x \sin \tau + y \cos \tau$, and the oscillation having the driving frequency $w \cos \nu\tau$ are considered because of their prime importance.[1] Following Mandelstam and Papalexi [72, p. 227], the amplitude w may further be approximated by

$$w = \frac{1}{1 - \nu^2} B \qquad (7.5)$$

This approximation is legitimate when the nonlinearity is small. However, as will be shown in Secs. 7.3c and 7.4, approximation is permissible even when the departure from linearity is large.

By substituting Eq. (7.4) into Eq. (7.1) and equating the coefficients of the terms containing $\sin \tau$ and $\cos \tau$ separately to zero, we obtain the following results according to the form of the nonlinear characteristic (7.2).

CASE 1.   WHEN THE NONLINEARITY IS GIVEN BY $f(v) = c_1 v + c_3 v^3$

In this case when the nonlinearity is symmetrical, that is, $f(v)$ is odd in v, the constant term z in Eq. (7.4) may usually be discarded; then, for $\nu = 2, 4, 5, \ldots$, the above-mentioned substitution leads to

$$\begin{aligned}
[1 - \tfrac{3}{4}(x^2 + y^2) - \tfrac{3}{2}w^2]x + ky &= 0 \\
[1 - \tfrac{3}{4}(x^2 + y^2) - \tfrac{3}{2}w^2]y - kx &= 0
\end{aligned} \qquad (7.6)$$

where $k = 2\delta/c_3$. By multiplying the first equation by y and the second by x and subtracting the second product from the first, we obtain

$$k(x^2 + y^2) = 0$$

This implies that the amplitude of the subharmonic oscillation is zero as long as the damping is present, that is, $k \neq 0$. Therefore, the subharmonic oscillations of orders $\frac{1}{2}, \frac{1}{4}, \frac{1}{5}, \ldots$ cannot occur in this case.[2]

[1] It is tacitly assumed that the damping coefficient 2δ is not so large that the term containing $\nu\tau$ may be omitted in Eq. (7.4).

[2] It must be mentioned that, by a closer investigation in which the constant term z in Eq. (7.4) is taken into account, the subharmonic oscillation of order $\frac{1}{2}$ may barely take place even in this case. This result is confirmed by experiments, though the range of occurrence is considerably restricted. Therefore, strictly speaking, the above conclusion should be modified in this respect.

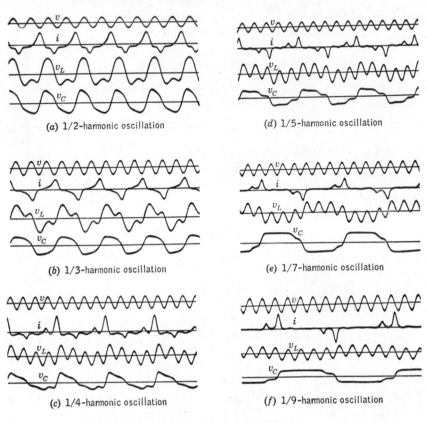

(a) 1/2-harmonic oscillation

(d) 1/5-harmonic oscillation

(b) 1/3-harmonic oscillation

(e) 1/7-harmonic oscillation

(c) 1/4-harmonic oscillation

(f) 1/9-harmonic oscillation

FIGURE 7.1   Waveforms of the subharmonic oscillations. v = applied voltage, i = oscillating current, v_L = terminal voltage across reactor, v_C = terminal voltage across capacitor.

However, as will be investigated in the following section, real roots of x and y which are not simultaneously zero may be obtained for $\nu = 3$. Thus we obtain a number of equilibrium states, and some of them are maintained in stable states.[1]   Hence, we conclude that the subharmonic oscillation of order $\frac{1}{3}$ may occur when the nonlinear term $c_3 v^3$ is contained in Eq. (7.2).

CASE 2.   WHEN THE NONLINEARITY IS GIVEN BY
$$f(v) = c_1 v + c_2 v^2 + c_3 v^3$$

The nonlinearity is unsymmetrical, and the constant term z in Eq. (7.4) must be considered.   The subharmonic oscillation of order $\frac{1}{2}$ may occur

[1] These equilibrium states are not necessarily maintained; they are able to last only if they are stable.

in this case.   The detailed investigation will be carried out in Secs. 7.6 to 7.8.[1]

CASE 3.   WHEN THE NONLINEARITY IS GIVEN BY $f(v) = c_1v + c_5v^5$

Although the cubic term c_3v^3 is absent in $f(v)$, the subharmonic oscillation of order ⅓ may still occur in this case.   The detailed investigation will be given in Sec. 7.4.   The subharmonic oscillation of order ⅕ is also maintained in this case.   It will be discussed in Sec. 9.4 in connection with the transient state of the oscillation.

From the foregoing consideration, it may be conjectured that the presence of the term $c_\nu v^\nu$ in Eq. (7.2) is desirable for the occurrence of the subharmonic oscillation of order $1/\nu$.   However, this is not a necessary condition, as we see in Case 3.   It may also be concluded that the subharmonic oscillation of order $1/\nu$ cannot occur if the highest degree of the power of the nonlinear terms in Eq. (7.2) is less than ν.

Finally, examples of the subharmonic oscillations of different orders are shown in Fig. 7.1.   These oscillograms are obtained in an electric circuit consisting of a saturable-core inductance and a capacitance in series.   Subharmonic oscillations occur in this circuit owing to the saturation of the core under the impression of an alternating voltage.

7.3 Subharmonic Oscillations of Order ⅓ with the Nonlinear Characteristic Represented by a Cubic Function [35]

We investigate the subharmonic oscillations of order ⅓ in a system described by the differential equation

$$\frac{d^2v}{d\tau^2} + 2\delta\frac{dv}{d\tau} + c_1v + c_3v^3 = B \cos 3\tau \tag{7.7}$$

(a) Nondissipative System

We begin with the system in which the damping coefficient $2\delta = 0$. Equation (7.7) becomes

$$\frac{d^2v}{d\tau^2} + c_1v + c_3v^3 = B \cos 3\tau \tag{7.8}$$

It is noted that $c_1 + c_3 = 1$ by virtue of Eq. (5.5) in Chap. 5.   By making

---

[1] The most appropriate form of the differential equation which has a subharmonic solution of order ½ is given by $f(v) = |v|v$ in Eq. (7.1) with an additional nonoscillatory term in the external force.

use of the approximation (7.5), a periodic solution of Eq. (7.8) is given by

$$v_0 = x \sin \tau + y \cos \tau + w \cos 3\tau$$

where $$w = \frac{1}{1 - 3^2} B = -\frac{1}{8} B$$ (7.9)

Substituting Eqs. (7.9) into (7.8) and equating the coefficients of the terms containing $\sin \tau$ and $\cos \tau$ separately to zero gives us

$$\left[1 - \frac{3}{4} (x^2 + y^2) - \frac{3}{2} w^2 \right] x = -\frac{3w}{4} 2xy$$

$$\left[1 - \frac{3}{4} (x^2 + y^2) - \frac{3}{2} w^2 \right] y = -\frac{3w}{4}(x^2 - y^2)$$ (7.10)

Multiplying the first equation by y and the second by x and subtracting the products so formed leads to

$$x = 0 \quad \text{or} \quad x = \pm \sqrt{3}\, y$$

For each value of x we have two pairs of the roots for x and y; hence, we obtain six equilibrium states in all. But we are interested only in the equilibrium states for which $x = 0$, because the other equilibrium states are obtained merely by shifting the phase of the oscillations by $2\pi/3$ or $4\pi/3$ radians in τ, that is, by one or two cycles of the external force. This is a natural consequence for the subharmonic oscillation of order $\frac{1}{3}$.

For $x = 0$, assuming that $y \neq 0$ in the second equation of (7.10), we obtain

$$y^2 + wy + 2w^2 - \frac{4}{3} = 0$$ (7.11)

This equation is plotted as a part of an ellipse in Fig. 7.2. In the figure is shown the relationship between the amplitude w (which is here assumed to be proportional to the external force B) and the amplitude y of the subharmonic oscillation.

We are now to determine the stability of the equilibrium states as found by Eq. (7.11). Proceeding in the usual manner, we consider a variation ξ from the equilibrium states which satisfies the equation

$$\frac{d^2 \xi}{d\tau^2} + (c_1 + 3c_3 v_0^2)\xi = 0$$

By substituting into this the equilibrium states represented by

$$v_0 = y \cos \tau + w \cos 3\tau$$

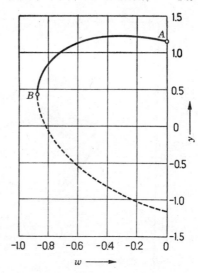

FIGURE 7.2 Amplitude characteristic of the $\frac{1}{3}$-harmonic oscillation; non-linearity by cubic function.

we obtain the following equation of Hill's type:

$$\frac{d^2\xi}{d\tau^2} + \left(\theta_0 + 2\sum_{n=1}^{3}\theta_n\cos 2n\tau\right)\xi = 0 \tag{7.12}$$

where
$$\theta_0 = c_1 + \tfrac{3}{2}c_3(y^2 + w^2) \qquad \theta_2 = \tfrac{3}{2}c_3wy$$
$$\theta_1 = \tfrac{3}{4}c_3(y^2 + 2wy) \qquad \theta_3 = \tfrac{3}{4}c_3w^2$$

Since w and y are determined by Eqs. (7.9) and (7.11), respectively, the parameters θ in Eqs. (7.12) may be readily calculated.

The stability condition in the nondissipative system is given by putting $\delta = 0$ in inequality (4.6) in Chap. 4, that is,

$$(\theta_0 - n^2 + \theta_n)(\theta_0 - n^2 - \theta_n) > 0$$

or
$$|\theta_0 - n^2| > |\theta_n| \qquad n = 1, 2, 3 \tag{7.13}$$

The condition for $n = 1$ ascertains the stability against buildup of an unstable oscillation whose principal frequency is the same as the sub-harmonic frequency. By the use of Eqs. (7.11) and (7.12), this condition leads to

$$2y > -w \tag{7.14}$$

Hence, the equilibrium states represented by dashed lines in Fig. 7.2 are unstable and do not actually exist. Figure 7.3 shows the loci of θ_n ($n = 1$, 2, 3) which are drawn by varying the value of w (or B) for the case of $c_1 = 0$ and $c_3 = 1$. As expected from the stability condition for $n = 1$, θ_1 enters into the first unstable region in the dashed interval ab. At the

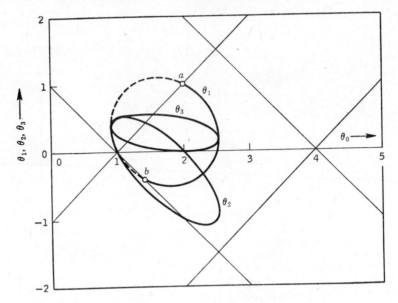

FIGURE 7.3   Loci of θ's in Eqs. (7.12) as obtained by varying w.

critical points a and b, we have

$$w = 0 \qquad \text{and} \qquad 2y = -w$$

These points respectively correspond to A and B in Fig. 7.2.   Figure 7.3 shows that neither θ_2 nor θ_3 enters into the corresponding unstable region.[1] Since the conditions (7.13) for $n = 2$ and $n = 3$ are thus satisfied, only the condition (7.14) will suffice to determine the stability.[2]   Hence, we may conclude that the full-line part of the characteristic curve in Fig. 7.2 represents the stable oscillations.

(b) *Dissipative System*

When the damping is present in a nonlinear system, the fundamental equation takes the form (7.7).   For convenience, we write this again,

---

[1] In Fig. 7.3 we see that the first and the second unstable regions partly overlap. This is due to the deficiency of approximation in computing the stability condition. Properly speaking, the unstable regions should not overlap.   In the above discussion, however, the periodic solution is obtained to a first approximation, so that the stability condition of the first approximation is also used.

[2] If the parameter θ_n ($n \geq 2$) enters into the nth unstable region, the unstable oscillation of order $n/3$ builds up.   If this oscillation grows to be predominant, the original subharmonic oscillation of order $\frac{1}{3}$ will no longer be maintained.   An example for such a case ($n = 2$) will be given in Sec. 7.4.

that is,

$$\frac{d^2v}{d\tau^2} + 2\delta \frac{dv}{d\tau} + c_1 v + c_3 v^3 = B \cos 3\tau \qquad (7.15)$$

The periodic solution is expressed by

$$v_0 = x \sin \tau + y \cos \tau + w \cos 3\tau$$

where

$$w = -\frac{B}{8} \qquad (7.16)$$

Substituting Eqs. (7.16) into (7.15) and equating the coefficients of the terms containing $\sin \tau$ and $\cos \tau$ separately to zero gives us

$$Ax + ky = -\frac{3w}{4} 2xy \qquad kx - Ay = \frac{3w}{4} (x^2 - y^2)$$

where

$$A = 1 - \frac{3}{4} (x^2 + y^2) - \frac{3}{2} w^2 \qquad k = \frac{2\delta}{c_3} \qquad (7.17)$$

By squaring and adding the first two equations, we obtain

$$A^2 + k^2 = \left(\frac{3w}{4}\right)^2 r^2 \qquad r^2 = x^2 + y^2 \qquad (7.18)$$

or, by putting $r^2 = R$ and $w^2 = W$,

$$\tfrac{9}{16}R^2 + (\tfrac{27}{16}W - \tfrac{3}{2})R + (\tfrac{9}{4}W^2 - 3W + k^2 + 1) = 0 \quad (7.19)$$

Hence the relationship between W and R, and consequently the amplitude of the subharmonic oscillation, may be determined. The result is illustrated in Fig. 7.4a and b for several values of k.

The components x and y of the amplitude r may be obtained as follows: From Eqs. (7.17) and (7.18), we obtain

$$x^3 - \frac{3}{4} Rx - \frac{kR}{3w} = 0 \qquad \text{and} \qquad y^3 - \frac{3}{4} Ry - \frac{AR}{3w} = 0$$

Hence the components x and y are given by

$$
\begin{array}{lll}
x = r \cos \theta & r \cos (\theta + 120°) & r \cos (\theta + 240°) \\
y = r \sin \theta & r \sin (\theta + 120°) & r \sin (\theta + 240°)
\end{array} \qquad (7.20)
$$

where

$$\sin 3\theta = -\frac{4A}{3wr} \qquad \cos 3\theta = \frac{4k}{3wr}$$

The stability problem may be treated in the same manner as before. The equation which characterizes a small variation ξ from the periodic solution is given by

$$\frac{d^2\xi}{d\tau^2} + 2\delta \frac{d\xi}{d\tau} + (c_1 + 3c_3 v_0^2)\xi = 0$$

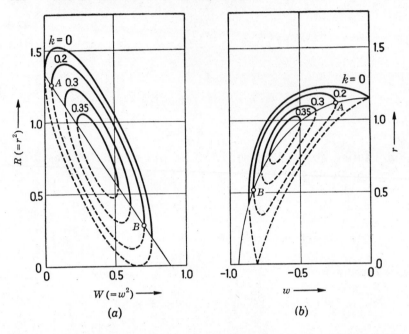

FIGURE 7.4   (a) Relationship between W and R in Eq. (7.19).   (b) Amplitude characteristic of the $\frac{1}{3}$-harmonic oscillation.

By the transformation $\xi = e^{-\delta\tau}\eta$, this leads to

$$\frac{d^2\eta}{d\tau^2} + (c_1 - \delta^2 + 3c_3v_0^2)\eta = 0$$

Substituting into this the periodic solution (7.16) gives the following Hill's equation:

$$\frac{d^2\eta}{d\tau^2} + \left[\theta_0 + 2\sum_{n=1}^{3}\theta_n\cos(2n\tau - \epsilon_n)\right]\eta = 0$$

where

$$\theta_0 = c_1 - \delta^2 + \tfrac{3}{2}c_3(x^2 + y^2 + w^2)$$

$$\theta_n{}^2 = \theta_{ns}{}^2 + \theta_{nc}{}^2 \qquad \epsilon_n = \tan^{-1}\frac{\theta_{ns}}{\theta_{nc}} \tag{7.21}$$

$$\theta_{1s} = \tfrac{3}{2}c_3x(y - w) \qquad \theta_{1c} = \tfrac{3}{4}c_3(-x^2 + y^2 + 2wy)$$

$$\theta_{2s} = \tfrac{3}{2}c_3wx \qquad\qquad \theta_{2c} = \tfrac{3}{2}c_3wy$$

$$\theta_{3s} = 0 \qquad\qquad\qquad \theta_{3c} = \tfrac{3}{4}c_3w^2$$

The stability condition in the dissipative system is given by inequality (4.6), that is,

$$(\theta_0 - n^2)^2 + 2(\theta_0 + n^2)\delta^2 + \delta^4 > \theta_n{}^2 \qquad n = 1, 2, 3 \tag{7.22}$$

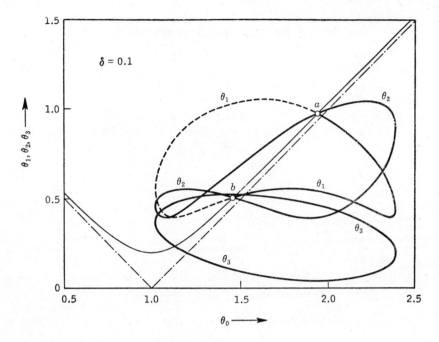

FIGURE 7.5   Loci of θ's in Eqs. (7.21) as obtained by varying w.

The condition for $n = 1$ is obtained by substituting θ_0 and θ_1 of Eqs. (7.21) into (7.22); further, by virtue of Eqs. (7.17) and (7.19), we ultimately find

$$R + \tfrac{3}{2}W - \tfrac{4}{9} > 0 \qquad (7.23)$$

Hence the equilibrium states represented by dashed lines in Fig. 7.4 are unstable and do not actually exist.   Figure 7.5 shows the loci of θ_n ($n = 1, 2, 3$) which are drawn by varying the value of w (or B) for the case of $c_1 = 0$, $c_3 = 1$, and $\delta = 0.1$.   As shown in the figure, the boundary curve of the first unstable region is given by a hyperbola,[1] and θ_1 enters into this region in the dashed interval ab.   At the critical points a and b, we have

$$R + \tfrac{3}{2}W - \tfrac{4}{9} = 0$$

These points respectively correspond to A and B in Fig. 7.4.   It is obvious from Fig. 7.5 that neither θ_2 nor θ_3 enters into the corresponding unstable region.   Therefore the condition (7.23) for $n = 1$ is sufficient to determine the stability in the case where the nonlinearity is characterized by a cubic function.

[1] If the damping $\delta = 0$, the hyperbola is reduced to straight lines (drawn by chain lines in the figure) which cross the abscissa at $(1,0)$.

(c) *Remarks on the Approximation in the Preceding Analysis*

As mentioned in Sec. 7.2, the approximation (7.5) is legitimate in the case where the nonlinearity is small, i.e., where the linear term c_1v in the right side of Eq. (7.2) is predominant over the nonlinear terms. However, this approximation is still permissible even when the linear term is absent in $f(v)$, as will be seen in the following example.

For the sake of simplicity, we consider the nondissipative case and put $c_1 = 0$ and $c_3 = 1$ in Eq. (7.8). Then we have

$$\frac{d^2v}{d\tau^2} + v^3 = B \cos 3\tau$$

By substituting a periodic solution

$$v_0 = y \cos \tau + w \cos 3\tau$$

and equating the coefficients of the terms containing $\cos \tau$ and $\cos 3\tau$ separately to zero, we obtain

$$y^2 + wy + 2w^2 - \tfrac{4}{3} = 0 \qquad \tfrac{1}{4}y^3 + \tfrac{3}{2}wy^2 + \tfrac{3}{4}w^3 - 9w = B \quad (7.24)$$

Figure 7.6 is obtained by plotting Eqs. (7.24). Thus we see that the approximation $w = -B/8$ is applicable to this case with acceptable

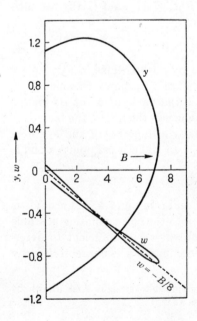

FIGURE 7.6 Relationship between w, y, and B in Eqs. (7.24).

Table 7.1   Values of y, w, and the θ's for B = 0 and B = 7.205

B	y	w	θ_0	θ_1	θ_2	θ_3
0	1.127	0.0510	1.9092	1.0392	0.0862	0.0019
7.205	0.285	−0.8655	1.2443	−0.3078	−0.3681	0.5619

accuracy.[1] It is also seen that, since w is not exactly proportional to B, the limiting values of B and w do not occur simultaneously.

We further show that the vertical tangency of the characteristic curve occurs at the stability limit where the parameters θ of the variational equation (7.12) are on the boundary of the first unstable region.[2] For the limiting values of B—that is, for $B = 0$ and $B = 7.205$ (maximum)—the amplitudes y and w are found by solving Eqs. (7.24). The parameters θ may also be calculated from Eqs. (7.12). They are shown in Table 7.1.

The boundary of the first unstable region is given, to the first approximation, by

$$\theta_0 = 1 \pm \theta_1 \tag{7.25}$$

Substituting the values of θ_1 in Table 7.1 gives us

$$\theta_0 = 2.0392 \quad \text{for} \quad B = 0$$
$$\theta_0 = 1.3078 \quad \text{for} \quad B = 7.205$$

These values differ from those given in Table 7.1 by 6.8 and 5.1 percent, respectively. This disagreement is due to the deficiency of approximation for Eq. (7.25). Therefore, for the values of y and w calculated by Eqs. (7.24), we have to apply a higher approximation: for example, as given by the development[3]

$$\theta_0 = 1 \pm \theta_1 - \tfrac{1}{8}\theta_1{}^2 - \tfrac{1}{6}\theta_2{}^2 - \tfrac{1}{16}\theta_3{}^2 \pm \tfrac{1}{4}\theta_1\theta_2 \pm \tfrac{1}{12}\theta_2\theta_3$$
$$\mp \tfrac{1}{64}\theta_1{}^3 + \tfrac{1}{48}\theta_1{}^2\theta_2 \pm \tfrac{5}{192}\theta_1{}^2\theta_3 \mp \tfrac{1}{144}\theta_1\theta_2{}^2$$
$$\mp \tfrac{1}{2304}\theta_1\theta_3{}^2 \pm \tfrac{1}{48}\theta_2{}^2\theta_3 - \tfrac{13}{288}\theta_1\theta_2\theta_3 + \cdots \tag{7.26}$$

Now, by substituting the values of $\theta_1 \sim \theta_3$ of Table 7.1 into Eq. (7.26), we obtain

$$\theta_0 = 1.9038 \quad \text{for} \quad B = 0$$
$$\theta_0 = 1.2352 \quad \text{for} \quad B = 7.205$$

---

[1] It is ascertained that the approximation is permissible in the dissipative case also.

[2] Since we are concerned with the nondissipative system, the characteristic exponent of the solution for Eq. (7.12) becomes zero at the stability limit.

[3] This relation is obtained by putting $\sigma = 0$ and $\sigma = -\pi/2$ in Eq. (II.4) in Appendix II.

which differ from the values given in Table 7.1 by 0.28 and 0.73 percent, respectively. Thus the discrepancy is much reduced; hence we may conclude that the stability limit is given by the condition that the amplitude B of the external force takes its limiting values.[1] This is a reasonable consequence from the physical point of view.

7.4 Subharmonic Oscillations of Order ⅓ with the Nonlinear Characteristic Represented by a Quintic Function [35]

For the brevity of calculation, we consider the nondissipative case only. Putting

$$\nu = 3 \qquad \delta = 0 \qquad \text{and} \qquad c_2 = c_3 = c_4 = c_6 = \cdots = 0$$

in Eqs. (7.1) and (7.2), we have

$$\frac{d^2v}{d\tau^2} + c_1 v + c_5 v^5 = B \cos 3\tau \tag{7.27}$$

where $c_1 + c_5 = 1$ by virtue of Eq. (7.3). By substituting the periodic solution

$$v_0 = y \cos \tau + w \cos 3\tau \qquad \text{with } w = -\frac{B}{8}$$

into (7.27) and equating to zero the coefficient of the term containing $\cos \tau$, we obtain

$$y^4 + \tfrac{5}{2}wy^3 + 6w^2y^2 + 3w^3y + 3w^4 - \tfrac{8}{5} = 0 \tag{7.28}$$

The relationship between y and w is illustrated in Fig. 7.7. Negative values of y are not shown in the figure because, as will be shown shortly, the equilibrium states in this part are unstable.

The stability problem may be treated analogously as in the preceding section. Thus, the variational equation leads to a Hill's equation of the form

$$\frac{d^2\xi}{d\tau^2} + \left(\theta_0 + 2 \sum_{n=1}^{6} \theta_n \cos 2n\tau \right) \xi = 0 \tag{7.29}$$

where
$$\begin{aligned}
\theta_0 &= c_1 + \tfrac{5}{8}c_5(3y^4 + 4wy^3 + 12w^2y^2 + 3w^4) \\
\theta_1 &= \tfrac{5}{4}c_5(y^4 + 3wy^3 + 3w^2y^2 + 3w^3y) \\
\theta_2 &= \tfrac{5}{16}c_5(y^4 + 12wy^3 + 6w^2y^2 + 12w^3y) \\
\theta_3 &= \tfrac{5}{4}c_5(wy^3 + 3w^2y^2 + w^4) \\
\theta_4 &= \tfrac{5}{8}c_5(3w^2y^2 + 2w^3y) \\
\theta_5 &= \tfrac{5}{4}c_5w^3y \\
\theta_6 &= \tfrac{5}{16}c_5w^4
\end{aligned} \tag{7.29a}$$

---

[1] It is noted that the amplitude w does not have its limiting values at the stability limit.

The stability condition for $n = 1$ is obtained by substituting θ_0 and θ_1 into inequality (7.13). Further, by virtue of Eq. (7.28), we ultimately find

$$y^3 + \tfrac{15}{8}wy^2 + 3w^2y + \tfrac{3}{4}w^3 > 0 \qquad \text{for } w < 0 \qquad (7.30)$$

Referring to Fig. 7.7, the equilibrium states in the interval between A and B satisfy the condition (7.30). The stability limits A and B are given by the condition that w takes its limiting values.

Figure 7.8 shows the loci of θ_n ($n = 1, 2, 3$) which are drawn by varying the value of w (or B) for the case of $c_1 = 0$ and $c_5 = 1$. As expected from the stability condition for $n = 1$, θ_1 enters into the first unstable region in the dashed interval ab. We see, further, that θ_2 enters into the second unstable region in the dashed interval cd. Hence the stability condition for $n = 2$ is not satisfied in the interval cd. In this case the unstable oscillation of order $\tfrac{2}{3}$ is excited, which disturbs the continuation of the original subharmonic oscillation.

As mentioned above, the curves in Fig. 7.8 are drawn for the case of $c_1 = 0$ and $c_5 = 1$. With increasing c_1 (or decreasing c_5), these curves move toward the point $(1,0)$, as we see from the expressions for the θ's in

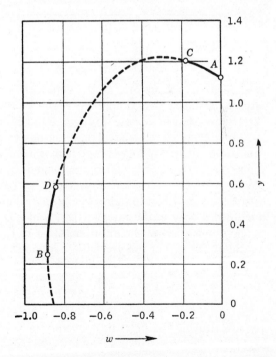

FIGURE 7.7 Amplitude characteristic of the $\tfrac{1}{3}$-harmonic oscillation; nonlinearity by quintic function.

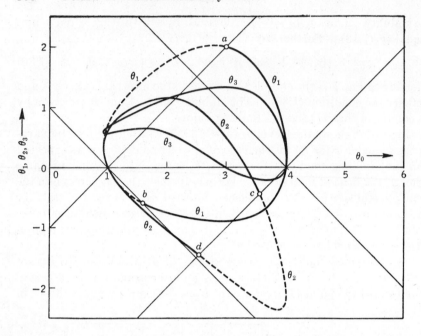

FIGURE 7.8   Loci of θ's in Eqs. (7.29a) as obtained by varying w.

Eqs. (7.29a).   Therefore, as the departure from linearity is reduced, the interval cd in the second unstable region contracts and finally disappears. We also see that in no case do the parameters θ_3 to θ_6 enter into the corresponding unstable regions.   Hence, it may be concluded that the stability conditions for $n = 2$ as well as for $n = 1$ must be considered in the case where the nonlinearity is characterized by a quintic function.   The points A to D in Fig. 7.7 respectively correspond to the critical points a to d in Fig. 7.8; therefore the subharmonic oscillation of order $\frac{1}{3}$ is maintained only in the intervals AC and BD.

We now discuss the approximation in the foregoing analysis just as we did at the end of the preceding section.   Let the differential equation be given by

$$\frac{d^2v}{d\tau^2} + v^5 = B \cos 3\tau$$

By substituting the periodic solution

$$v_0 = y \cos \tau + w \cos 3\tau$$

and equating the coefficients of the terms containing $\cos \tau$ and $\cos 3\tau$

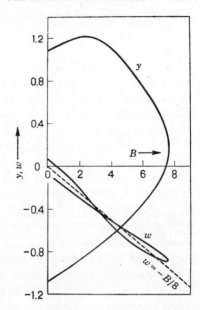

FIGURE 7.9 Relationship between w, y, and B in Eqs. (7.31).

separately to zero, we obtain

$$y^4 + \tfrac{5}{2}wy^3 + 6w^2y^2 + 3w^3y + 3w^4 - \tfrac{8}{5} = 0$$
$$\tfrac{5}{16}y^5 + \tfrac{15}{8}wy^4 + \tfrac{15}{8}w^2y^3 + \tfrac{15}{4}w^3y^2 + \tfrac{5}{8}w^5 - 9w = B \quad (7.31)$$

Figure 7.9 is obtained by plotting Eqs. (7.31). Thus we see that the approximation $w = -B/8$ may also be applicable with acceptable accuracy.

For the limiting values of B—that is, for $B = 0$ and $B = 7.546$ (maximum)—the amplitudes y and w are determined by solving Eqs. (7.31). The parameters θ are also calculated from Eqs. (7.29a). They are shown in Table 7.2.

Substituting the values of θ_1 into Eq. (7.25) gives us

$$\theta_0 = 3.0327 \quad \text{for} \quad B = 0$$
$$\theta_0 = 1.3385 \quad \text{for} \quad B = 7.546$$

These values differ from those given in Table 7.2 by 9.1 and 5.8 percent, respectively. By again applying a higher approximation (7.26) instead

Table 7.2   Values of y, w, and the θ's for $B = 0$ and $B = 7.546$

B	y	w	θ_0	θ_1	θ_2	θ_3
0	1.0759	0.0716	2.7795	2.0327	0.7655	0.1337
7.546	0.1542	-0.8818	1.2651	-0.3385	-0.3736	0.8210

of (7.25), we obtain

$$\theta_0 = 2.7234 \quad \text{for} \quad B = 0$$
$$\theta_0 = 1.2415 \quad \text{for} \quad B = 7.546$$

which differ from the values given in Table 7.2 by 1.9 and 2.0 percent, respectively. Hence we may conclude that the stability limit is given by the condition that the amplitude B of the external force takes its limiting values.

We have thus far discussed the stability limit of the first unstable region. A similar investigation may be carried out for the second unstable region. But in this case the stability limit has no significant relation to the external force, and besides, as we shall see in discussion later of another experiment, a small parasitic oscillation of order $2/3$ may coexist with the original subharmonic oscillation in the neighborhood of the stability limit. Hence, the points c and d in Fig. 7.8, however accurately they might be determined, would not represent the critical points at which the original subharmonic oscillation ceases to exist. Therefore, a further investigation into the stability limit of the second unstable region is not important and hence is omitted here.

7.5 Experimental Investigation [35]

In this section we compare the theoretical results of the preceding sections with experiments conducted for an electric oscillatory circuit containing a saturable-core inductor and a capacitor in series. As mentioned in Sec. 5.1a, the circuit equation takes the form of Eq. (7.1) when an alternating voltage (60 cps in our case) is applied to the circuit. If an initial condition is prescribed appropriately, a subharmonic oscillation of order $1/3$, that is, of 20 cps, may easily be started in the circuit.

By making use of an ordinary transformer-core inductor as the nonlinear element, we first determine the region in which the subharmonic oscillation of order $1/3$ is sustained. This region is shown in Fig. 7.10a by shaded area. The appearance of the blank part inside the sustaining region is an arresting feature previously reported by McCrumm [68] and the author, but no theoretical consideration was given at that time.

From the preceding analysis, however, it will be deduced that the blank part corresponds to the unstable regions of order $n \geq 2$, because the nonlinear characteristic of the ordinary transformer core is expressed by[1]

$$f(v) = c_1 v + c_3 v^3 + c_5 v^5 + c_7 v^7 + \cdots \tag{7.32}$$

---

[1] Physically, Eq. (7.32) represents the magnetization curve of the core, i.e., the relationship between the magnetic flux v and the magnetizing current $f(v)$.

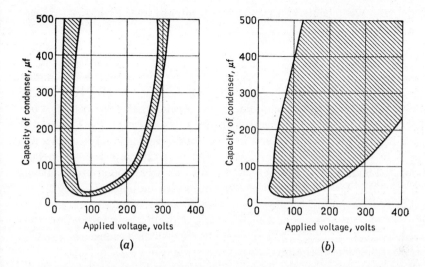

FIGURE 7.10 Regions in which the ⅓-harmonic oscillation is sustained. (*a*) Magnetization curve by Eq. (7.32). (*b*) Magnetization curve by Eq. (7.33).

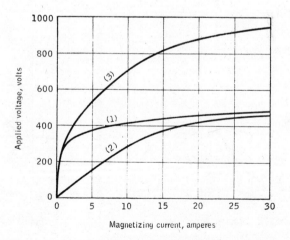

FIGURE 7.11 Combined characteristic approximated to a cubic curve. (1) Without air gap, (2) with air gap, (3) combined characteristic of (1) and (2).

in which the coefficients c_5, c_7, ... predominate over c_1 and c_3. If we used a core whose characteristic is expressed by

$$f(v) = c_1v + c_3v^3 \tag{7.33}$$

the blank part inside the shaded region would be eliminated, as one might expect from the theoretical consideration in Sec. 7.3. Such a core is not available in practice, but we can obtain the characteristic of Eq. (7.33) by connecting a number of inductance coils in series and adjusting the length of the air gap which is interposed in each core. In Fig. 7.11 is shown an example in which two cores are used, one with air gap and the other without. The resultant characteristic shows a fairly good approximation to Eq. (7.33). By making use of this composite inductor, the region in which the subharmonic oscillation of order ⅓ occurs is sought and plotted in Fig. 7.10*b*. We see that the self-excited oscillations related to the unstable regions of order $n \geq 2$ are completely excluded; hence the experimental verification of the preceding analysis is quite satisfactory.

We further measure the harmonic contents in these oscillations with a heterodyne harmonic analyzer. The result in the cases in which the nonlinearities are given by Eqs. (7.32) and (7.33), respectively, is shown in Fig. 7.12. In Fig. 7.12*a*, we observe the higher harmonics of orders ⅔, ⅗, ⅞, ... The oscillation of order ⅔ is particularly significant, because it is this oscillation (related to the second unstable region) that grows up rapidly and interrupts the original subharmonic oscillation (see

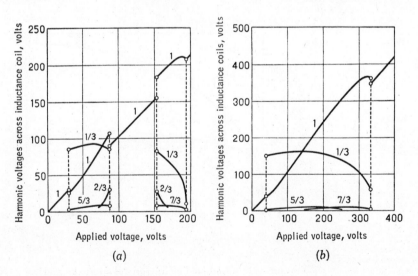

FIGURE 7.12 Harmonic analysis of the ⅓-harmonic oscillations. (*a*) Magnetization curve by Eq. (7.32). (*b*) Magnetization curve by Eq. (7.33).

Figs. 7.8 and 7.10*a*).   On the other hand, in Fig. 7.12*b*, no such obstructive oscillation is observed; the subharmonic oscillation of order $\frac{1}{3}$ is sustained in the whole region (see Figs. 7.3 and 7.10*b*).

Finally, it is to be noted that though the subharmonic oscillation of order $\frac{1}{5}$ may occur when the nonlinearity is given by Eq. (7.32), this oscillation has not been observed when the nonlinearity is given by Eq. (7.33).   This result also agrees with the investigation in Sec. 7.2.

7.6 *Subharmonic Oscillations of Order* $\frac{1}{2}$ *with the Nonlinear Characteristic Represented by a Cubic Function*

This section deals with the subharmonic oscillation of order $\frac{1}{2}$, the least period of which is twice the period of the external force.   As mentioned in Sec. 7.2, the oscillation of order $\frac{1}{2}$ is apt to occur when the nonlinearity is unsymmetrical; therefore, we consider the differential equation

$$\frac{d^2v}{d\tau^2} + k\frac{dv}{d\tau} + c_1v + c_2v^2 + c_3v^3 = B\cos 2\tau \tag{7.34}$$

in which the nonlinear restoring force is made unsymmetrical by the addition of a quadratic term.   Analogously as for the case of Sec. 5.2, this equation is readily transformed to the alternative form

$$\frac{d^2v'}{d\tau^2} + k\frac{dv'}{d\tau} + c_1'v' + c_3'v'^3 = B\cos 2\tau + B_0 \tag{7.35}$$

in which the restoring force is symmetrical, but the external force is unsymmetrical since it contains a unidirectional component B_0.   For the convenience of analysis, we shall take the form of Eq. (7.35) and investigate the subharmonic oscillations of order $\frac{1}{2}$ in what follows.

(a) *Periodic States of Equilibrium*

Since there will be no confusion in discarding the primes in Eq. (7.35), we write the fundamental equation as[1]

$$\frac{d^2v}{d\tau^2} + k\frac{dv}{d\tau} + c_1v + c_3v^3 = B\cos 2\tau + B_0 \tag{7.36}$$

and assume a periodic solution of the form

$$v_0(\tau) = z + x\sin\tau + y\cos\tau + w\cos 2\tau \tag{7.37}$$

in which the constant term z, the subharmonic oscillation $x\sin\tau + y\cos\tau$, and the oscillation having the driving frequency $w\cos 2\tau$ are considered to be of prime importance.   Following Mandelstam and Papalexi [72,

---

[1] It is noted that $c_1 + c_3 = 1$.

p. 227], the amplitude w may further be approximated by

$$w = \frac{1}{1 - 2^2} B = -\frac{1}{3} B \tag{7.38}$$

We shall show in part (c) that this approximation is permissible.

Substituting Eq. (7.37) into (7.36) and equating the coefficients of the terms containing $\sin \tau$ and $\cos \tau$ and of the nonoscillatory term separately to zero yields

$$Ax + ky = -3c_3wxz \qquad kx - Ay = -3c_3wyz$$
$$c_1z + c_3[(\tfrac{3}{2}r^2 + z^2 + \tfrac{3}{2}w^2)z - \tfrac{3}{4}w(x^2 - y^2)] = B_0 \tag{7.39}$$
where $\quad A = (1 - c_1) - c_3(\tfrac{3}{4}r^2 + 3z^2 + \tfrac{3}{2}w^2) \qquad r^2 = x^2 + y^2$

Elimination of x and y in Eqs. (7.39) gives us

$$A^2 + k^2 = (3c_3wz)^2$$
$$c_1z + c_3\left[\left(\frac{3}{2}r^2 + z^2 + \frac{3}{2}w^2\right)z + \frac{Ar^2}{4c_3z}\right] = B_0 \tag{7.40}$$

and the components x and y of the amplitude r are found to be

$$\begin{array}{ll} x = r\cos\theta & r\cos(\theta + 180°) \\ y = r\sin\theta & r\sin(\theta + 180°) \end{array} \tag{7.41}$$
where $\quad \sin 2\theta = -\dfrac{k}{3c_3wz} \qquad \cos 2\theta = -\dfrac{A}{3c_3wz}$

An example of the amplitude characteristics calculated by using Eqs. (7.40) is shown in Fig. 7.13, where the system parameters are given by[1]

$$c_1 = 0 \qquad c_3 = 1.0 \qquad \text{and} \qquad k = 0.1$$

(b) Stability Condition of the Periodic Solutions

We investigate the stability of the equilibrium states as determined by Eqs. (7.39). Proceeding in the usual manner, we consider a small variation ξ from the equilibrium state. Then the variational equation is given by

$$\frac{d^2\xi}{d\tau^2} + k\frac{d\xi}{d\tau} + c_1\xi + 3c_3v_0^2\xi = 0 \tag{7.42}$$

---

[1] We see from Eqs. (7.40) and (7.41) that, if the sign of B_0 is reversed, the signs of z and consequently of $\sin 2\theta$ and $\cos 2\theta$ are also reversed, resulting in the shift in θ by 90 degrees. Hence, if the sign of B_0 is reversed, the components x and y are given by

$$\begin{array}{ll} x = r\cos(\theta + 90°) & r\cos(\theta + 270°) \\ y = r\sin(\theta + 90°) & r\sin(\theta + 270°) \end{array}$$

When $B_0 = 0$ in particular, four types of the $\tfrac{1}{2}$-harmonic oscillations exist, each differing in phase by 90 degrees from the other.

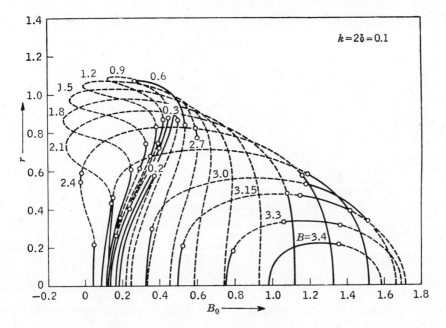

FIGURE 7.13   Amplitude characteristics of the $\frac{1}{2}$-harmonic oscillation; nonlinearity by cubic function.

and, by use of the transformation

$$\xi = e^{-\delta\tau}\eta \qquad \text{with} \qquad 2\delta = k$$

Eq. (7.42) leads to

$$\frac{d^2\eta}{d\tau^2} + (c_1 - \delta^2 + 3c_3v_0^2)\eta = 0 \qquad (7.43)$$

Since the periodic solution $v_0(\tau)$ has been determined in the preceding part, substitution of Eq. (7.37) into (7.43) gives a Hill's equation of the form

$$\frac{d^2\eta}{d\tau^2} + \left[\theta_0 + 2 \sum_{n=1}^{4} \theta_n \cos (n\tau - \epsilon_n) \right] \eta = 0$$

where $\theta_0 = c_1 + \frac{3}{2}c_3(r^2 + 2z^2 + w^2) - \delta^2$

$$\theta_n{}^2 = \theta_{ns}{}^2 + \theta_{nc}{}^2 \qquad \epsilon_n = \tan^{-1}\frac{\theta_{ns}}{\theta_{nc}} \qquad (7.44)$$

$$\theta_{1s} = 3c_3x(z - \tfrac{1}{2}w) \qquad \theta_{1c} = 3c_3y(z + \tfrac{1}{2}w)$$
$$\theta_{2s} = \tfrac{3}{2}c_3xy \qquad \theta_{2c} = \tfrac{3}{4}c_3(-x^2 + y^2 + 4wz)$$
$$\theta_{3s} = \tfrac{3}{2}c_3wx \qquad \theta_{3c} = \tfrac{3}{2}c_3wy$$
$$\theta_{4s} = 0 \qquad \theta_{4c} = \tfrac{3}{4}c_3w^2$$

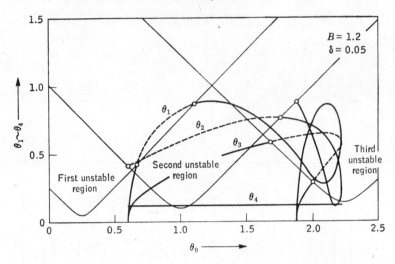

FIGURE 7.14   Loci of θ's in Eqs. (7.44) as obtained by varying B_0.

Following the investigation in Sec. 4.2, the equilibrium states will be stable provided that

$$\left[\theta_0 - \left(\frac{n}{2}\right)^2\right]^2 + 2\left[\theta_0 + \left(\frac{n}{2}\right)^2\right]\delta^2 + \delta^4 > \theta_n{}^2 \qquad n = 1, 2, 3, 4 \quad (7.45)$$

This is the stability condition (of the first approximation) for the nth unstable region. In order that the periodic states of equilibrium be stable, the condition (7.45) must be satisfied for all values of n simultaneously.[1] Once the periodic solution (7.37) is known, the values of θ_0 to θ_4 may readily be calculated from Eqs. (7.44). Figure 7.14 shows the loci of θ_1 to θ_4, which are drawn by varying B_0 in the case in which $2\delta = 0.1$ and $B = 1.2$. The unstable regions in this plane are readily obtained by the condition (7.45). The boundaries between the stable and the unstable regions are given by hyperbolic curves, the upper sides of which are unstable. Hence, the loci of θ_1 to θ_3 enter into the unstable regions in the respective dashed intervals, and consequently the corresponding equilibrium states are unstable.

Thus far the discussion of stability refers to the first approximation as given by inequality (7.45). However, the analytical result obtained in this way does not agree very well with an experiment which will be described in Sec. 7.8. Particularly, the stability condition for $n = 2$ fails to verify that the vertical tangency of the characteristic curves (see Fig.

[1] If θ_n enters into the nth unstable region, the $n/2$th harmonic of the subharmonic oscillation, i.e., the oscillation of order $n/4$, is excited with negative damping (see Sec. 4.4c).

7.13) occurs at the stability limit of the second unstable region. This is due to the deficiency of approximation in the derivation of the stability condition (7.45). Therefore, it is preferable to use the improved stability condition (4.15) in Sec. 4.3 instead of (7.45). Thus we once again write[1]

$$4(\delta^2 - \mu^2) = (\theta_0 + \delta^2 - 1)^2 - \theta_2{}^2 - 2\,\frac{\theta_0 + \delta^2 - 1}{\theta_0 + \delta^2}\,\theta_1{}^2$$

$$+ \frac{2}{\theta_0 + \delta^2}\,[(\theta_{1c}{}^2 - \theta_{1s}{}^2)\theta_{2c} + 2\theta_{1c}\theta_{1s}\theta_{2s}] + 4\delta^2 > 0 \quad (7.46)$$

Substituting the parameters θ as given by Eqs. (7.44) gives

$$\frac{1}{c_1 + \tfrac{3}{2}c_3(r^2 + 2z^2 + w^2)}\left[2c_3{}^2\left(A + \frac{3}{2}\,c_3 w^2\right)\left(1 - \frac{3}{2}\,r^2 + 3z^2 - \frac{3}{2}\,w^2\right)\right.$$

$$\left. + \frac{1}{2}\,c_3 A r^2\left(\frac{A}{2z^2} + 3c_3\right)\right] - c_3 A > 0 \quad (7.47)$$

Furthermore, by virtue of Eqs. (7.39), this condition leads to

$$\left(1 - \frac{3}{4}\,r^2 - 3z^2\right)z\,\frac{dB_0}{dr} > 0 \qquad (7.48)$$

where we have considered the case in which B_0 varies but B is kept constant. Hence it is clear that the characteristic curves of Fig. 7.13 have vertical tangents at the stability limit of the second unstable region.

Now, taking again the case of $k = 2\delta = 0.1$ and $B = 1.2$, we show in Fig. 7.15a the amplitude r of the subharmonic oscillation against B_0. The curve is extended on both sides by fine lines to obtain a single value of r for a given B_0. The value of μ_2 (the subscript designating the order of the corresponding unstable region) is calculated by Eq. (7.46), that is,[2]

$$4\mu_2{}^2 = -\,(\theta_0 + \delta^2 - 1)^2 + \theta_2{}^2 + 2\,\frac{\theta_0 + \delta^2 - 1}{\theta_0 + \delta^2}\,\theta_1{}^2$$

$$- \frac{2}{\theta_0 + \delta^2}\,[(\theta_{1c}{}^2 - \theta_{1s}{}^2)\theta_{2c} + 2\theta_{1c}\theta_{1s}\theta_{2s}] \quad (7.49)$$

and is plotted against B_0 in Fig. 7.15b. The curves of μ_1 and μ_3 calculated from

$$\mu_n{}^2 = -\left[\theta_0 + \left(\frac{n}{2}\right)^2\right] + \sqrt{n^2\theta_0 + \theta_n{}^2} \qquad \text{[cf. Eq. (3.86) in Chap. 3]}$$

$$(7.50)$$

---

[1] This particular condition for $n = 2$ is most important, because it tests the stability against the buildup of an unstable oscillation having the same frequency as the subharmonic frequency.

[2] As noted in Sec. 4.3a, Eq. (7.49) is valid only for $|\mu_2| = \delta$; therefore, it does not give a correct value of μ_2 when $|\mu_2| \neq \delta$. However, this is not important when we are concerned only with the stability problem.

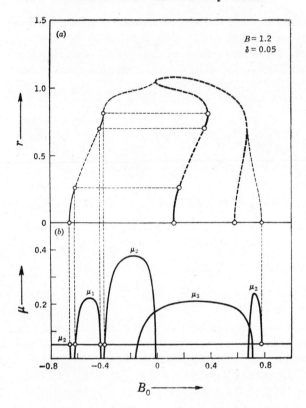

FIGURE 7.15   Stability of the equilibrium states. (*a*)
Amplitude of the ½-harmonic oscilla-
tion. (*b*) Characteristic   exponent   related to the stability.

are also shown in the figure.[1]   The equilibrium states are stable when
$|\mu_n| < \delta$.   On the other hand, if $|\mu_n| > \delta$, the $n/2$th harmonic of the subharmonic oscillation, i.e., the $n/4$th harmonic of the external force, builds
up and the original subharmonic oscillation ceases to exist.   The stable
states are shown by full lines in Fig. 7.15*a*.

By making similar calculations for various values of B, we find that
the full-line portions of the characteristic curves in Fig. 7.13 represent

---

[1] In Fig. 7.15*b* we see that the curves of μ_2 and μ_3 intersect each other, so that real
values of μ_2 and μ_3 exist simultaneously for certain values of B_0.   This is due to the
deficiency of approximation in computing the characteristic exponent.   Properly
speaking, these curves should not intersect, because an intersection would lead to the
contradiction that there are more than two linearly independent solutions for the
second-order differential equation of Eqs. (7.44).

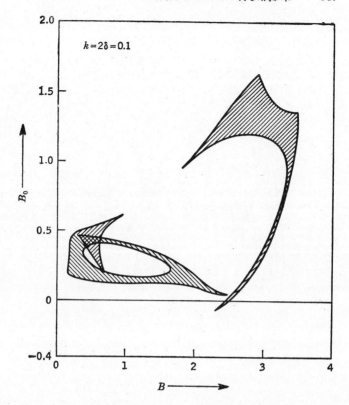

FIGURE 7.16   Region in which the ½-harmonic oscillation is sustained (calculated).

the stable states of equilibrium. Further plotted in the BB_0 plane of Fig. 7.16 is the region in which the subharmonic oscillation of order ½ is sustained. The subharmonic oscillation does not occur in the blank areas surrounded by the shaded region. As one sees from the result of Fig. 7.15, this is due to the buildup of an unstable oscillation having the ¼-harmonic or the ¾-harmonic frequency. It is also noticed that two different types of ½-harmonic oscillation may occur in the doubly shaded portion of the region; which one will occur depends on the initial condition.[1] Finally, the complete amplitude characteristics of the ½-harmonic oscillation are shown in Fig. 7.17, in which the stable states are indicated by full lines.

[1] Further details concerning the possible types of ½-harmonic oscillation will be discussed in Sec. 9.5 in connection with the transient state of the oscillation.

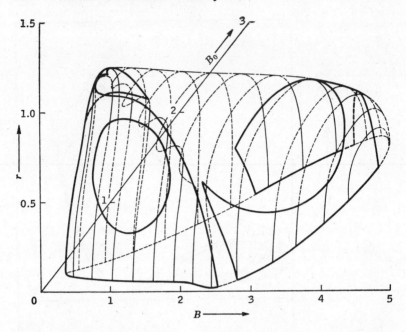

FIGURE 7.17   Complete amplitude characteristics of the ½-harmonic oscillation (calculated).

(c) On the Approximation of Eq. (7.38)

As we have previously mentioned in Sec. 7.3c, the relation

$$w = \frac{1}{1 - \nu^2} B = -\frac{1}{8} B \qquad \text{for } \nu = 3$$

is a fairly good approximation even when the departure from linearity is large. This is also the case when we deal with the subharmonic oscillation of order ½ (that is, $\nu = 2$). We shall confirm this for the nondissipative case. We assume the periodic solution of the form

$$v_0 = z + y \cos \tau + w \cos 2\tau \qquad (7.51)$$

where z, y, and w are unknown quantities to be determined shortly.

By substituting Eq. (7.51) into the original Eq. (7.36), thereby putting $k = 0$, $c_1 = 0$, and $c_3 = 1$, and equating the constant terms, the coefficients of $\cos \tau$ and $\cos 2\tau$ on both sides, respectively, we obtain

$$\begin{aligned}
z^3 + \tfrac{3}{2}zy^2 + \tfrac{3}{2}zw^2 + \tfrac{3}{4}y^2w &= B_0 \\
3z^2 + 3zw + \tfrac{3}{4}y^2 + \tfrac{3}{2}w^2 - 1 &= 0 \\
3z^2w + \tfrac{3}{2}zy^2 + \tfrac{3}{2}zy^2w + \tfrac{3}{4}w^3 - 4w &= B
\end{aligned} \qquad (7.52)$$

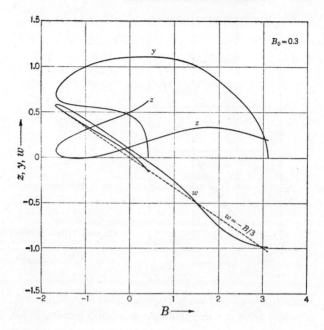

FIGURE 7.18 Relationship between w, y, z, and B in Eqs. (7.52).

The coefficients z, y, and w in Eq. (7.51) can be determined by solving Eqs. (7.52) for given values of B and B_0. When the calculation is carried out, it is seen that the approximation (7.38) may be used with acceptable accuracy. Taking the case of $B_0 = 0.3$ by way of example, the values of z, y, and w are plotted against B in Fig. 7.18. It is apparent that the exact values of w calculated from Eqs. (7.52) are closely approximated by $w = -B/3$.

7.7 Subharmonic Oscillations of Order $\frac{1}{2}$ with the Nonlinear Characteristic Represented by a Symmetrically Quadratic Function [38]

(a) Suppression of the Undesirable Self-excited Oscillations

As mentioned in Secs. 7.3 and 7.5, the subharmonic oscillation of order $\frac{1}{3}$ is most stably maintained when the nonlinear characteristic is given by a cubic function. From the analogy of this fact, one may infer that an appropriate form of the nonlinearity to produce the subharmonic oscillation of order $\frac{1}{2}$ will be

$$f(v) = |v|v \tag{7.53}$$

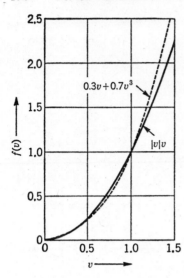

FIGURE 7.19   Nonlinear characteristic $|v|v$ and its approximation by a power-series expansion.

and thus the function is quadratic for $v \gtrless 0$ but is odd in v.*   For lack of a better term we call it a *symmetrically quadratic function*.   Accordingly, we write the fundamental equation for the subharmonic oscillation of order $\frac{1}{2}$ as

$$\frac{d^2v}{d\tau^2} + k\frac{dv}{d\tau} + |v|v = B \cos 2\tau + B_0 \qquad (7.54)$$

However, the expression $|v|v$ is difficult to handle analytically; therefore, we expand this into a power series as given by Eq. (7.2).   If we take only the first two terms for simplicity, we have the same equation (7.36) under the assumption that $|v|v$ is permissibly approximated by $c_1v + c_3v^3$.   For instance, the nonlinearity in Eq. (7.36) may be fixed by

$$f(v) = c_1v + c_3v^3 = 0.3v + 0.7v^3 \qquad (7.55)$$

The constants c_1 and c_3 are so chosen that the difference between $|v|v$ and $c_1v + c_3v^3$ is small enough for the range of the variable v in which the $\frac{1}{2}$-harmonic oscillation occurs.   These characteristics are compared in Fig. 7.19.

In the preceding section we have shown the amplitude characteristic of the $\frac{1}{2}$-harmonic oscillation in the case where $c_1 = 0$ and $c_3 = 1.0$.   In this case the subharmonic oscillation may become unstable, owing to the buildup of an unstable oscillation related to the first or the third unstable region of the stability chart for Hill's equation.   We shall show in what follows that such unstable oscillations are suppressed if the nonlinearity is given by Eq. (7.53) or (7.55).   Figure 7.20a shows an amplitude charac-

* Analogously, an appropriate form of $f(v)$ to produce the subharmonic oscillation of order $1/\nu$ will be $|v^{\nu-1}|v$.

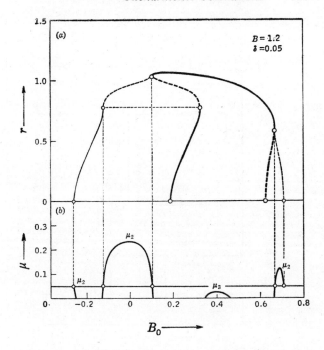

FIGURE 7.20   Stability of the equilibrium states. (a) Amplitude of the ½-harmonic oscillation. (b) Characteristic exponent related to the stability.

teristic of the ½-harmonic oscillation where the system parameters in Eq. (7.36) are given by

$$c_1 = 0.3 \qquad c_3 = 0.7 \qquad k = 0.1 \qquad \text{and} \qquad B = 1.2$$

Through use of Eqs. (7.49) and (7.50), the characteristic exponents μ_1 to μ_3 are calculated; they are shown in Fig. 7.20b. Thus we see that the unstable oscillation is caused only by the condition $|\mu_2| > \delta$ and that the buildup of the ¼-harmonic and the ¾-harmonic is eliminated (cf. Fig. 7.15).

(b) Analog-computer Analysis

By making use of an analog computer, we seek the region in which the subharmonic oscillation of order ½ is sustained. The systems under consideration are described by

$$\frac{d^2v}{d\tau^2} + 2\delta \frac{dv}{d\tau} + v^3 = B \cos 2\tau + B_0 \qquad (7.56)$$

and

$$\frac{d^2v}{d\tau^2} + 2\delta \frac{dv}{d\tau} + |v|v = B \cos 2\tau + B_0 \qquad (7.57)$$

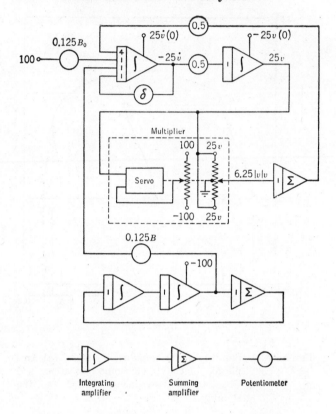

FIGURE 7.21   Computer block diagram for Eq. (7.57).

Figure 7.21 shows the block diagram of an analog-computer setup for the solution of Eq. (7.57).   The symbols in the figure follow the conventional notation.[1]   The nonlinear characteristic $|v|v$ is exactly obtained by a servomultiplier connection as indicated in the figure.   For the solution of Eq. (7.56), this servomultiplier may be replaced by an ordinary multiplier which produces a cubic function.

The subharmonic oscillation of order $\frac{1}{2}$ may be started if the initial conditions, $v(0)$ and $\dot{v}(0)$, on the integrating amplifiers are appropriately chosen.   Then, by slowly varying the values of B and B_0, we obtain the region in which the subharmonic oscillation is sustained.   Figure 7.22 shows such a region (shaded) for the system described by Eq. (7.56), and Fig. 7.23 is obtained for Eq. (7.57).   As expected from the theoretical

---

[1] The integrating amplifiers in the block diagram integrate the inputs with respect to the machine time t (in seconds), which is 2 times the nondimensional time τ, that is, $t = 2\tau$.

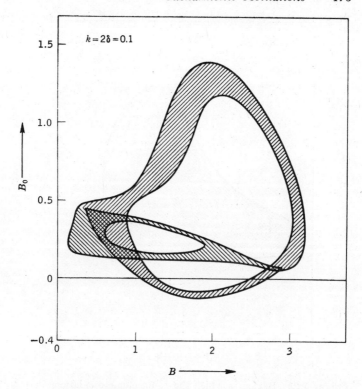

FIGURE 7.22 Region in which the ½-harmonic oscillation is sustained; nonlinearity by cubic function (analog-computer analysis).

consideration in Sec. 7.6, the occurrence of the blank areas inside the shaded region of Fig. 7.22 is due to the buildup of unstable oscillations having the ¼-harmonic and the ¾-harmonic frequencies (Fig. 7.16). On the other hand, no such obstructive oscillations are observed in Fig. 7.23. Thus we see that the buildup of the ¼-harmonic and the ¾-harmonic are completely eliminated when the nonlinearity is given by $f(v) = |v|v$.

7.8 *Experimental Investigation*

In this section we observe the subharmonic oscillation of order ½ governed by the original equation (7.36), and we compare the results obtained from experiment with those of the foregoing analysis.

The schematic diagram in Fig. 7.24 shows an electric circuit in which the subharmonic oscillation takes place owing to the presence of a saturable-core inductor L under the impression of an alternating voltage

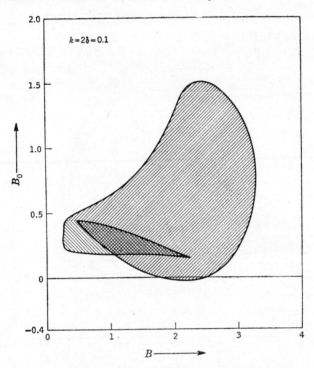

FIGURE 7.23   Region in which the ½-harmonic oscillation
is sustained; nonlinearity by symmetrically
quadratic function (analog-computer analysis).

$E \sin 2\omega t$.   As shown in the figure, the secondary winding is provided on
the core in order to afford the asymmetry to the nonlinear characteristic
by forcing a constant direct-current flow through it.   With the notation
of Fig. 7.24, we have

$$n\frac{d\phi}{dt} + Ri_R = E \sin 2\omega t \qquad Ri_R = \frac{1}{C}\int i_C \, dt \qquad i = i_R + i_C \quad (7.58)$$

where n is the number of turns of the primary winding and ϕ is the magnetic flux in the core.

By proceeding in the same manner as described in Sec. 5.1a, we
introduce the nondimensional variables u, u_0, and v in place of i, i_0, and
ϕ by the relations

$$i = I_n u \qquad i_0 = I_n u_0 \qquad \text{and} \qquad \phi = \Phi_n v \qquad (7.59)$$

where I_n and Φ_n are appropriate base quantities of the current and the

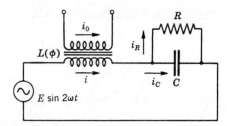

FIGURE 7.24   Oscillatory circuit containing reactor with direct current superposed.

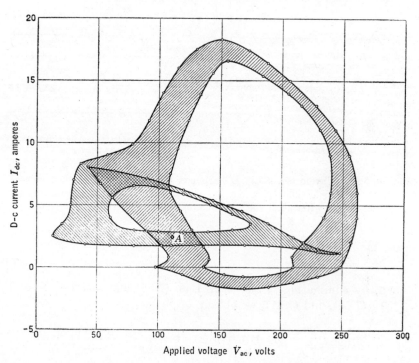

FIGURE 7.25   Region in which the ½-harmonic oscillation is sustained (experimental).

flux, respectively.   Then, neglecting hysteresis, we may take the saturation curve of the form[1]

$$u + u_0 = c_1 v + c_3 v^3 + c_5 v^5 + \cdots \tag{7.60}$$

where c_1, c_3, c_5, . . . are constants characterizing the core.   Further, as we have done before, the base quantities I_n and Φ_n may expediently be

[1] It is tacitly assumed that the secondary winding of the reactor has the same number of turns as the primary winding.

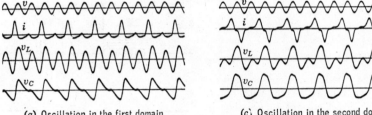

(a) Oscillation in the first domain

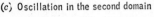

(c) Oscillation in the second domain

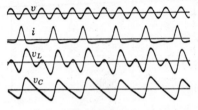

(b) Similar to (a), but accompanied
with 1/4-harmonic

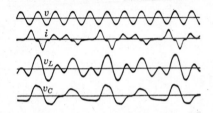

(d) Similar to (c), but accompanied
with 3/4-harmonic

Figure 7.26   Waveforms of the ½-harmonic oscillation. v = applied voltage,
i = oscillating current, v_L = terminal voltage across reactor, and
v_C = terminal voltage across capacitor.

fixed by

$$n\omega^2 C\Phi_n = I_n \qquad c_1 + c_3 + c_5 + \cdots = 1 \qquad (7.61)$$

Then, by eliminating i_R and i_C in Eqs. (7.58) and making use of Eqs.
(7.59), (7.60), and (7.61), we obtain

$$\frac{d^2v}{d\tau^2} + k\frac{dv}{d\tau} + c_1v + c_3v^3 + c_5v^5 + \cdots = B\cos 2\tau + B_0$$

where
$$\tau = \omega t - \frac{1}{2}\tan^{-1}\frac{k}{2} \qquad k = \frac{1}{\omega CR} \qquad (7.62)$$

$$B = \frac{E}{n\omega\Phi_n}\sqrt{4 + k^2} \qquad B_0 = u_0$$

Furthermore, in order to secure a better agreement with the analysis in
Sec. 7.6, we use a composite reactor[1] the saturation curve of which is
given by a cubic function

$$u + u_0 = c_1v + c_3v^3$$

Then Eq. (7.62) leads·to the original form (7.36).

[1] As already explained in Sec. 7.5, a number of magnetic cores with air gaps of
appropriate length are used to obtain the desired saturation curve.

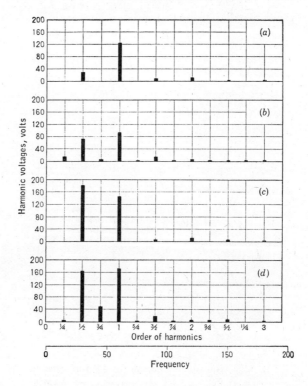

FIGURE 7.27 Harmonic analysis of v_L in the oscillograms of Fig. 7.26.

We now proceed to describe the result of experiments which are effective in verifying the foregoing analysis. We first determine the region in which the subharmonic oscillation of order $\frac{1}{2}$ is sustained. This region is plotted in the $V_{ac}I_{dc}$ plane of Fig. 7.25, where V_{ac} is the applied a-c voltage and I_{dc} is the constant direct current in the secondary winding (Fig. 7.24). Since the coordinates V_{ac} and I_{dc} respectively correspond to B and B_0 in the preceding sections, a considerable agreement is found between the results in Figs. 7.16 and 7.25.[1] In Fig. 7.25 we see that the region of the $\frac{1}{2}$-harmonic oscillation may be divided into two domains as indicated by changing the direction of the hatching. For conven'ence we call them the first and the second domains, the former being extended from $I_{dc} \cong 2$ to 8 amp and the latter from $I_{dc} \cong -2$ to 18 amp. In the particular areas common to these domains, for instance, at the point A, two types of the $\frac{1}{2}$-harmonic oscillation exist for a given

---

[1] The region of the $\frac{1}{2}$-harmonic oscillation in Fig. 7.16 extends down to the negative side of B_0 as the damping coefficient k (or 2δ) decreases.

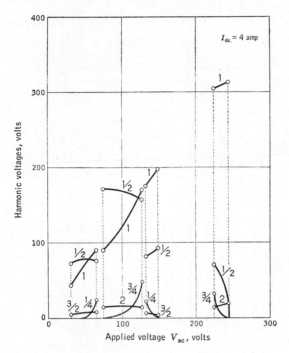

FIGURE 7.28   Harmonic analysis of v_L for varying V_{ac}.

set of V_{ac} and I_{dc}. The waveforms of these oscillations will be shown shortly.

As expected from the result of Fig. 7.15, the blank area inside the first domain is due to the self-excitation of an unstable oscillation of order $\frac{1}{4}$, because in this case the characteristic exponent μ_1 associated with the first unstable region of Hill's equation grows up; in other words, the stability condition $|\mu_1| < \delta$ fails. Similarly, in the blank area surrounded by the second domain, an unstable oscillation of order $\frac{3}{4}$ disturbs the original $\frac{1}{2}$-harmonic oscillation, because $|\mu_3|$ is greater than δ in this case.

The above discussion is confirmed experimentally by observing the waveform of the $\frac{1}{2}$-harmonic oscillation. Oscillograms (a) and (c) in Fig. 7.26 are obtained for the values of V_{ac} and I_{dc} as given by the coordinates of the point A in Fig. 7.25, (a) being associated with the first domain and (c) with the second domain. Oscillograms (b) and (d) are obtained in the cases where the point A approaches the blank areas of the first and the second domains, respectively. The voltage v_L across the reactor is analyzed by making use of a harmonic analyzer. The

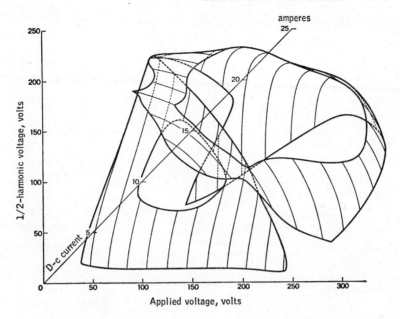

FIGURE 7.29 Complete amplitude characteristics of the ½-harmonic oscillation (experimental).

result is shown in Fig. 7.27, in which (a) to (d) respectively correspond to oscillograms (a) to (d) in Fig. 7.26. Thus, as expected, the oscillation of order ¼ is detected in (b) and the oscillation of order ¾ is noticeable in (d).[1]

We further measure the harmonic voltages across the reactor for varying V_{ac} while holding I_{dc} at a constant value (4 amp), and we obtain Fig. 7.28. It is evidently shown that the buildup of the unstable oscillations of orders ¼ and ¾ causes the collapse of the ½-harmonic oscillation in the blank areas of the first and the second domains, respectively. Finally, the complete amplitude characteristics of the ½-harmonic oscillation measured by using a harmonic analyzer are plotted in Fig. 7.29. The figure shows a satisfactory agreement with the theoretical result obtained in Fig. 7.17.

From the experimental results and the preceding analysis, the salient features of the ½-harmonic oscillation are summarized as follows:

1. The ½-harmonic oscillation is likely to occur in the case in which

---

[1] It is noted that, in (a) and (c) of Fig. 7.27, the oscillations of orders ½ and 1 predominate over all others. Since the reactor voltage is proportional to the time derivative of the flux in the core, the approximation made for the periodic solution of the flux, that is, Eq. (7.37), is quite satisfactory.

either the nonlinear restoring force or the external force is unsymmetrical. However, it is noted that the $\frac{1}{2}$-harmonic oscillation may still occur in a symmetrical system if the damping is sufficiently small.[1]

2. Contrary to the result obtained for the $\frac{1}{3}$-harmonic oscillation, two different types of the $\frac{1}{2}$-harmonic oscillation are sustained, under appropriate conditions, in the same system but with different initial conditions (see Sec. 9.5 for further details).

3. The $\frac{1}{2}$-harmonic oscillation is frequently accompanied by the self-excitation of an unstable oscillation of order $\frac{1}{4}$ or $\frac{3}{4}$, which will interrupt the continuation of the original oscillation. The generalized stability condition as given in Chap. 4 is useful in detecting this kind of instability.

4. An appropriate form of the nonlinearity to produce a $\frac{1}{2}$-harmonic oscillation is given by a symmetrically quadratic function. The buildup of unstable oscillations of orders $\frac{1}{4}$ and $\frac{3}{4}$ is completely suppressed in this case.

[1] Subharmonic oscillations of order $\frac{1}{2}$ occur when $B_0 = 0$ in Fig. 7.16 or when $I_{dc} = 0$ in Fig. 7.25.

Part III

Forced Oscillations
in Transient States

Chapter 8

Harmonic Oscillations

8.1 Introduction

It was mentioned in Sec. 4.1 that various types of periodic solution may exist for a given nonlinear differential equation, depending on different values of the initial condition. In Part II we have concentrated our attention on the periodic solutions and discussed their stability. We shall now investigate the transient state concerned with the oscillations until they get to the steady state. When we have done this, the relationship between the initial conditions and the resulting periodic oscillations will be made clear. But it is usually not possible to solve nonlinear differential equations with arbitrary initial conditions, except for special cases in which exact solutions are known in analytic form [18, p. 55]. Fortunately, however, we can to some extent succeed in investigating the transient state of the oscillations by means of the topological method of analysis [32, 33, 36].

Thus, the method of solution used in the following sections is, first, to transform second-order nonlinear differential equations of the non-autonomous type to the form

$$\frac{dy}{dx} = \frac{Y(x,y)}{X(x,y)} \tag{8.1}$$

and then to find its integral curves through the basic idea that the singular points of Eq. (8.1) are correlated with the steady state of the oscillations and the integral curves with the transient state.[1]

By using this method of analysis, we first study the harmonic oscillations in the present chapter and then the subharmonic oscillations in Chap. 9.

8.2 Periodic Solutions and Their Stability

We have discussed this problem in Sec. 5.1. Our object in this section is to review it with attention to the transient state rather than to confine attention to the steady state.

---

[1] This method of analysis has been used by Andronow and Witt [4] for the study of entrainment in a vacuum-tube oscillator.

We consider a system described by the differential equation

$$\frac{d^2v}{d\tau^2} + k\frac{dv}{d\tau} + v^3 = B\cos\tau \qquad (8.2)$$

where the restoring force is expressed by a cubic function in v and the damping coefficient k is assumed to be constant.[1]   In the case of harmonic oscillations in which the fundamental component having the period 2π predominates over the higher harmonics, the solution of Eq. (8.2) may be approximated by the form

$$v(\tau) = x(\tau)\sin\tau + y(\tau)\cos\tau \qquad (8.3)$$

in which the amplitudes $x(\tau)$ and $y(\tau)$ are both functions of τ and ultimately reduce to constants after passage through the transient state.[2]

Substituting Eq. (8.3) into (8.2) and equating the coefficients of the terms containing $\sin\tau$ and $\cos\tau$ separately to zero gives us

$$\frac{dx}{d\tau} = \frac{1}{2}(-kx + Ay + B) \equiv X(x,y)$$
$$\frac{dy}{d\tau} = \frac{1}{2}(-Ax - ky) \equiv Y(x,y) \qquad (8.4)$$

with $\qquad A = 1 - \tfrac{3}{4}r^2 \qquad r^2 = x^2 + y^2$

under the assumptions that the amplitudes $x(\tau)$ and $y(\tau)$ are such slowly varying functions of τ that $d^2x/d\tau^2$ and $d^2y/d\tau^2$ may be neglected and that the damping coefficient k is a sufficiently small quantity that $k\,dx/d\tau$ and $k\,dy/d\tau$ may also be discarded.   The result which will be obtained from Eqs. (8.4) may not account for the occurrence of pronounced higher-harmonic oscillations.   But as far as we deal with harmonic oscillations, Eqs. (8.4) may be considered to be legitimate.[3]

Equations (8.4) play a significant role in the following investigation, since they serve as the fundamental equations in studying the transient state as well as the steady state.   We shall, for the time being, consider the steady state where the amplitudes $x(\tau)$ and $y(\tau)$ in Eq. (8.3) are

---

[1] A linear term in v is discarded in Eq. (8.2), since it results in no significant change in the subsequent analysis.

[2] The periodic solution has been given by Eq. (5.8).

[3] The autonomous system (8.4) may also be obtained by making use of the averaging method described in Sec. 1.4b.   Strictly speaking, the assumption (originally due to Appleton and van der Pol [5, 86]) which is required for the derivation of the autonomous system (8.4) should be examined in order to expect a correct description of the oscillations governed by Eq. (8.2).   Some attempts in this line have been made by Kryloff and Bogoliuboff [55, p. 28] and Lefschetz [3, p. 341] in the case when the nonlinearity is sufficiently small.

constant, that is,

$$\frac{dx}{d\tau} = X(x,y) = 0 \qquad \frac{dy}{d\tau} = Y(x,y) = 0 \tag{8.5}$$

Substituting these conditions in Eqs. (8.4) leads to determination of the steady-state amplitude $r_0 \; (= \sqrt{x_0{}^2 + y_0{}^2})$ of the periodic solution $v_0(\tau)$ by

$$(A^2 + k^2)r_0{}^2 = B^2 \tag{8.6}$$

and the components x_0, y_0 of the amplitude r_0 are found to be

$$x_0{}^2 = \frac{r_0{}^2}{1 + (A/k)^2} \qquad y_0{}^2 = \frac{r_0{}^2}{1 + (k/A)^2} \tag{8.7}$$

These results agree with the investigation in Sec. 5.1b.

We have already discussed the stability of the periodic solution in Sec. 5.1c. As mentioned in Sec. 4.4a, the stability condition for the first unstable region, i.e., the condition (4.6) for $n = 1$, may also be obtained by making use of the Routh-Hurwitz criterion. Along this line we derive the stability condition of the periodic solution as follows.

Let ξ and η be small variations from the amplitudes x_0 and y_0, respectively, and determine whether these variations approach zero or not with increase of the time τ. From Eqs. (8.4) we obtain[1]

$$\frac{d\xi}{d\tau} = a_1\xi + a_2\eta \qquad \frac{d\eta}{d\tau} = b_1\xi + b_2\eta \tag{8.8}$$

where

$$a_1 = \left(\frac{\partial X}{\partial x}\right)_{\substack{x=x_0,\\ y=y_0}} = \frac{1}{2}\left(-k - \frac{3}{2}x_0 y_0\right)$$

$$a_2 = \left(\frac{\partial X}{\partial y}\right)_{\substack{x=x_0,\\ y=y_0}} = \frac{1}{2}\left(A - \frac{3}{2}y_0{}^2\right)$$

$$b_1 = \left(\frac{\partial Y}{\partial x}\right)_{\substack{x=x_0,\\ y=y_0}} = \frac{1}{2}\left(-A + \frac{3}{2}x_0{}^2\right)$$

$$b_2 = \left(\frac{\partial Y}{\partial y}\right)_{\substack{x=x_0,\\ y=y_0}} = \frac{1}{2}\left(-k + \frac{3}{2}x_0 y_0\right)$$

$$\tag{8.8a}$$

The characteristic equation of the system defined by Eqs. (8.8) is given by

$$\begin{vmatrix} a_1 - \lambda & a_2 \\ b_1 & b_2 - \lambda \end{vmatrix} = 0$$

or

$$\lambda^2 - (a_1 + b_2)\lambda + a_1 b_2 - a_2 b_1 = 0 \tag{8.9}$$

The variations ξ and η approach zero with the time τ, provided that the real part of λ is negative. In this case the corresponding periodic solution is stable. This stability condition is given by the Routh-Hurwitz cri-

---

[1] Terms of degree higher than the first in ξ and η are neglected.

terion (Sec. 3.3), that is,

$$-a_1 - b_2 > 0 \qquad \text{and} \qquad a_1 b_2 - a_2 b_1 > 0 \qquad (8.10)$$

Substituting Eqs. (8.8a) into (8.10) and using Eq. (8.6) leads to

$$k > 0$$

$$\text{and} \qquad \frac{27}{16} r_0{}^4 - 3r_0{}^2 + k^2 + 1 > 0 \qquad \text{or} \qquad \frac{dB^2}{dr_0{}^2} > 0 \qquad (8.11)$$

Evidently the first condition is fulfilled from the outset, since we are concerned with the positive damping. The second condition is the same as we have previously obtained in Sec. 5.1c.

8.3 Analysis of Harmonic Oscillations by Means of Integral Curves

As mentioned before, our object is to study the solution of Eq. (8.2) in the transient state, which, with the lapse of time, yields ultimately the periodic solution. For this purpose it is useful to investigate, following Poincaré [81, 82, 84] and Bendixson [6],[1] the integral curves of the following equation derived from Eqs. (8.4), that is,

$$\frac{dy}{dx} = \frac{Y(x,y)}{X(x,y)} \qquad (8.12)$$

Since the time τ does not appear explicitly in this equation, we can draw the integral curves in the xy plane with the aid of the isocline method or otherwise.[2] As mentioned in Sec. 8.1, periodic solutions are correlated with $x(\tau) = $ constant, $y(\tau) = $ constant of Eqs. (8.4), that is, with the singular points of Eq. (8.12) where both $X(x,y)$ and $Y(x,y)$ vanish.

Now suppose that we fix a point $(x(0),y(0))$ in the xy plane as an initial condition. Then the representative point $(x(\tau),y(\tau))$ moves, with the increase of time τ, along the integral curve which passes through the initial point $(x(0),y(0))$ and leads ultimately to a stable singular point.[3] Thus the transient solutions are correlated with the integral curves of Eq. (8.12) and the steady (periodic) solutions with the singular points in the xy plane. The time-varying response of $v(\tau)$ in the transient state

---

[1] See also Refs. 16 and 89.

[2] See Secs. 2.5 to 2.7 for other graphical methods of solution.

[3] Generally speaking, the representative point may not always lead to a singular point, but may tend to a limit cycle along which the representative point travels permanently as τ increases. However, by making use of Bendixson's criterion (Sec. 2.2e), one readily sees that no limit cycle exists for Eq. (8.12), because

$$\frac{\partial X}{\partial x} + \frac{\partial Y}{\partial y} = -k$$

by virtue of Eqs. (8.8a). When $k = 0$, there exists a continuum of closed trajectories around a center, but no limit cycle exists (Sec. 8.5).

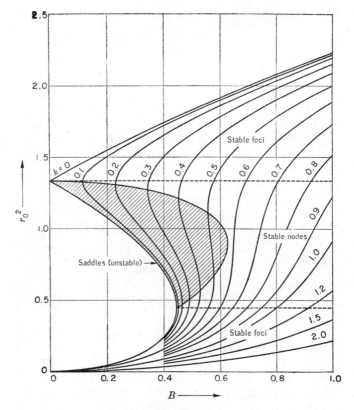

FIGURE 8.1   Amplitude characteristic of the harmonic oscillation
and the corresponding singularities.

may be obtained by the line integral

$$\tau = \int \frac{ds}{\sqrt{X^2(x,y) + Y^2(x,y)}} \qquad ds = \sqrt{(dx)^2 + (dy)^2} \qquad (8.13)$$

where $X(x,y)$ and $Y(x,y)$ are given by Eqs. (8.4) and ds is the line element
along the integral curve.

We discuss, for the time being, the types of the singular points
of Eq. (8.12) which are correlated with the periodic solutions of Eq. (8.2).
As mentioned in Sec. 2.2a, these singularities are classified according to
the nature of the roots of the characteristic equation (8.9), that is,

$$\lambda_{1,2} = \frac{a_1 + b_2 \pm \sqrt{(a_1 - b_2)^2 + 4a_2b_1}}{2} \qquad (8.14)$$

where a_1, a_2, b_1, and b_2 are given by Eqs. (8.8a).

The relationships between B and r_0^2 are first calculated from Eq. (8.6)
and plotted in Fig. 8.1 for several values of k.   We now distinguish these

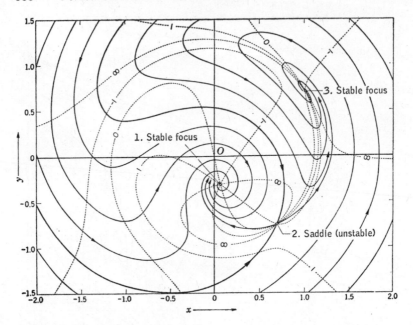

FIGURE 8.2   Integral curves for the harmonic oscillation with three singularities.   Full line: integral curves.   Dotted line: isoclines.

periodic states of equilibrium[1] according to the types of singularities.
The boundary line between nodes and saddles is to be found from Eqs.
(2.12) and (2.13) in Sec. 2.2a, that is,

$$a_1 b_2 - a_2 b_1 = 0 \tag{8.15}$$

which, upon substitution of Eqs. (8.8a), leads to

$$27\!/_{16} r_0{}^4 - 3 r_0{}^2 + k^2 + 1 = 0 \tag{8.16}$$

This equation is identical with $dB^2/dr_0{}^2 = 0$ [see the second condition of
(8.11)].   The region of the saddle points is marked by hatching where
$dB^2/dr_0{}^2$ is negative.   Since one of the characteristic roots is always positive, the periodic states in this región are unstable.   This agrees with the
result obtained in the foregoing section.

Similarly, the boundary lines between nodes and foci are determined
by Eqs. (2.12) and (2.14) in Sec. 2.2a, that is,

$$(a_1 - b_2)^2 + 4 a_2 b_1 = 0 \tag{8.17}$$

which, upon substitution of Eqs. (8.8a), leads to

$$27\!/_{16} r_0{}^4 - 3 r_0{}^2 + 1 = 0 \qquad \text{that is} \qquad r_0{}^2 = 4\!/_3,\ 4\!/_9 \tag{8.18}$$

[1] These periodic states of equilibrium actually exist when they are correlated
with stable singularities.

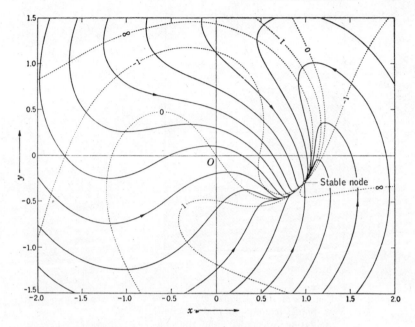

FIGURE 8.3 Integral curves for the harmonic oscillation with one singularity.
Full line: integral curves. Dotted line: isoclines.

These boundaries are shown by dashed lines in Fig. 8.1. Since, by Eqs.
(8.8*a*), $a_1 + b_2 = -k$, it is readily seen that the periodic states of equilibrium in the regions of nodes and foci are stable.

8.4 Phase-plane Analysis of the Harmonic Oscillations [32, 33, 36]

In the preceding section we have briefly referred to the transient
solutions which are correlated with the integral curves of Eq. (8.12). It
is, however, useful and illuminating to consider the integral curves for
certain typical cases. The special cases we consider are those stemming
from use of the following values of k and B in Eqs. (8.4), namely,

Case 1: $\qquad\qquad k = 0.2 \qquad B = 0.3$

and

Case 2: $\qquad\qquad k = 0.7 \qquad B = 0.75$

As observed in Fig. 8.1, there are three different states of equilibrium
in the first case, whereas there is only one in the second case. The
integral curves for these cases are shown in Figs. 8.2 and 8.3, respectively.
The singularities are determined by Eqs. (8.5); the corresponding details
are listed in Table 8.1.

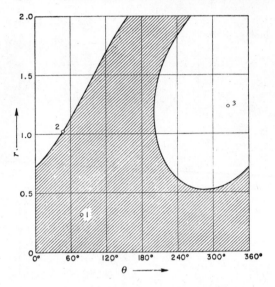

FIGURE 8.4   Regions of initial conditions in the
$r\theta$ plane leading to the resonant
oscillation (unshaded) and the nonresonant oscillation (shaded).

The integral curves in Figs. 8.2 and 8.3 are drawn with the aid of the
isoclines represented by dotted lines, the numbers on which indicate the
values of dy/dx for the respective isoclines.   As one sees from Eqs. (8.4),
a representative point $(x(\tau),y(\tau))$ moves, with the increase of time τ, along
the integral curve in the direction of the arrows and tends ultimately to a
stable singular point.

In Fig. 8.2 there are three singularities, points 1 and 3 are stable,
and the corresponding periodic states are realized.   The remaining
singularity, point 2, is a saddle point, which is intrinsically unstable.

Table 8.1   Singular Points in Figs. 8.2 and 8.3

Singular point	x_0	y_0	λ_1, λ_2	μ_1, μ_2*	Classification
Fig. 8.2, 1	0.067	-0.310	$-0.100 \pm 0.423i$		Stable focus
Fig. 8.2, 2	0.699	-0.748	$0.170, -0.370$	$0.392, \quad 2.113$	Saddle (unstable)
Fig. 8.2, 3	1.012	0.703	$-0.100 \pm 0.289i$		Stable focus
Fig. 8.3	0.983	-0.295	$-0.082, -0.618$	$1.275, -12.21$	Stable node

* μ_1, μ_2 are the tangential directions of the integral curves at the singular points
(Sec. 2.2a).

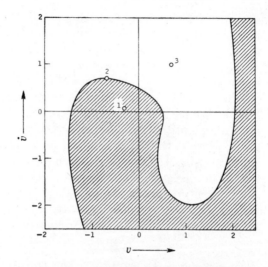

FIGURE 8.5   Regions of initial conditions in the $v\dot{v}$ plane leading to the resonant oscillation (unshaded) and the nonresonant oscillation (shaded).

The corresponding periodic state cannot be sustained, because any slight deviation from the point 2 will result in passage to the stable state represented by either the point 1 or point 3.   It is worth noting that one of the integral curves (drawn in thick line in the figure) that contains the saddle point divides the whole plane into two regions, in one of which all integral curves tend to the singularity of point 1 and in the other to the singularity of point 3.   This particular integral curve is referred to as a separatrix.

Figure 8.4 shows these regions in the $r\theta$ plane, where $r(0)$ is the amplitude and $\theta(0)$ is the phase angle of $v(\tau)$ at $\tau = 0$, that is,

$$r(0) = \sqrt{x^2(0) + y^2(0)} \qquad \theta(0) = \tan^{-1}\frac{-y(0)}{x(0)} \qquad (8.19)$$

Further, from Eqs. (8.3) and (8.4), we obtain

$$v(0) = y(0)$$
$$\dot{v}(0) = x(0) + \dot{y}(0) = x(0) - \tfrac{1}{2}[x(0) + ky(0) - \tfrac{3}{4}r^2(0)x(0)] \qquad (8.20)$$

where the dots denote differentiation with respect to τ.   By making use of these relations, the above regions are reproduced in the $v\dot{v}$ plane, as shown in Fig. 8.5.   From these figures the relationship between the initial conditions and the resulting oscillations is apparent: an oscillation started with any initial conditions in the shaded region tends ultimately to the singularity of point 1, whereas an oscillation started from the unshaded region tends to the singularity of point 3. It is also seen that the singular point 2 is situated on the boundary curve between these two regions.

In Fig. 8.3 we have only one stable node; in other words, we have a single periodic solution for any initial conditions prescribed.

The amplitude and the phase angle of an oscillation in the neighborhood of the periodic state approach, with the increase of time, the final state of damped sinusoids when the corresponding singularity is a stable focus, whereas they approach the final state with damped exponentials when the singularity is a stable node. When the initial value of the amplitude r is large, the transient oscillation has a higher frequency than the external frequency, because the representative point $(x(\tau),y(\tau))$ correlated with the transient oscillation moves along the integral curve in the counterclockwise direction around the origin O.

8.5 Geometrical Discussion of the Integral Curves for Conservative Systems

Although dissipation exists in all natural systems, the study of conservative systems is yet of some value, first, for the interesting character of the integral curves and, second, as an introduction to the following study of dissipative systems.

Putting $k = 0$ in Eqs. (8.4), we have

$$\frac{dx}{d\tau} = \frac{1}{2}\left(-\frac{3}{4}r^2y + y + B\right) \equiv X(x,y)$$
$$\frac{dy}{d\tau} = \frac{1}{2}\left(\frac{3}{4}r^2x - x\right) \equiv Y(x,y)$$

(8.21)

from which we obtain

$$Y(x,y)\,dx - X(x,y)\,dy = 0 \qquad (8.22)$$

Since, by Eqs. (8.21), $\partial X/\partial x + \partial Y/\partial y = 0$, Eq. (8.22) becomes an exact differential equation, and hence the complete integral is

$$\tfrac{3}{16}r^4 - \tfrac{1}{2}r^2 - By = C \qquad (8.23)$$

where C is a constant of integration. The integral curves of Eqs. (8.21) are readily obtained by plotting Eq. (8.23); an example will be given shortly.

To investigate the integral curves in the neighborhood of a singular point, we now transfer the origin to the singular point (x_0,y_0) by introducing the new variables ξ and η defined by

$$x = x_0 + \xi \qquad y = y_0 + \eta$$

Thereby the fundamental equations (8.4) become

$$\frac{d\xi}{d\tau} = a_1\xi + a_2\eta - \frac{3}{8}(y_0\xi^2 + 2x_0\xi\eta + 3y_0\eta^2 + \xi^2\eta + \eta^3)$$

$$\frac{d\eta}{d\tau} = b_1\xi + b_2\eta + \frac{3}{8}(3x_0\xi^2 + 2y_0\xi\eta + x_0\eta^2 + \xi^3 + \xi\eta^2)$$

where

$$a_1 = \left(\frac{\partial X}{\partial x}\right)_{\substack{x=x_0,\\y=y_0}} = \frac{1}{2}\left(-k - \frac{3}{2}x_0y_0\right)$$

$$a_2 = \left(\frac{\partial X}{\partial y}\right)_{\substack{x=x_0,\\y=y_0}} = \frac{1}{2}\left[1 - \frac{3}{4}(x_0^2 + 3y_0^2)\right] \qquad (8.24)$$

$$b_1 = \left(\frac{\partial Y}{\partial x}\right)_{\substack{x=x_0,\\y=y_0}} = \frac{1}{2}\left[-1 + \frac{3}{4}(3x_0^2 + y_0^2)\right]$$

$$b_2 = \left(\frac{\partial Y}{\partial y}\right)_{\substack{x=x_0,\\y=y_0}} = \frac{1}{2}\left(-k + \frac{3}{2}x_0y_0\right)$$

These equations are applicable to systems with and without damping, and they reduce to Eqs. (8.8) if we neglect the terms of degree higher than the first in ξ and η.

Now, in the present case, putting $k = 0$ in Eqs. (8.24), and remembering that $x_0 = 0$ from Eqs. (8.7), we have

$$\frac{d\xi}{d\tau} = a_2\eta - \frac{3}{8}(y_0\xi^2 + 3y_0\eta^2 + \xi^2\eta + \eta^3)$$

$$\frac{d\eta}{d\tau} = b_1\xi + \frac{3}{8}(2y_0\xi\eta + \xi^3 + \xi\eta^2) \qquad (8.25)$$

where

$$a_2 = \tfrac{1}{2}(1 - \tfrac{9}{4}y_0^2) \qquad b_1 = \tfrac{1}{2}(-1 + \tfrac{3}{4}y_0^2)$$

Equations (8.25) may also be integrated as in the case of Eqs. (8.21). The result is

$$b_1\xi^2 - a_2\eta^2 + \tfrac{3}{4}[y_0\eta(\xi^2 + \eta^2) + \tfrac{1}{4}(\xi^2 + \eta^2)^2] = C \qquad (8.26)$$

where C is a constant of integration.

In order to classify the types of singularities, we calculate the characteristic roots from Eq. (8.14); they are

$$\lambda_1, \lambda_2 = \pm\sqrt{a_2b_1} \qquad a_1 + b_2 = 0 \qquad (8.27)$$

The Br_0^2 curve for $k = 0$ is reproduced in Fig. 8.6 (cf. Fig. 8.1). We divide this curve into sections I to III (as indicated in the figure), whose boundaries are given by the points A and B at which $r_0^2 = \tfrac{4}{3}$ and $\tfrac{4}{9}$,

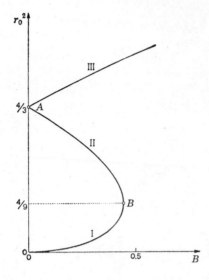

FIGURE 8.6   Amplitude characteristic of the harmonic oscillation in a conservative system.

respectively.   We then have the following singularities for the respective sections:

Section I: Center[1] $a_2 b_1 < 0$
Section II: Saddle point $a_2 b_1 > 0$
Section III: Center[1] $a_2 b_1 < 0$

NUMERICAL EXAMPLE

We consider a case in which $B = 0.2$.   There are three states of equilibrium (Fig. 8.6), and the corresponding singularities are listed in Table 8.2.

Table 8.2   Singular Points in Fig. 8.7

Singular point	y_0	λ_1, λ_2	μ_1, μ_2	Classification
1	-0.207	$\pm 0.468i$		Center (neutral)
2	-1.037	± 0.262	∓ 0.369	Saddle (unstable)
3	1.244	$\pm 0.316i$		Center (neutral)

The integral curves of Eqs. (8.21) are readily obtained by plotting Eq. (8.23) for various values of C.   The result is shown in Fig. 8.7.   We see that, in a conservative system, each integral curve forms a closed trajectory and does not tend to a stable singularity.   Hence, with the

[1] As pointed out by Poincaré [84, p. 95], the condition that the characteristic roots are imaginary is not sufficient to distinguish a center from a focal point.   However, in the present case the singularity is a center, as verified in Sec. V.2.

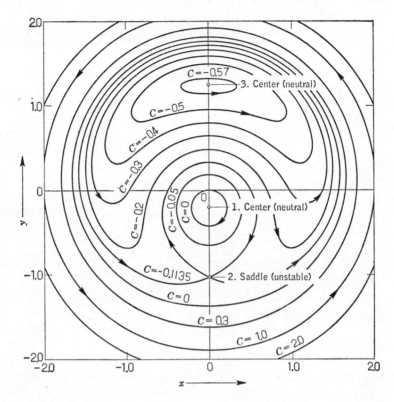

FIGURE 8.7   Integral curves for the harmonic oscillation in a conservative system.

increase of time, a representative point $(x(\tau),y(\tau))$ which starts from a given initial condition $(x(0),y(0))$ moves along the closed trajectory in the direction of the arrows, and retraces the same trajectory repeatedly. This implies that the amplitude and the phase angle of the oscillation are no longer constant but vary periodically in the steady state. Thus the phase of the oscillation may lead or lag the external force. When the closed trajectory does not encircle the origin, the leading and lagging angles cancel each other in every cycle in which the representative point completes one revolution along the closed trajectory; in this case the oscillation is synchronized with the external force. On the other hand, when the closed trajectory contains the origin in its interior, a phase difference of 2π radians would result in every cycle; so the oscillation is no longer synchronized with the external force.

Finally, we discuss the cases in which the systems are specified by the points A and B of Fig. 8.6.

(a) *Integral Curves for the System Represented by the Point A of Fig. 8.6*

Putting $B = 0$ in Eqs. (8.21) gives us

$$\frac{dy}{dx} = -\frac{x}{y}$$

or by integrating this,

$$x^2 + y^2 = \text{const}$$

Hence the integral curves are represented by a family of concentric circles centered at the origin, so the singularity (i.e., the origin in this case) is a center. The period T, the time required for the representative point $(x(\tau), y(\tau))$ to complete one revolution along the closed trajectories, is given by

$$T = \oint \frac{ds}{\sqrt{X^2(x,y) + Y^2(x,y)}} = \oint \frac{ds}{\frac{1}{2}(1 - \frac{3}{4}r^2)r} = \frac{4\pi}{1 - \frac{3}{4}r^2} \quad (8.28)$$

Now, if we prescribe, as an initial condition, a point $(x(0), y(0))$ on that circle the radius of which is given by $r = \sqrt{4/3}$, the period T will tend to infinity. As one sees from Eqs. (8.21), the representative point stays in this case at the initial position. This means that the frequency of the oscillation is the same as the external frequency. Further, from Eqs. (8.21), one also sees that the representative point moves in the counterclockwise direction when $r^2 > 4/3$ and in the clockwise direction when $r^2 < 4/3$. In the former case, the oscillation has a higher frequency than the external force; in the latter case, the situation is reversed. Thus, in the end, we may conclude that the frequency of the oscillation changes with varying r and coincides with the external frequency only when $r^2 = 4/3$.

(b) *Integral Curves for the System Represented by the Point B of Fig. 8.6*

In this case $k = 0$ and $B = 4/9$. From Eqs. (8.6) and (8.7) we obtain

$$x_0 = 0 \qquad y_0 = -2/3 \qquad \text{and} \qquad r_0^2 = 4/9 \qquad (8.29)$$

As illustrated in Fig. 8.1, the regions of nodes, foci, and saddles come together at the point B. We shall investigate the character of this singular point, as follows.[1]

From Eqs. (8.25),

$$a_1 = a_2 = b_2 = 0 \qquad b_1 = -1/3$$

[1] The following analysis refers to Sec. V.1 and the contribution due to Bendixson [6, pp. 58, 62, 74].

and so we find that $\lambda_1 = \lambda_2 = 0$. Equations (8.25) become

$$\frac{d\xi}{d\tau} = \frac{1}{4}(\xi^2 + 3\eta^2) - \frac{3}{8}(\xi^2\eta + \eta^3)$$

$$\frac{d\eta}{d\tau} = -\frac{1}{3}\xi - \frac{1}{2}\xi\eta + \frac{3}{8}(\xi^3 + \xi\eta^2)$$

(8.30)

or, substituting $dz = b_1\,d\tau = -\frac{1}{3}\,d\tau$ into (8.30), we obtain

$$\frac{d\xi}{dz} = -\frac{3}{4}(\xi^2 + 3\eta^2) + \frac{9}{8}(\xi^2\eta + \eta^3)$$

$$\frac{d\eta}{dz} = \xi + \frac{3}{2}\xi\eta - \frac{9}{8}(\xi^3 + \xi\eta^2)$$

(8.31)

The integral curves in the $\xi\eta$ plane tend to the origin with the tangent $\xi = 0$. Hence, by making use of the transformation $\xi = x_1\eta$, we have

$$\frac{dx_1}{dz} = -\frac{9}{4}\eta - x_1^2 + \frac{9}{8}\eta^2 - \frac{9}{4}x_1^2\eta + \frac{9}{4}x_1^2\eta^2 + \frac{9}{8}x_1^4\eta^2$$

$$\frac{d\eta}{dz} = x_1\eta + \frac{3}{2}x_1\eta^2 - \frac{9}{8}x_1\eta^3 - \frac{9}{8}x_1^3\eta^3$$

(8.32)

Now the integral curves in the $x_1\eta$ plane tend to the origin, with the tangent $\eta = 0$; and further, by the transformation $\eta = x_1y_1$, Eqs. (8.32) lead to

$$\frac{dx_1}{dz} = -x_1^2 - \frac{9}{4}x_1y_1 + \left(-\frac{9}{4}x_1^3y_1 + \frac{9}{8}x_1^2y_1^2 + \frac{9}{4}x_1^4y_1^2 + \frac{9}{8}x_1^6y_1^2\right)$$

$$\frac{dy_1}{dz} = 2x_1y_1 + \frac{9}{4}y_1^2 + \left(\frac{15}{4}x_1^2y_1^2 - \frac{9}{8}x_1y_1^3 - \frac{27}{8}x_1^3y_1^3 - \frac{9}{4}x_1^5y_1^3\right)$$

(8.33)

The tangents of the integral curves at the origin of the x_1y_1 plane are determined by

$$x_1y_1(x_1 + \tfrac{3}{2}y_1) = 0 \qquad \text{(see Sec. V.1)} \qquad (8.34)$$

However, in this equation, the tangents $x_1 = 0$ and $y_1 = 0$ are reduced to the origin of the $\xi\eta$ plane; therefore, we shall be interested only in the integral curves which have the tangent $x_1 + \tfrac{3}{2}y_1 = 0$ at the origin. To investigate them, we make use of the transformation

$$y_1 = (y_2 - \tfrac{2}{3})x_1$$

Then Eqs. (8.33) become

$$x_1\frac{dy_2}{dx_1} = \frac{-3y_2 + \tfrac{9}{2}y_2^2 + x_1^2\phi(x_1,y_2)}{\tfrac{1}{2} - \tfrac{9}{4}y_2 + x_1\psi(x_1,y_2)}$$

(8.35)

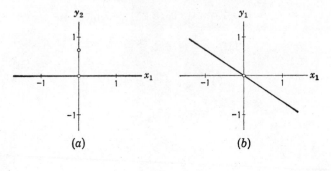

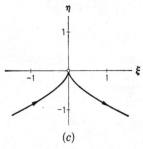

FIGURE 8.8   Integral curves in the respective coordinates, the
singularity being correlated with the point B of
Fig. 8.6.

where $\phi(x_1,y_2)$ and $\psi(x_1,y_2)$ are polynomials in x_1 and y_2.   As mentioned
in Sec. V.1, Eq. (8.35) takes the form

$$x^m \frac{dy}{dx} = ay + bx + B(x,y)$$

where $B(x,y)$ contains terms of degree higher than the first in x and y.
For Eq. (8.35), we have

$$m = 1 \quad \text{odd} \quad \text{and} \quad a = -6 < 0$$

so that the singularity $(x_1 = 0, y_2 = 0)$ is a saddle point[1] and the integral
curves tend to it, with tangents $x_1 = 0$ and $y_2 = 0$.

Now, from the foregoing transformations, we have

$$\xi = x_1\eta = x_1{}^2 y_1 = x_1{}^3 \left(y_2 - \tfrac{2}{3}\right) \qquad \eta = x_1 y_1 = x_1{}^2 \left(y_2 - \tfrac{2}{3}\right)$$

As mentioned before, the tangent $x_1 = 0$ is reduced to the origin of the

---

[1] There is another singularity at $x_1 = 0$ and $y_2 = \tfrac{2}{3}$ for Eq. (8.35), but this
singularity need not be investigated when we consider the integral curves in the
$\xi\eta$ plane.

$\xi\eta$ plane, but the tangent $y_2 = 0$ is transformed to

$$\xi = -\tfrac{2}{3}x_1^3 \qquad \eta = -\tfrac{2}{3}x_1^2 \tag{8.36}$$

which may be considered to represent the integral curves in the neighborhood of the origin of the $\xi\eta$ plane. The tangent $y_2 = 0$ in the respective coordinate planes is illustrated in Fig. 8.8. There are two and only two branches of integral curves which tend to the origin of the $\xi\eta$ plane with the tangent $\xi = 0$, because, if there were more than two, there would be more than two branches of the integral curves tending to the origin of the $x_1 y_2$ plane with the tangent $y_2 = 0$, which contradicts the nature of the saddle point.

In conclusion, the singularity is a cusp, and, as we see from Eqs. (8.30), the representative point $(\xi(\tau), \eta(\tau))$ moves, with the increase of time, along the integral curves in the direction of the arrows (Fig. 8.8c). Therefore, the equilibrium state corresponding to this singular point is unstable. We may notice that, with increasing B, the center 1 and the saddle point 2 of Fig. 8.7 approach each other and finally coalesce, resulting in a cusp.

8.6 Geometrical Discussion of the Integral Curves for Dissipative Systems

We have considered the integral curves and the singular points correlated with the harmonic oscillations in Secs. 8.3 and 8.4. As mentioned there, the types of singularities are determined once the roots (different from zero) of the characteristic equation (8.9) are known. However, there still remain special cases to discuss in which one or both roots are zero. The corresponding singular points are of higher order (Sec. 2.2d). An example of such a singularity is that of the cusp of the foregoing section. In this section we shall consider some special cases and investigate the nature of the corresponding singular points.

(a) Singular Points on the Boundaries between the Regions of Nodes and Foci (Fig. 8.1)

As mentioned in Sec. 8.3, these boundaries are given by

$$r_0^2 = \tfrac{4}{3} \quad \text{and} \quad \tfrac{4}{9} \quad \text{[see Eqs. (8.18)]}$$

Since Eq. (8.17) holds in this case, the characteristic roots λ_1 and λ_2 are readily calculated from Eqs. (8.8a) and (8.14); they are

$$\lambda_1 = \lambda_2 = -\tfrac{1}{2}k \tag{8.37}$$

Hence the singularity is simply a stable node, and the tangential direction of the integral curves at the singular point is uniquely determined.[1]

[1] Refer to Eq. (2.16).

(b) *Singular Points on the Boundary between the Regions of Nodes and Saddles (Fig. 8.1)*

This boundary is given by the condition $dB^2/dr_0^2 = 0$. Solving Eq. (8.16) for r_0^2 gives

$$r_0^2 = \frac{8 \pm 4\sqrt{1 - 3k^2}}{9} \tag{8.38}$$

By calculating x_0^2 and y_0^2 by Eqs. (8.7) and substituting them into Eqs. (8.24), we obtain

$$a_1 = a_2 = 0$$
$$b_1 = \frac{1}{3}\left[\frac{6k^2(2 \pm \sqrt{1 - 3k^2})}{1 + 3k^2 \mp \sqrt{1 - 3k^2}} - 1\right] \tag{8.39}$$
$$b_2 = -k$$

so that the characteristic roots are

$$\lambda_1 = 0 \qquad \lambda_2 = -k \tag{8.40}$$

Thus the singular point with which we are dealing is not simple. We investigate the stability in detail, as follows.

Equations (8.24) may be written in this case as

$$\frac{d\xi}{d\tau} = -\frac{3}{8}(y_0\xi^2 + 2x_0\xi\eta + 3y_0\eta^2 + \xi^2\eta + \eta^3)$$
$$\frac{d\eta}{d\tau} = b_1\xi - k\eta + \frac{3}{8}(3x_0\xi^2 + 2y_0\xi\eta + x_0\eta^2 + \xi^3 + \xi\eta^2) \tag{8.41}$$

Substituting $dz = -k\,d\tau$ into Eqs. (8.41) gives us

$$\frac{d\xi}{dz} = X_1 \qquad \frac{d\eta}{dz} = \eta - \alpha\xi + Y_1$$

where

$$X_1 = \frac{3}{8k}(y_0\xi^2 + 2x_0\xi\eta + 3y_0\eta^2 + \xi^2\eta + \eta^3)$$
$$Y_1 = -\frac{3}{8k}(3x_0\xi^2 + 2y_0\xi\eta + x_0\eta^2 + \xi^3 + \xi\eta^2) \tag{8.42}$$
$$\alpha = \frac{b_1}{k}$$

The tangents of the integral curves at the origin of the $\xi\eta$ plane are determined by (Sec. V.1)

$$\xi(\eta - \alpha\xi) = 0 \tag{8.43}$$

We first show that there are two, and only two, branches of the integral curves which tend to the origin with the tangent $\xi = 0$. By making use of the transformation $\xi = \phi\eta$, we obtain, from Eqs. (8.42),

$$\frac{d\phi}{dz} = -\phi + \frac{9}{8k}\, y_0\eta + X_2 \qquad \frac{d\eta}{dz} = \eta + Y_2 \qquad (8.44)$$

where X_2 and Y_2 are polynomials containing terms of degree higher than the first in ϕ and η. The characteristic equation becomes

$$(\lambda + 1)(\lambda - 1) = 0$$

so that the singularity, i.e., the origin of the $\phi\eta$ plane, is a saddle point. Hence, there are four branches of integral curves tending to the origin. Two of these are represented by $\eta = 0$, but they reduce to the origin of the $\xi\eta$ plane. We have, therefore, two and only two branches of the integral curves which tend to the origin of the $\xi\eta$ plane with the tangent $\xi = 0$, one of them being situated above and the other below the ξ axis.

Now we may conclude that all the other integral curves which tend to the origin have the tangent $\eta - \alpha\xi = 0$. In order, therefore, to investigate them, we apply the transformation

$$\eta = (\alpha + \psi)\xi$$

to Eqs. (8.42) and obtain

$$\xi^2 \frac{d\psi}{d\xi} = \frac{\psi + Y_1/\xi}{X_1/\xi^2} - \xi(\alpha + \psi) \qquad (8.45)$$

or

$$\xi^2 \frac{d\psi}{d\xi} = a\psi + b\xi + B_1(\xi,\psi)$$

with

$$(8.46)$$

$$a = \frac{8k}{3}\, \frac{1}{2\alpha x_0 + (3\alpha^2 + 1)y_0} \qquad b = -\frac{3(1 + \alpha^2)(x_0 + \alpha y_0)}{2\alpha x_0 + (3\alpha^2 + 1)y_0}$$

where $B_1(\xi,\psi)$ is a series containing terms of degree higher than the first in ξ and ψ. This takes the form of Eq. (V.3) in Sec. V.1 with $m = 2$ (even).[1] Hence, on dividing the $\xi\psi$ plane into two regions along the ψ axis, we see that all the integral curves tend to the origin on one side of the ψ axis (which side it is will depend on the sign of a) and that, on the other side, one and only one branch of the integral curves tends to the origin, while all the others veer away from the origin. Therefore, we may conclude that the equilibrium state corresponding to the singularity is unstable.

In order to derive an approximate equation of the integral curves in the neighborhood of the singularity, we make use of the transformation

$$p = \frac{1}{a}\, \xi \qquad q = a\psi + b\xi$$

---

[1] It is here assumed that $2\alpha x_0 + (3\alpha^2 + 1)y_0$ is not zero. We shall shortly discuss the case in which this is zero.

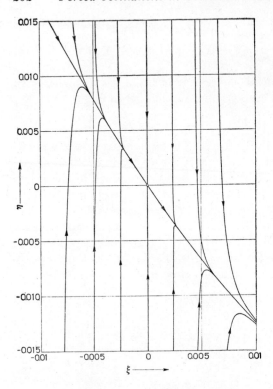

FIGURE 8.9   Coalescence of node and saddle, resulting in a node-saddle distribution
of the integral curves.

Then Eq. (8.46) becomes

$$p^2 \frac{dq}{dp} = q + B_1'(p,q) \qquad (8.47)$$

Since $B_1'(p,q)$ contains terms of degree higher than the first in p and q, it
may be neglected as compared with q, since we confine our discussion to
the singular point and its vicinity alone.   By integrating Eq. (8.47)
under this condition, we have

$$q = Ce^{-1/p} \qquad (8.48)$$

where C is a constant of integration.   Turning back to the original
coordinate plane, we finally obtain

$$\eta = \left(\alpha - \frac{b}{a} \xi + C'e^{-a/\xi} \right) \xi \qquad C' = \frac{C}{a} \qquad (8.49)$$

NUMERICAL EXAMPLE

In order to illustrate the foregoing analysis, we consider a system with $k = 0.2$. Then, from Eq. (8.38), we obtain

$$r_0{}^2 = 1.3058 \qquad \text{and} \qquad 0.4720$$

For the latter value of $r_0{}^2$ the following quantities are readily obtained:

$$
\begin{aligned}
x_0 &= 0.2032 & y_0 &= -0.6563 \\
B &= 0.4645 & \alpha &= -1.460 \\
a &= -0.0979 & b &= 2.004
\end{aligned}
$$

By substituting these values into Eq. (8.49), we may draw the integral curves for several values of C'. The result is shown in Fig. 8.9. From Eqs. (8.41) we see that the representative point $(\xi(\tau), \eta(\tau))$ moves along the integral curves in the direction of the arrows. Thus a point $(x(\tau), y(\tau))$ tends to the origin for $\xi < 0$ but leaves the origin for $\xi > 0$. Hence, as already mentioned, the equilibrium state corresponding to the singularity is unstable.

Now, in order to complete our discussion, we must consider the exceptional case in which

$$2\alpha x_0 + (3\alpha^2 + 1) y_0 = 0 \tag{8.50}$$

This takes place when Eq. (8.38) has two equal roots. In this case we have

$$1 - 3k^2 = 0 \qquad \text{or} \qquad k = \frac{1}{\sqrt{3}}$$

and the following values which satisfy the condition (8.50) are readily obtained:

$$
r_0{}^2 = \frac{8}{9} \qquad x_0 = \frac{\sqrt{2}}{\sqrt{3}} \qquad y_0 = -\frac{\sqrt{2}}{3} \qquad B = \frac{4\sqrt{2}}{9}
$$

$$
a_1 = a_2 = 0 \qquad b_1 = \frac{1}{3} \qquad b_2 = -\frac{1}{\sqrt{3}}
$$

$$
\lambda_1 = -\frac{1}{\sqrt{3}} \qquad \lambda_2 = 0
$$

Substituting these values into Eq. (8.45) gives us

$$
\xi^2 \frac{d\psi}{d\xi} = \frac{\psi - \sqrt{2}\,\xi - \tfrac{3}{2}\xi^2 - \tfrac{3}{4}\xi^3\psi - (3\sqrt{2}/8)\xi\psi^2 - (3\sqrt{3}/8)\xi^2\psi^2}{\tfrac{1}{2}\xi + (3\sqrt{3}/4)\xi\psi - (3\sqrt{6}/8)\psi^2 + \tfrac{9}{8}\xi\psi^2 + (3\sqrt{3}/8)\xi\psi^3}
$$
$$
-\frac{1}{\sqrt{3}}\,\xi - \xi\psi
$$

However, this may not be transformed to Eq. (8.46); and so, applying further the transformation $\psi = (\sqrt{2} + \phi)\xi$, we obtain

$$\xi^3 \frac{d\phi}{d\xi} =$$

$$\frac{\phi - \dfrac{\sqrt{3}}{2}\xi - \dfrac{3}{4}\xi^2(\sqrt{2}+\phi) - \dfrac{3\sqrt{2}}{8}\xi^2(\sqrt{2}+\phi)^2 - \dfrac{3\sqrt{3}}{8}\xi^3(\sqrt{2}+\phi)^2}{\dfrac{1}{2} + \dfrac{3\sqrt{3}}{4}\xi(\sqrt{2}+\phi) - \dfrac{3\sqrt{6}}{8}\xi(\sqrt{2}+\phi)^2 + \dfrac{9}{8}\xi^2(\sqrt{2}+\phi)^2 + \dfrac{3\sqrt{3}}{8}\xi^3(\sqrt{2}+\phi)^3}$$

$$-\frac{1}{\sqrt{3}}\xi - 2\xi^2(\sqrt{2}+\phi)$$

or
$$\xi^3 \frac{d\phi}{d\xi} = 2\phi - \frac{4}{\sqrt{3}}\xi + B_2(\xi,\phi) \tag{8.51}$$

where $B_2(\xi,\phi)$ contains terms of degree higher than the first. This takes the form of Eq. (V.3) with $m = 3$ (odd) and $a = 2 > 0$. Hence the singularity is a nodal point. In order to obtain the integral curves in the neighborhood of the singularity, we put

$$p = \frac{1}{\sqrt{2}}\xi \qquad q = 2\phi - \frac{4}{\sqrt{3}}\xi$$

Then Eq. (8.51) may be written as

$$p^3 \frac{dq}{dp} = q + B_2'(p,q) \tag{8.52}$$

where $B_2'(p,q)$ contains terms of degree higher than the first. Hence, neglecting this, we integrate Eq. (8.52) to obtain

$$q = Ce^{-1/2p^2}$$

C being a constant of integration. In the original $\xi\eta$ plane, we obtain

$$\eta = \left(\frac{1}{\sqrt{3}} + \sqrt{2}\,\xi + \frac{2}{\sqrt{3}}\xi^2 + \frac{C}{2}\xi e^{-1/\xi^2}\right)\xi \tag{8.53}$$

The integral curves are calculated for several values of C; they are plotted in Fig. 8.10. From Eqs. (8.41), we see that the representative point which has started from any initial conditions moves along the integral curve in the direction of the arrows and tends ultimately to the origin. Hence the singular point and the corresponding equilibrium state are stable.

In the end, we may conclude that an equilibrium state corresponding to the singular point on the boundary of the regions of nodes and saddles is usually unstable. Hence, the oscillation is transferred to the other

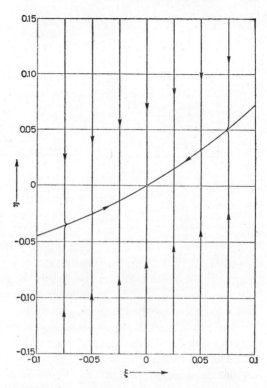

FIGURE 8.10   Coalescence of node and saddle, resulting in a nodal distribution of
the integral curves.

equilibrium state, which is stable.   However, in the case where $k = 1/\sqrt{3}$
and $r_0{}^2 = \frac{8}{9}$, the equilibrium state becomes stable.   From the physical
point of view this is a plausible result, since there is no other equilibrium
state in this particular case (Fig. 8.1).

8.7 Experimental Investigation [33, 36, 96]

As we mentioned in Sec. 4.1, it is a salient feature of nonlinear oscillations that different types of response may be obtained for a given system,
depending on the values of the initial conditions.   In Fig. 4.1, for
example, we see three types of oscillation; namely, two kinds of harmonic
oscillation and a subharmonic oscillation of order $\frac{1}{3}$.   They are obtained
in an electric circuit containing a nonlinear inductance.   We here discuss
experiments on a similar circuit (the details to be illustrated presently)
and determine the regions of initial conditions which give rise to the

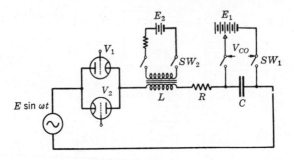

FIGURE 8.11   Experimental circuit with means provided for prescribing the initial conditions.

respective types of periodic oscillation.   However, we here confine our
attention to the harmonic oscillation only; the occurrence of the subharmonic oscillation will be treated in the following chapter.

We now proceed to give some experimental results which will be
useful in verifying the foregoing analysis.   As we have shown in Fig. 8.2,
there are two kinds of stable oscillation associated with the singularities of
points 1 and 3.   In Sec. 5.1c we have distinguished those as the resonant
and the nonresonant oscillations, according as the amplitude of the
oscillation is larger or smaller.   Consequently, it is clear that the singularity of point 3 in Fig. 8.2 is correlated with the resonant oscillation and
the singularity of point 1 with the nonresonant oscillation.

In the foregoing analysis the initial conditions were prescribed with
the sine and cosine components of $v(0)$, that is, the coordinates $x(0)$ and
$y(0)$ of the initial point in the xy plane.   This is not practical for the
present experiment, however, so that we prescribe the initial conditions,
instead, with the initial charging voltage across the capacitor and the
phase angle of the applied voltage at the instant when it is impressed on
the circuit.   The schematic diagram of Fig. 8.11 shows an electric circuit
used for our experiment.   The initial charging voltage V_{C0} across the
capacitor C is supplied from the battery E_1 by momentarily closing the
switch SW_1.   As regards the other initial condition, i.e., the phase angle
at which the oscillation starts, an ordinary mechanical switch is not
adequate to effect the accurate timing; therefore, we use an electronic
switch consisting of two thyratrons V_1 and V_2 connected in inverse-
parallel.   These tubes are made conductive by impressing a positive
potential on their grids at the desired phase angle of the applied voltage.[1]

It is a noteworthy fact that the residual magnetism in the saturable
core has a serious effect on the experiment.   Therefore, the core must be

[1] Refer to Appendix VI for further details of the control circuit.

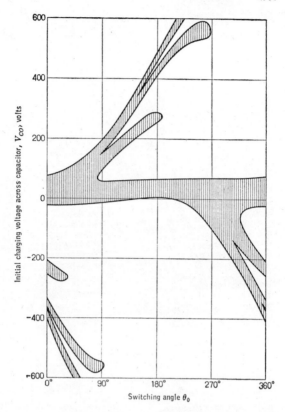

FIGURE 8.12   Regions of initial conditions leading
to the resonant oscillation (un-
shaded) and the nonresonant oscillation (shaded).

demagnetized, or, at least, the residual magnetism should be kept constant.   If this precaution is neglected, a random result will be obtained
in every experiment.   Hence, as shown in Fig. 8.11, the inductor core is
provided with a secondary winding through which a direct current flows
from the battery E_2 upon closing the switch SW_2.   Thus, prior to each
experiment, the iron core is premagnetized in a certain direction by
letting an ample current flow through the secondary winding; thus a
constant residual magnetism is obtained after the opening of the switch
SW_2.

Following this procedure, we perform the experiment by starting the
oscillations with various combinations of the initial conditions V_{C0} and
θ_0 (θ_0 being the switching angle of the applied voltage) and obtain Fig.
8.12.   In the figure the shaded area is the region of initial conditions

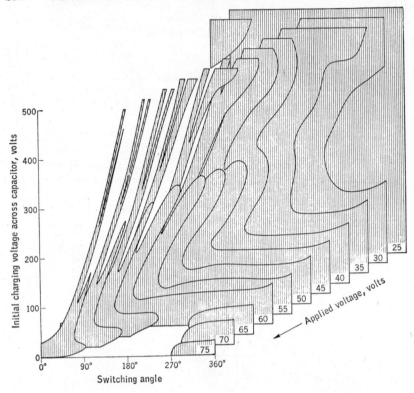

FIGURE 8.13   Regions of initial conditions leading to the resonant oscillation
(unshaded) and the nonresonant oscillation (shaded) obtained for
different values of the applied voltage.

which give rise to the nonresonant state, and the blank area corresponds
to the resonant state.   It will be noticed that the shaded region is slightly
shifted upward owing to the unidirectional premagnetization.   If the
core is completely demagnetized before each experiment, the shaded area
should appear symmetrically with respect to the abscissa when the figure
for $V_{C0} < 0$ is shifted horizontally by 180 degrees in the direction of the
θ_0 axis.

Similar experiments are carried out for several values of the applied
voltage; the result is plotted in Fig. 8.13.   As may be expected from the
preceding analysis (Fig. 8.1), the shaded regions (related to the nonresonant state) are contracted with the increase of the applied voltage.
However, even with an exceedingly high value of V_{C0}, the nonresonant
state may still occur if we appropriately choose the initial switching
angle θ_0.   These experimental results agree with the results of the preceding analysis, except for slight discrepancies such as the branching of
the shaded regions (cf. Fig. 8.4).

Chapter 9

Subharmonic Oscillations

9.1 Analysis of Subharmonic Oscillations by Means of Integral Curves

In the preceding chapter we have considered the transient state of the harmonic oscillation and discussed in detail the stability of equilibrium states correlated with singular points. We are now to deal with the subharmonic oscillation whose fundamental frequency is a fraction $1/\nu$ (ν being a positive integer) of the driving frequency. As mentioned in Chap. 7, such oscillations may occur in nonlinear systems, particularly in the case where the system is described by Duffing's equation

$$\frac{d^2v}{d\tau^2} + k\frac{dv}{d\tau} + f(v) = B \cos \nu\tau \tag{9.1}$$

and the periodic solutions have been assumed, to a first approximation, by [see Eq. (7.4)]

$$v = z + x \sin \tau + y \cos \tau + w \cos \nu\tau$$

where
$$w = \frac{1}{1 - \nu^2} B \tag{9.2}$$

In this chapter we consider that the transient solution of Eq. (9.1) also takes the form (9.2), but we assume that the coefficients z, x, and y are slowly varying functions of the time τ. The amplitude w is fixed by the second equation of (9.2). This implies that, when a subharmonic oscillation occurs, the accompanying oscillation is a harmonic oscillation of the nonresonant type (Sec. 5.1c).[1] Following the method of analysis described in Chap. 8, we shall investigate the integral curves and singularities corresponding to the subharmonic oscillations of orders ⅓, ⅕, and ½ in the following sections.

Before closing these introductory remarks, it should be mentioned that the present analysis answers only the question whether a certain type of subharmonic oscillation occurs. If a number of different types of

[1] It is confirmed by experiment that no subharmonic oscillation exists along with a resonant oscillation.

response are to be expected (e.g., subharmonic, nonresonant, and resonant oscillations), this method of analysis is practically inapplicable, since analysis must be effected by graphical solution in a higher-dimensional phase space. Such cases have to be attacked by a different method, which will be studied in Chap. 10.

9.2 Phase-plane Analysis of the Subharmonic Oscillations of Order ⅓ [33, 36]

We first treat the subharmonic oscillations of order ⅓ (that is, $\nu = 3$), and consider the differential equation

$$\frac{d^2v}{d\tau^2} + k\frac{dv}{d\tau} + v^3 = B \cos 3\tau \tag{9.3}$$

Since the nonlinearity is symmetrical, the nonoscillatory term z in the solution (9.2) may be discarded. Then the solution of Eq. (9.3) becomes

$$v(\tau) = x(\tau) \sin \tau + y(\tau) \cos \tau + w \cos 3\tau \tag{9.4}$$

where

$$w = \frac{1}{1 - 3^2} B = -\frac{1}{8} B$$

By substituting this solution into Eq. (9.3) and equating the terms containing $\sin \tau$ and $\cos \tau$ separately to zero, we obtain

$$\frac{dx}{d\tau} = \frac{1}{2}\left[-kx + Ay + \frac{3}{4} w(x^2 - y^2) \right] \equiv X(x,y)$$

$$\frac{dy}{d\tau} = \frac{1}{2}\left(-Ax - ky - \frac{3}{2} wxy \right) \equiv Y(x,y) \tag{9.5}$$

where

$$A = 1 - \tfrac{3}{4}r^2 - \tfrac{3}{2}w^2 \qquad r^2 = x^2 + y^2$$

It should, however, be remembered that the assumptions mentioned in Sec. 8.2 must be made for the derivation of Eqs. (9.5).

The periodic solutions are determined by

$$X(x,y) = 0 \qquad Y(x,y) = 0 \tag{9.6}$$

From Eqs. (9.5) and (9.6) we see that the steady-state amplitude r_0 satisfies the equation

$$A^2 + k^2 = \left(\frac{3w}{4}\right)^2 r_0^2 \tag{9.7}$$

which is identical with Eq. (7.18). The steady-state amplitudes x_0 and y_0 are also given by Eqs. (7.20). It was mentioned in Sec. 7.3b that the stability of the periodic solution may be determined only by the condition (7.22) for $n = 1$. This stability condition is also derived from the Routh-Hurwitz criterion (as we derived it in Sec. 8.2), or from the consideration of stability of the singular points of

$$\frac{dy}{dx} = \frac{Y(x,y)}{X(x,y)} \tag{9.8}$$

where $X(x,y)$ and $Y(x,y)$ are given by Eqs. (9.5). As before, the singularities of this equation are correlated with the periodic solutions of Eq. (9.3) and the integral curves with the transient solutions. The character of the singular point (x_0,y_0) is determined from the characteristic equation

$$\lambda^2 - (a_1 + b_2)\lambda + a_1 b_2 - a_2 b_1 = 0$$

where
$$a_1 = \left(\frac{\partial X}{\partial x}\right)_{\substack{x=x_0, \\ y=y_0}} = \frac{1}{2}\left[-k + \frac{3}{2}(wx_0 - x_0 y_0)\right]$$

$$a_2 = \left(\frac{\partial X}{\partial y}\right)_{\substack{x=x_0, \\ y=y_0}} = \frac{1}{2}\left[A - \frac{3}{2}(wy_0 + y_0{}^2)\right]$$

$$b_1 = \left(\frac{\partial Y}{\partial x}\right)_{\substack{x=x_0, \\ y=y_0}} = \frac{1}{2}\left[-A - \frac{3}{2}(wy_0 - x_0{}^2)\right] \tag{9.9}$$

$$b_2 = \left(\frac{\partial Y}{\partial y}\right)_{\substack{x=x_0, \\ y=y_0}} = \frac{1}{2}\left[-k - \frac{3}{2}(wx_0 - x_0 y_0)\right]$$

In order to illustrate the geometrical configuration of integral curves, a system with

$$k = 0.2 \quad \text{and} \quad B = 3.2$$

is investigated. Then, from the second equation of (9.4), $w = -0.4$. Upon substitution of these values into Eqs. (9.5), the integral curves of Eq. (9.8) are plotted in Fig. 9.1 by using the isocline method. In the figure we see seven singular points, which may also be determined from Eqs. (7.18) and (7.20) in Sec. 7.3b. The details are listed in Table 9.1.

Table 9.1   Singular Points of Fig. 9.1

Singular point	x_0	y_0	λ_1, λ_2	$\mu_1, \mu_2{}^*$	Classification
1	-1.123	-0.378	$-0.100 \pm 0.599i$		Stable focus
2	0.889	-0.784	$-0.100 \pm 0.599i$		Stable focus
3	0.234	1.162	$-0.100 \pm 0.599i$		Stable focus
4	-0.858	0.209	$0.365, -0.565$	$0.613, -7.274$	Saddle (unstable)
5	0.610	0.639	$0.365, -0.565$	$-35.24, -0.407$	Saddle (unstable)
6	0.248	-0.848	$0.365, -0.565$	$-0.543, 0.767$	Saddle (unstable)
7	0	0	$-0.100 \pm 0.500i$		Stable focus

* μ_1, μ_2 are the tangential directions of the integral curves at the singular points.

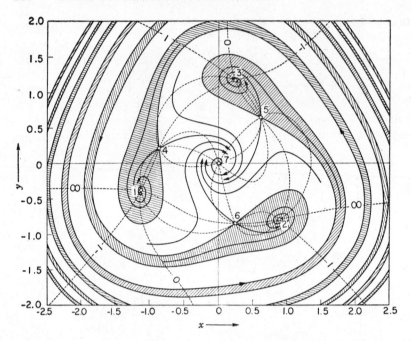

FIGURE 9.1   Integral curves for the subharmonic oscillation of order ⅓.
Full line: integral curves.   Dashed line: isoclines.

In Fig. 9.1 we see that the singular points 1 to 3 represent the stable states of subharmonic oscillation and that they are equidistant from and equiangular about the origin.   The angular distance between any two of these singular points corresponds to one complete cycle of the external force.   This is a plausible result for subharmonic oscillations of order ⅓.

As we also see from the integral curves of Fig. 9.1, any oscillation started from a point (which prescribes the initial condition) in the shaded regions leads ultimately to one of the singularities of points 1 to 3, resulting in subharmonic response, whereas any oscillation started from the unshaded region leads ultimately to the origin, resulting in no subharmonic oscillation.   Analogously to the case of harmonic oscillations, the $v\tau$ relation in the transient state is determined by the line integral of Eq. (8.13) in Sec. 8.3.   When the initial condition is given by a point far from the origin, the transient oscillation has a higher frequency than the subharmonic frequency, because the representative point $(x(\tau),y(\tau))$ associated with the transient oscillation moves along the integral curve in the counterclockwise direction around the origin.[1]

[1] A detailed investigation of the stability of equilibrium states which are associated with the singular points of higher order may also be carried out in the same manner as in Secs. 8.5 and 8.6.

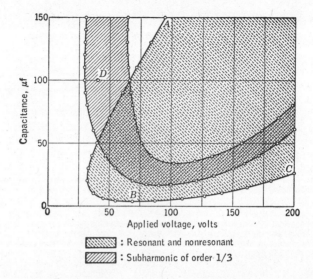

: Resonant and nonresonant

: Subharmonic of order 1/3

FIGURE 9.2   Regions in which different types of oscillation are sustained.

9.3 *Experimental Investigation*

By making use of the electric circuit of Fig. 8.11, we shall describe
some experiments concerned with the subharmonic oscillation of order $\frac{1}{3}$
and compare their results with the foregoing analysis.

By varying the applied voltage V and the capacitor C of Fig. 8.11, we
first determine the region in which the subharmonic oscillations of order
$\frac{1}{3}$ are sustained.   An experimental result is shown in Fig. 9.2, where this
region is indicated by hatching.   We also determine the region in which
both the resonant and the nonresonant oscillations of harmonic frequency
are sustained.   In Fig. 9.2 is drawn the curve ABC, which shows the
boundary of this region.   An oscillation which is first in the nonresonant
state changes into the resonant oscillation when the voltage is raised up
to the boundary BC.   On the contrary, a transition takes place from the
resonant to the nonresonant state when the voltage is reduced to that on
the boundary AB.   Hence, within the boundary curve ABC shaded by
dashed lines, either the resonant or the nonresonant oscillation may occur
according to the initial condition prescribed.   We shall, in what follows,
observe the transient state of a subharmonic oscillation of order $\frac{1}{3}$ and
seek the regions of initial conditions which give rise to the subharmonic
response.

As illustrated in Fig. 9.2, there are two kinds of periodic oscillation,
i.e., the nonresonant and the subharmonic oscillations at the point D for

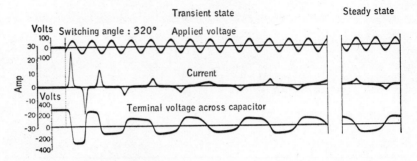

FIGURE 9.3   Oscillation leading to the subharmonic response of order ⅓.

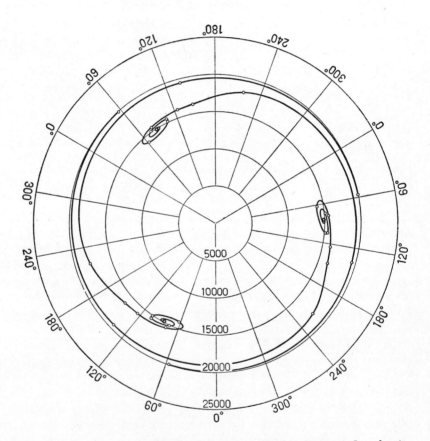

FIGURE 9.4   Polar diagram showing the relationship between flux density
and phase angle of the applied voltage. Radial scale: flux density (crest value in gauss). Angular scale. phase angle of the
applied voltage.

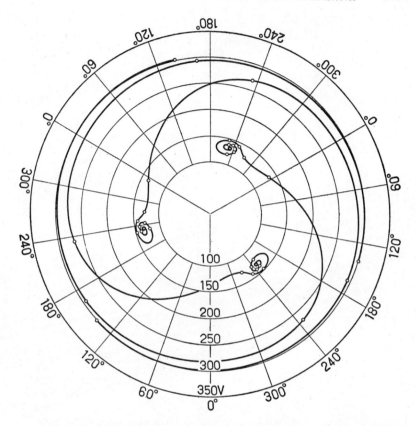

FIGURE 9.5   Polar diagram showing the relationship between capacitor voltage and phase angle of the applied voltage. Radial scale:
voltage across capacitor (crest value). Angular scale: phase
angle of the applied voltage.

which

$$C = 100 \ \mu f \qquad \text{and} \qquad V = 40 \text{ volts}$$

We shall now describe the experiments with this system.   As mentioned
in Sec. 8.7, we prescribe the initial conditions with the initial charging
voltage V_{C0} across the capacitor and the phase angle θ_0 of the applied
voltage from which the oscillation starts.

Figure 9.3 shows the transient state of an oscillation which gives rise
to the subharmonic oscillation of order $\frac{1}{3}$ in the steady state, where the
initial conditions are given by

$$V_{C0} = 312 \text{ volts} \qquad \text{and} \qquad \theta_0 = 320°$$

Since the initial charging voltage is comparatively high, the transient

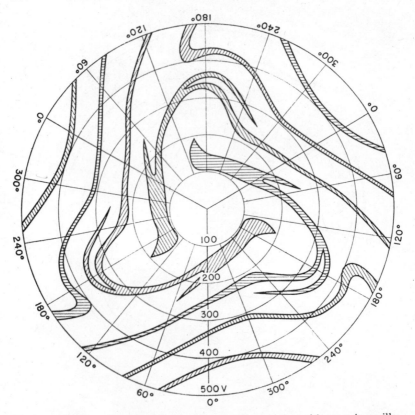

FIGURE 9.6   Regions of initial conditions leading to the subharmonic oscillation of order ⅓.   Radial scale: initial charging voltage across capacitor.   Angular scale: switching angle.

oscillation has a higher frequency than the subharmonic frequency.   In order to compare this result more clearly with the preceding analysis, we measure the flux density in the core and the phase angle of the applied voltage at every instant when the oscillating current takes its extreme value.[1]   These values are plotted in polar coordinates, as shown in Fig. 9.4.   The trajectories in the figure correspond to the integral curves in the shaded regions of Fig. 9.1; thus we see that the trajectories spiral to the ultimate points with the increase of time.

Similarly, crest values of the capacitor voltage and the corresponding phase angles of the applied voltage are plotted in Fig. 9.5.   Since the oscillating current becomes zero at the instant when the capacitor is

---

[1] Inasmuch as the core loss is negligible, the exciting current and the magnetic flux in the core are considered to be in phase.   Therefore, the flux density takes an extreme value at the above-mentioned phase angle.

charged up to a crest voltage (Fig. 9.3), a subharmonic oscillation occurs if these crest voltages and phase angles are chosen as the initial conditions. Hence, the points in Fig. 9.5 are the ones that prescribe the initial conditions which give rise to the subharmonic response.   By making use of an electronic synchronous switch as explained in Sec. 8.7, we further determine (by varying the initial conditions V_{C0} and θ_0) the regions of initial conditions which lead to the nonresonant state and the subharmonic response.   The result is shown in Fig. 9.6, where the shaded regions entail those which lead to the subharmonic response and the unshaded regions entail those which lead to the nonresonant state.[1]   The figure agrees with that analytically obtained in Fig. 9.1.

9.4 *Subharmonic Oscillations of Order* $\frac{1}{5}$ *[36]*

It was mentioned in Sec. 7.2 that the subharmonic oscillations of order $\frac{1}{5}$ may occur when the nonlinearity $f(v)$ in Eq. (9.1) is given by a quintic function in v.   So we write

$$\frac{d^2v}{d\tau^2} + k\frac{dv}{d\tau} + v^5 = B \cos 5\tau \tag{9.10}$$

and assume the periodic solution of the form

$$v(\tau) = x(\tau) \sin \tau + y(\tau) \cos \tau + w \cos 5\tau$$

where
$$w = \frac{1}{1 - 5^2}B = -\frac{1}{24}B \tag{9.11}$$

By a procedure analogous to that of Sec. 9.2, we obtain

$$\frac{dx}{d\tau} = \frac{1}{2}\left[-kx + Ay - \frac{5}{16}w(x^4 - 6x^2y^2 + y^4) \right] \equiv X(x,y)$$

$$\frac{dy}{d\tau} = \frac{1}{2}\left[-Ax - ky + \frac{5}{4}wxy(x^2 - y^2) \right] \equiv Y(x,y) \tag{9.12}$$

where $A = 1 - \frac{5}{8}r^4 - \frac{15}{4}w^2r^2 - \frac{15}{8}w^4$ $r^2 = x^2 + y^2$

The steady states of the oscillation are given by

$$X(x,y) = 0 \qquad Y(x,y) = 0 \tag{9.13}$$

Upon substitution of Eqs. (9.12) into (9.13), the steady-state amplitude r_0 is determined by

$$A^2 + k^2 = \left(\frac{5w}{16}\right)^2 r_0^6 \tag{9.14}$$

---

[1] The regions of initial conditions which give rise to the resonant, nonresonant, and subharmonic oscillations will be shown in Sec. 10.6a.

and the components x_0, y_0 are given by

$$x_0 = r_0 \cos (\theta + 72° \times n)$$
$$y_0 = r_0 \sin (\theta + 72° \times n) \qquad n = 0, 1, 2, 3, 4 \qquad (9.15)$$

where $\qquad \cos 5\theta = -\dfrac{16k}{5wr_0{}^3} \qquad \sin 5\theta = \dfrac{16A}{5wr_0{}^3}$

(a) Phase-plane Analysis

The transient state is defined by the integral curves of

$$\frac{dy}{dx} = \frac{Y(x,y)}{X(x,y)} \qquad (9.16)$$

where $X(x,y)$ and $Y(x,y)$ are given by Eqs. (9.12). The character of the singular point (x_0,y_0) is determined from the characteristic equation

$$\lambda^2 - (a_1 + b_2)\lambda + a_1 b_2 - a_2 b_1 = 0 \qquad (9.17)$$

where

$$a_1 = \left(\frac{\partial X}{\partial x}\right)_{\substack{x=x_0,\\y=y_0}} = \frac{1}{2}\left\{-k - \frac{5}{4}[2x_0 y_0(r_0{}^2 + 3w^2) + wx_0(x_0{}^2 - 3y_0{}^2)]\right\}$$

$$a_2 = \left(\frac{\partial X}{\partial y}\right)_{\substack{x=x_0,\\y=y_0}} = \frac{1}{2}\left\{A - \frac{5}{4}[2y_0{}^2(r_0{}^2 + 3w^2) - wy_0(3x_0{}^2 - y_0{}^2)]\right\}$$

$$b_1 = \left(\frac{\partial Y}{\partial x}\right)_{\substack{x=x_0,\\y=y_0}} = \frac{1}{2}\left\{-A + \frac{5}{4}[2x_0{}^2(r_0{}^2 + 3w^2) + wy_0(3x_0{}^2 - y_0{}^2)]\right\}$$

$$b_2 = \left(\frac{\partial Y}{\partial y}\right)_{\substack{x=x_0,\\y=y_0}} = \frac{1}{2}\left\{-k + \frac{5}{4}[2x_0 y_0(r_0{}^2 + 3w^2) + wx_0(x_0{}^2 - 3y_0{}^2)]\right\}$$

$$(9.17a)$$

In order to illustrate the geometrical configuration of the integral curves, a system with

$$k = 0.04 \qquad \text{and} \qquad B = 9.6$$

is investigated. Then, from the second equation of (9.11), $w = -0.4$. After substitution of these values into Eqs. (9.12), the corresponding integral curves of Eq. (9.16) are plotted in Fig. 9.7 by using the isocline method. There are eleven singular points; the details are listed in Table 9.2.

In the figure the singular points 1 to 5 are stable foci and correspond to the stable states of subharmonic oscillations. The integral curves, which tend to the saddle points 6 to 10 with the increase of time τ, separate the whole plane into two groups, i.e., the shaded and unshaded regions. An oscillation started from a point $(x(0),y(0))$ in the shaded regions tends to one of the singularities of points 1 to 5 corresponding to

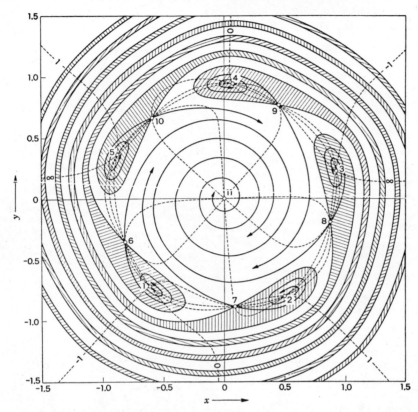

FIGURE 9.7   Integral curves for the subharmonic oscillation of order ⅕.   Full
line: integral curves.   Dashed line: isoclines.

Table 9.2   Singular Points of Fig. 9.7

Singular point	x_0	y_0	λ_1, λ_2	μ_1, μ_2*	Classification
1	-0.616	-0.724	$-0.020 \pm 0.618i$		Stable focus
2	0.499	-0.810	$-0.020 \pm 0.618i$		Stable focus
3	0.924	0.225	$-0.020 \pm 0.618i$		Stable focus
4	0.072	0.948	$-0.020 \pm 0.618i$		Stable focus
5	-0.876	0.362	$-0.020 \pm 0.618i$		Stable focus
6	-0.819	-0.354	$0.512, -0.552$	$150.9, \quad -1.143$	Saddle (unstable)
7	0.083	-0.888	$0.512, -0.552$	$-0.333, \quad 0.432$	Saddle (unstable)
8	0.870	-0.195	$0.512, -0.552$	$1.375, -10.70$	Saddle (unstable)
9	0.454	0.767	$0.512, -0.552$	$-1.403, -0.223$	Saddle (unstable)
10	-0.589	0.669	$0.512, -0.552$	$0.314, \quad 1.693$	Saddle (unstable)
11	0	0	$-0.020 \pm 0.500i$		Stable focus

* μ_1, μ_2 are the tangential directions of the integral curves at the singular points.

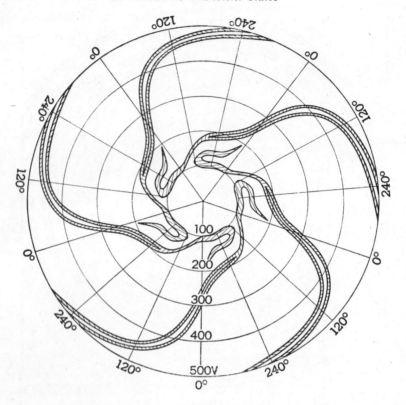

FIGURE 9.8   Regions of initial conditions leading to the subharmonic
oscillation of order ⅓. Radial scale: initial charging voltage
across capacitor. Angular scale: switching angle.

subharmonic response, whereas any oscillation started from the unshaded
region tends to the singularity of point 11. Since $x = 0$ and $y = 0$ at
the point 11, no subharmonic oscillation results in this case.

(b) Experimental Investigation

By a procedure analogous to that of Sec. 9.3, we perform an experiment to determine the regions of initial conditions which lead to subharmonic oscillations of order ⅓ and harmonic oscillations of the nonresonant type. These regions, shown in Fig. 9.8, agree with the analytic
results of Fig. 9.7.

9.5 Subharmonic Oscillations of Order ½ [36, 38]

(a) Fundamental Equations

In Chap. 7 we have discussed the subharmonic oscillations of order
½ in the steady state. We have particularly concentrated our attention

on the stability of periodic oscillations by making use of the stability theory of Hill's equation. As mentioned in Sec. 7.7a, an appropriate form of the nonlinearity to produce a subharmonic oscillation of order $\frac{1}{2}$ is given by $f(v) = |v|v$; so we write the differential equation for the subharmonic oscillations of order $\frac{1}{2}$ as

$$\frac{d^2v}{d\tau^2} + k\frac{dv}{d\tau} + |v|v = B\cos 2\tau + B_0 \qquad (9.18)$$

or as its equivalent (which is easier to treat analytically),

$$\frac{d^2v}{d\tau^2} + k\frac{dv}{d\tau} + c_1v + c_3v^3 = B\cos 2\tau + B_0 \qquad (9.19)$$

We take the periodic solution as of the form

$$v(\tau) = z(\tau) + x(\tau)\sin\tau + y(\tau)\cos\tau + w\cos 2\tau$$

where $\qquad\qquad w = \dfrac{1}{1-2^2}B = -\dfrac{1}{3}B \qquad\qquad (9.20)$

In this case the stability of the periodic solution may be examined by using the condition (4.15) (see also Sec. 7.6b).[1] This stability condition is also derived from the Routh-Hurwitz criterion, as one sees shortly.

By substituting Eqs. (9.20) into (9.19) and equating the coefficients of the terms containing $\cos\tau$ and $\sin\tau$ and of the nonoscillatory term separately to zero, we obtain

$$\frac{dx}{d\tau} = \frac{1}{2}(-kx + Ay - 3c_3wyz) \equiv X(x,y,z)$$

$$\frac{dy}{d\tau} = \frac{1}{2}(-Ax - ky - 3c_3wxz) \equiv Y(x,y,z) \qquad (9.21)$$

$$B_0 = c_1z + c_3[(\tfrac{3}{2}r^2 + z^2 + \tfrac{3}{2}w^2)z - \tfrac{3}{4}w(x^2 - y^2)] \equiv Z(x,y,z)$$

where $\quad A = (1 - c_1) - c_3(\tfrac{3}{4}r^2 + 3z^2 + \tfrac{3}{2}w^2) \qquad r^2 = x^2 + y^2$

It is noted that the same assumptions as those mentioned in Sec. 8.2 must be made for the derivation of Eqs. (9.21).

(b) *Periodic Solutions and Their Stability*

In order to obtain the relationship between the initial condition and the resulting response, we must first know the types of the steady-state oscillations under various combinations of the system parameters. Therefore, we shall be concerned, for the time being, with the $\frac{1}{2}$-harmonic oscillations in the steady state.

---

[1] The stability condition for the nth unstable region ($n = 1, 3, 4, \ldots$) is generally satisfied when the nonlinearity is given by $f(v) = |v|v$ (refer to Sec. 7.7).

Since $x(\tau)$, $y(\tau)$, and $z(\tau)$ in Eqs. (9.20) become constant in the steady state, we have[1]

$$\frac{dx}{d\tau} = 0 \qquad \frac{dy}{d\tau} = 0 \qquad \text{and} \qquad \frac{dz}{d\tau} = 0 \qquad (9.22)$$

Substituting these conditions into Eqs. (9.21) gives us the same relation as already given by Eqs. (7.39) in Sec. 7.6a. Hence the steady-state components x_0, y_0, z_0, and r_0 of the solution (9.20) are determined from Eqs. (7.40) and (7.41).

We now investigate the stability of the periodic solutions by making use of the Routh-Hurwitz criterion. To this end, we consider sufficiently small variations ξ, η, and ζ from the equilibrium state defined by

$$\xi = x - x_0 \qquad \eta = y - y_0 \qquad \zeta = z - z_0 \qquad (9.23)$$

Then, if these variations ξ, η, and ζ tend to zero with increasing τ, the solution is stable. By substituting Eqs. (9.23) into (9.21), we obtain

$$\frac{d\xi}{d\tau} = a_{11}\xi + a_{12}\eta + a_{13}\zeta$$

$$\frac{d\eta}{d\tau} = a_{21}\xi + a_{22}\eta + a_{23}\zeta \qquad (9.24)$$

$$0 = a_{31}\xi + a_{32}\eta + a_{33}\zeta$$

with

$$a_{11} = \left(\frac{\partial X}{\partial x}\right)_0 = -\frac{1}{2}\left(\frac{3}{2}c_3 x_0 y_0 + k\right)$$

$$a_{12} = \left(\frac{\partial X}{\partial y}\right)_0 = \frac{1}{2}\left(A - \frac{3}{2}c_3 y_0^2 - 3c_3 w z_0\right)$$

$$a_{13} = \left(\frac{\partial X}{\partial z}\right)_0 = -\frac{3}{2}c_3 y_0(2z_0 + w)$$

$$a_{21} = \left(\frac{\partial Y}{\partial x}\right)_0 = -\frac{1}{2}\left(A - \frac{3}{2}c_3 x_0^2 + 3c_3 w z_0\right)$$

$$a_{22} = \left(\frac{\partial Y}{\partial y}\right)_0 = \frac{1}{2}\left(\frac{3}{2}c_3 x_0 y_0 - k\right) \qquad (9.24a)$$

$$a_{23} = \left(\frac{\partial Y}{\partial z}\right)_0 = \frac{3}{2}c_3 x_0(2z_0 - w)$$

$$a_{31} = \left(\frac{\partial Z}{\partial x}\right)_0 = \frac{3}{2}c_3 x_0(2z_0 - w)$$

$$a_{32} = \left(\frac{\partial Z}{\partial y}\right)_0 = \frac{3}{2}c_3 y_0(2z_0 + w)$$

$$a_{33} = \left(\frac{\partial Z}{\partial z}\right)_0 = c_1 + c_3\left(\frac{3}{2}r_0^2 + 3z_0^2 + \frac{3}{2}w^2\right)$$

[1] By virtue of Eqs. (9.21), the third condition results from the first two.

where $\left(\dfrac{\partial X}{\partial x}\right)_0, \cdots, \left(\dfrac{\partial Z}{\partial z}\right)_0$ denote the values of $\dfrac{\partial X}{\partial x}, \cdots, \dfrac{\partial Z}{\partial z}$ at $x = x_0$,
$y = y_0$, and $z = z_0$. The characteristic equation of the system (9.24) is

$$\begin{vmatrix} a_{11} - \lambda & a_{12} & a_{13} \\ a_{21} & a_{22} - \lambda & a_{23} \\ a_{31} & a_{32} & a_{33} \end{vmatrix} = 0 \qquad (9.25)$$

or

$$a_{33}\lambda^2 - \left\{ \begin{vmatrix} a_{11} & a_{13} \\ a_{31} & a_{33} \end{vmatrix} + \begin{vmatrix} a_{22} & a_{23} \\ a_{32} & a_{33} \end{vmatrix} \right\} \lambda + \begin{vmatrix} a_{11} & a_{12} & a_{13} \\ a_{21} & a_{22} & a_{23} \\ a_{31} & a_{32} & a_{33} \end{vmatrix} = 0 \quad (9.26)$$

The Routh-Hurwitz criterion states that the system (9.24) and, consequently, the periodic solution are stable provided that

$$a_{33} > 0 \qquad \begin{vmatrix} a_{11} & a_{13} \\ a_{31} & a_{33} \end{vmatrix} + \begin{vmatrix} a_{22} & a_{23} \\ a_{32} & a_{33} \end{vmatrix} < 0$$

$$\begin{vmatrix} a_{11} & a_{12} & a_{13} \\ a_{21} & a_{22} & a_{23} \\ a_{31} & a_{32} & a_{33} \end{vmatrix} \equiv \Delta > 0 \qquad (9.27)$$

The first and the second conditions of (9.27) are fulfilled from the outset because, by Eqs. (9.24a),

$$a_{33} = c_1 + c_3(\tfrac{3}{2}r_0^2 + 3z_0^2 + \tfrac{3}{2}w^2) > 0$$

$$\begin{vmatrix} a_{11} & a_{13} \\ a_{31} & a_{33} \end{vmatrix} + \begin{vmatrix} a_{22} & a_{23} \\ a_{32} & a_{33} \end{vmatrix} = -k[c_1 + c_3(\tfrac{3}{2}r_0^2 + 3z_0^2 + \tfrac{3}{2}w^2)] < 0 \qquad (9.28)$$

Hence the third inequality $\Delta > 0$ is the essential condition for the stability of the periodic solution. Substitution of Eqs. (9.24a) in the determinant Δ leads to a lengthy expression; however, by virtue of Eqs. (7.40) and (7.41) in Sec. 7.6a, the stability condition ultimately leads to

$$\Delta = 6c_3r_0^2z_0\left(1 - c_1 - \frac{3}{4}c_3r_0^2 - 3c_3z_0^2\right)\frac{dB_0}{dr_0^2} > 0 \qquad (9.29)$$

Since $c_1 + c_3 = 1$ [see Eq. (7.3)], this condition is identical with (7.48).

NUMERICAL ANALYSIS

In order to present a concrete discussion of the foregoing analysis, we consider some representative combinations of the system parameters and investigate the types of $\frac{1}{2}$-harmonic oscillations in what follows.

As mentioned in Sec. 7.7a, the nonlinearity $|v|v$ is approximated by

$$|v|v \cong c_1v + c_3v^3 = 0.3v + 0.7v^3 \qquad (9.30)$$

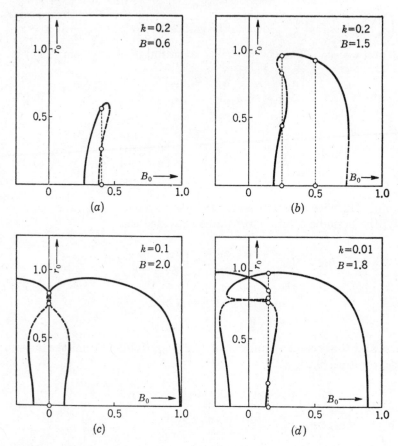

FIGURE 9.9   Amplitude characteristics of the ½-harmonic oscillation.

This approximation is considered to be permissible within the range of v in which the ½-harmonic oscillation occurs. By making use of Eqs. (7.40) in Sec. 7.6a, the amplitude characteristics (r_0 versus B_0) are calculated for several values of k and B and plotted in Fig. 9.9. The stability of the periodic solutions is investigated by using the condition (9.29). The result is shown in the figure by distinguishing the characteristic curves with full lines and dashed lines, corresponding to the stable and the unstable states, respectively. It will be noticed that, since $x = 0$ and $y = 0$ satisfy Eqs. (9.22), $v(\tau) = z_0 + w \cos 2\tau$ is another periodic solution. We see in Fig. 9.9 that various types of ½-harmonic oscillations exist according to the different values of the system parameters. They are as follows:

CASE 1. $k = 0.2$, $B = 0.6$, AND $B_0 = 0.4$ (FIG. 9.9a)

There are two $\frac{1}{2}$-harmonic oscillations, differing only in phase by π radians.   The periodic solution without $\frac{1}{2}$-harmonic (that is, $r_0 = 0$) is readily found to be stable.   Therefore, the $\frac{1}{2}$-harmonic oscillation occurs only when the initial condition is properly chosen.

CASE 2. $k = 0.2$, $B = 1.5$, AND $B_0 = 0.5$ (FIG. 9.9b)

As regards the $\frac{1}{2}$-harmonic oscillations, the situation is the same as in the first case.   However, the periodic solution $r_0 = 0$ is unstable; therefore all initial conditions lead to the $\frac{1}{2}$-harmonic response.

CASE 3. $k = 0.2$, $B = 1.5$, AND $B_0 = 0.25$ (FIG. 9.9b)

There are two different values for r_0, and for each of them there are two $\frac{1}{2}$-harmonic oscillations that differ in phase by π radians.   The periodic solution with $r_0 = 0$ is unstable; therefore, all initial conditions lead to the $\frac{1}{2}$-harmonic response.

CASE 4. $k = 0.1$, $B = 2.0$, AND $B_0 = 0$ (FIG. 9.9c)

There are, as mentioned in Sec. 7.6a, four $\frac{1}{2}$-harmonic oscillations, each differing in phase by $\pi/2$ radians from the other.   The periodic solution with $r_0 = 0$ is stable; therefore, the $\frac{1}{2}$-harmonic oscillation occurs only when the initial condition is properly chosen.

CASE 5. $k = 0.01$, $B = 1.8$, AND $B_0 = 0.15$ (FIG. 9.9d)

There are three different values for r_0, and for each of them there are two $\frac{1}{2}$-harmonic oscillations that differ in phase by π radians.   The periodic solution with $r_0 = 0$ is unstable; therefore, all initial conditions lead to the $\frac{1}{2}$-harmonic response.

9.6 Phase-plane Analysis of the Subharmonic Oscillations of Order $\frac{1}{2}$

As mentioned before, our object is to study the solution of Eq. (9.18) or (9.19) in the transient state, which, with the increase of time, ultimately yields the periodic solution.   For this purpose we investigate the integral curves of

$$\frac{dy}{dx} = \frac{Y(x,y,z)}{X(x,y,z)} \tag{9.31}$$

with $Z(x,y,z) = B_0$

where $X(x,y,z)$, $Y(x,y,z)$, and $Z(x,y,z)$ are given by Eqs. (9.21).   One will readily see from the third equation of (9.21) that z is uniquely determined once the values of x and y are known.   The periodic solutions satisfy the conditions (9.22), and they are therefore expressed by the singular points of Eqs. (9.31), that is, by the points at which both $X(x,y,z)$ and $Y(x,y,z)$ vanish.

The types of singular points are classified according to the roots λ of the characteristic equation (9.26). By use of Eqs. (9.28), the discriminant D of Eq. (9.26) becomes

$$D \equiv \left\{ \begin{vmatrix} a_{11} & a_{13} \\ a_{31} & a_{33} \end{vmatrix} + \begin{vmatrix} a_{22} & a_{23} \\ a_{32} & a_{33} \end{vmatrix} \right\}^2 - 4a_{33}\Delta = a_{33}(a_{33}k^2 - 4\Delta) \quad (9.32)$$

It is also noted, from Eqs. (9.28), that

$$\begin{vmatrix} a_{11} & a_{13} \\ a_{31} & a_{33} \end{vmatrix} + \begin{vmatrix} a_{22} & a_{23} \\ a_{32} & a_{33} \end{vmatrix} < 0 \quad \text{and} \quad a_{33} > 0 \quad (9.33)$$

Hence the singular points of Eqs. (9.31) are classified as follows:

1. If $D \geqq 0$ and $\Delta > 0$, the characteristic roots λ are both real and of the negative sign, so that the singularity is a stable node.

2. If $D > 0$ and $\Delta < 0$, the characteristic roots λ are real but of opposite sign, so that the singularity is a saddle point which is intrinsically unstable.

3. If $D < 0$, the characteristic roots λ are conjugate complex, and the real part is negative, so that the singularity is a stable focus.

NUMERICAL ANALYSIS

First, let us consider an example corresponding to Case 3 in Sec. 9.5b, where the system parameters are given by

$$k = 0.2 \qquad B = 1.5 \qquad \text{and} \qquad B_0 = 0.25$$

As illustrated in Fig. 9.9b, there are two kinds of the ½-harmonic oscillations with different amplitudes. The integral curves for this particular case are plotted in Fig. 9.10. As expected, there are seven singular points 1, 2, . . . , 7; the details are listed in Table 9.3.

Table 9.3   Singular Points of Fig. 9.10

Singular point	x_0	y_0	z_0	λ_1, λ_2	μ_1, μ_2*	Classification
1	0.376	0.883	0.265	$-0.100 \pm 0.090i$		Stable focus
2	-0.376	-0.883	0.265	$-0.100 \pm 0.090i$		Stable focus
3	0.586	0.586	0.190	$0.098, -0.298$	0.156, 5.322	Saddle (unstable)
4	-0.586	-0.586	0.190	$0.098, -0.298$	0.156, 5.322	Saddle (unstable)
5	0.407	0.176	0.263	$-0.100 \pm 0.055i$		Stable focus
6	-0.407	-0.176	0.263	$-0.100 \pm 0.055i$		Stable focus
7	0	0	0.375	$0.086, -0.286$	$0.696, -0.696$	Saddle (unstable)

* μ_1, μ_2 are the tangential directions of the integral curves at the singular points.

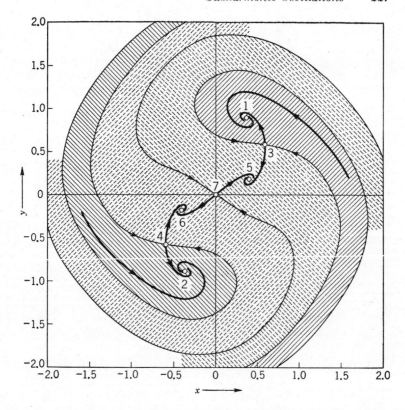

FIGURE 9.10   Integral curves of Eqs. (9.31) in the xy plane, the system
parameters being $k = 0.2$, $B = 1.5$, and $B_0 = 0.25$.

By Eqs. (9.21) the representative point $(x(\tau),y(\tau))$ moves, with
increasing τ, along the integral curve in the direction of the arrows and
tends ultimately to one of the stable singularities 1, 2, 5, and 6.   Since
the distance between the singular point and the origin shows the amplitude r_0, the singularities 1 and 2 represent the ½-harmonic oscillations
that have the same amplitude but are of opposite phase; the same thing
is true for the singularities of points 5 and 6.   The singularities of points
3, 4, and 7 are saddle points which are intrinsically unstable; the corresponding periodic states cannot be sustained, because any slight deviation from the saddle point will lead the oscillation to one of the stable
foci.   The separatrices, i.e., the integral curves which enter the saddle
points, divide the whole plane into four regions as indicated with different
hatching.   All integral curves in one of these regions tend to the stable
singularity which is contained in that region.   Hence the relationship
existing between the initial condition $(x(0),y(0))$ and the resulting

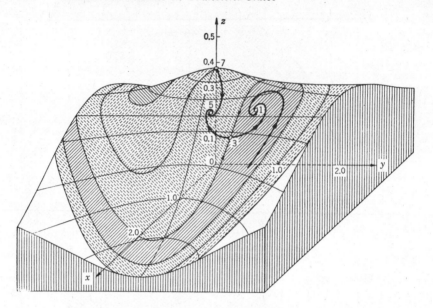

FIGURE 9.11   Integral curves of Eqs. (9.31) in the xyz space, the system parameters being $k = 0.2$, $B = 1.5$, and $B_0 = 0.25$.

$\frac{1}{2}$-harmonic oscillation will be made clear. Since the origin is an unstable singularity, all initial conditions lead to the $\frac{1}{2}$-harmonic response.

Thus far, the integral curves are considered in the xy plane. Since the nonoscillatory component $z(\tau)$ also varies as the values of $x(\tau)$ and $y(\tau)$, the integral curves are really on the surface which is determined by the third equation of (9.21). Figure 9.11 shows the geometrical configuration of the integral curves in the xyz space.

Now suppose that an initial condition for the solution of Eq. (9.19) is prescribed by $v(0)$ and $\dot{v}(0)$, where the dot refers to differentiation with respect to τ; then $x(0)$, $y(0)$, and $z(0)$ corresponding to this initial condition are determined by Eqs. (9.20) and (9.21), that is,

$$
\begin{aligned}
v(0) &= z(0) + y(0) + w \\
\dot{v}(0) &= \dot{z}(0) + x(0) + \dot{y}(0) \cong x(0) \\
c_1 z(0) + c_3 \{ [\tfrac{3}{2} r^2(0) + z^2(0) &+ \tfrac{3}{2} w^2] z(0) - \tfrac{3}{4} w[x^2(0) - y^2(0)] \} \\
&= B_0
\end{aligned}
\tag{9.34}
$$

By making use of these relations, the regions of initial conditions in Fig. 9.10 are reproduced on the $v\dot{v}$ plane as illustrated in Fig. 9.12. Since, in the steady state,

$$
\begin{aligned}
v(\tau) &= z_0 + x_0 \sin \tau + y_0 \cos \tau + w \cos 2\tau \\
\dot{v}(\tau) &= x_0 \cos \tau - y_0 \sin \tau - 2w \sin 2\tau
\end{aligned}
\tag{9.35}
$$

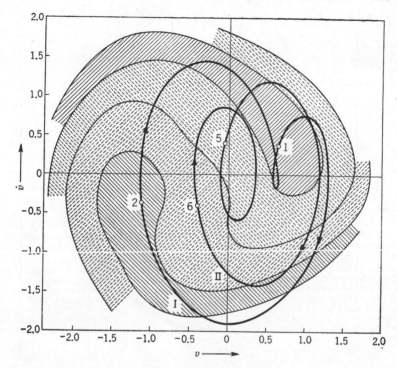

FIGURE 9.12   Regions of initial conditions leading to the ½-harmonic response and the trajectories of the periodic solutions correlated with the stable singularities in Fig. 9.10.

the periodic solutions correlated with the stable singularities of points 1, 2 and 5, 6 in Fig. 9.10 are shown by the closed curves I and II, respectively. The period required for the representative point $(v(\tau), \dot{v}(\tau))$ to complete one cycle on the curve I (or II) is 2π, or twice the period of the external force. A trajectory which starts from an initial point $(v(0), \dot{v}(0))$ in one of these regions, say, the region containing the point 1 (or 2), will tend to the closed curve I; the representative point $(v(\tau), \dot{v}(\tau))$ in the steady state will then pass through the point 1 (or 2) when $\tau = 2n\pi$, n being a sufficiently large positive integer. Similarly, initial conditions in the region containing the point 5 (or 6) will lead the oscillation to the steady state represented by the closed curve II, and the representative point $(v(\tau), \dot{v}(\tau))$ in the steady state will pass through the point 5 (or 6) when $\tau = 2n\pi$.

Second, let us consider an example corresponding to Case 4 in Sec. 9.5*b*, where the system parameters are given by

$$k = 0.1 \qquad B = 2.0 \qquad \text{and} \qquad B_0 = 0$$

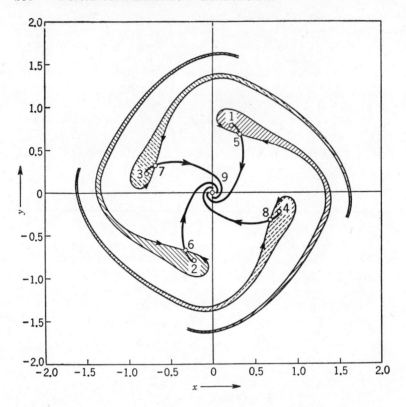

FIGURE 9.13   Integral curves of Eqs. (9.31) in the xy plane, the system
parameters being $k = 0.1$, $B = 2.0$, and $B_0 = 0$.

As illustrated in Fig. 9.9c, there are four $\frac{1}{2}$-harmonic oscillations, each
having the same amplitude as but differing in phase by $\pi/2$ radians from
the other.   The integral curves for this particular case are plotted in
Fig. 9.13.   As expected, there are nine singularities, points 1, 2, . . . , 9,
the details of which are listed in Table 9.4.

In Fig. 9.13 we see that the singular points 1 to 4 represent the
stable states of the $\frac{1}{2}$-harmonic oscillations and that they are equidistant
from and equiangular about the origin.   The angular distance between
the adjacent singular points corresponds to one-half cycle of the external
force.   The singular points 5 to 8 are saddle points; therefore, the corresponding periodic solutions are unstable.   Contrary to the previous
case, the singular point 9, that is, the origin, is a stable focus.   Hence
the conclusion follows that any oscillation starting from a point in the
shaded regions leads ultimately to one of the singularities of points 1 to 4,
resulting in the $\frac{1}{2}$-harmonic response, whereas any oscillation which

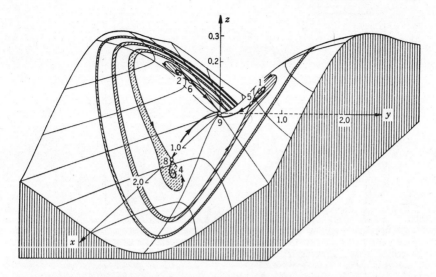

FIGURE 9.14 Integral curves of Eqs. (9.31) in the xyz space, the system parameters being $k = 0.1$, $B = 2.0$, and $B_0 = 0$.

starts from the unshaded region leads ultimately to the origin, resulting in no $\frac{1}{2}$-harmonic response. By making use of the third equation of (9.21), the integral curves in the xyz space are calculated and shown in Fig. 9.14.

By a procedure analogous to that for the previous example, the regions of initial conditions and the stable singularities in Fig. 9.13 are reproduced on the $v\dot{v}$ plane, as illustrated in Fig. 9.15. The periodic solutions corresponding to the singularities of points 1, 2 and 3, 4 are shown by the closed curves I and II, respectively. Since these oscillations are

Table 9.4 Singular Points of Fig. 9.13

Singular point	x_0	y_0	z_0	λ_1, λ_2	μ_1, μ_2^*	Classification
1	0.229	0.789	0.134	$-0.050 \pm 0.140i$		Stable focus
2	-0.229	-0.789	0.134	$-0.050 \pm 0.140i$		Stable focus
3	-0.789	0.229	-0.134	$-0.050 \pm 0.140i$		Stable focus
4	0.789	-0.229	-0.134	$-0.050 \pm 0.140i$		Stable focus
5	0.320	0.683	0.093	$0.095, -0.195$	0.301, 2.434	Saddle (unstable)
6	-0.320	-0.683	0.093	$0.095, -0.195$	0.301, 2.434	Saddle (unstable)
7	-0.683	0.320	-0.093	$0.095, -0.195$	$-3.322, -0.411$	Saddle (unstable)
8	0.683	-0.320	-0.093	$0.095, -0.195$	$-3.322, -0.411$	Saddle (unstable)
9	0	0	0	$-0.050 \pm 0.117i$		Stable focus

* μ_1, μ_2 are the tangential directions of the integral curves at the singular points.

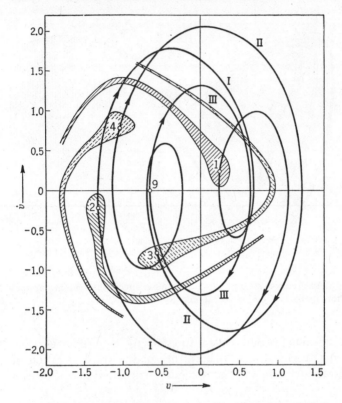

FIGURE 9.15   Regions of initial conditions leading to the ½-
harmonic response and the trajectories of the
periodic solutions correlated with the stable
singularities in Fig. 9.13.

the ½-harmonics, the time required for the representative point $(v(\tau),\dot{v}(\tau))$
to complete one cycle on the curve I (or II) is 2π.   The singularity 9,
that is, the origin of Fig. 9.13, is correlated with the oscillation without
½-harmonic response; the periodic solution corresponding to this singularity is represented by the closed curve III.   A representative point
$(v(\tau),\dot{v}(\tau))$ passes through the point 9 when $\tau = n\pi$, where n is a sufficiently large positive integer; thus the time required for a point $(v(\tau),\dot{v}(\tau))$
to complete one cycle on the curve III is π, or equal to the period of the
external force.

Finally, Fig. 9.16 shows a list of representative patterns of the initial
conditions which lead to the ½-harmonic responses.   These patterns are
explained as follows:

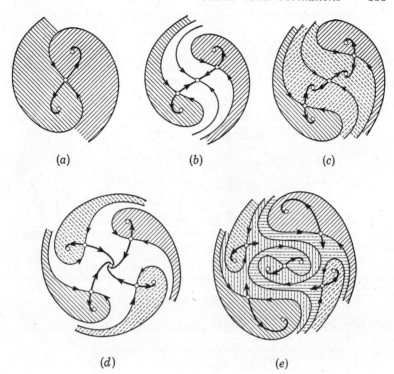

(a)　　　　　(b)　　　　　(c)

(d)　　　　　(e)

FIGURE 9.16 Patterns of initial conditions leading to the ½-harmonic response.

a. All initial conditions lead to one of the two ½-harmonic oscillations that have the same amplitude but differ in phase by π radians.

b. Initial conditions lead either to the ½-harmonic response or to the oscillation without ½-harmonic. The ½-harmonic oscillations have the same amplitude but differ in phase by π radians.

c. All initial conditions lead to the ½-harmonic response. The ½-harmonics have two different amplitudes, and for each of these there are two oscillations that differ in phase by π radians.

d. Initial conditions lead either to the ½-harmonic response or to the oscillation without ½-harmonic. The ½-harmonic oscillations have the same amplitude, but each differs in phase by $\pi/2$ radians from the other.

e. All initial conditions lead to the ½-harmonic response. The ½-harmonics have three different amplitudes, and for each of these there are two oscillations that differ in phase by π radians.

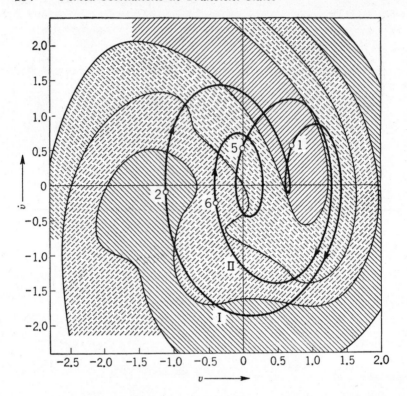

FIGURE 9.17   Regions of initial conditions leading to the ½-harmonic response and the trajectories of the periodic solutions, both
obtained by analog-computer analysis (see Fig. 9.12).

9.7 Analog-computer Analysis

An analog-computer setup to solve the equation

$$\frac{d^2v}{d\tau^2} + 2\delta\frac{dv}{d\tau} + |v|v = B\cos 2\tau + B_0 \tag{9.36}$$

has already been given in Fig. 7.21.[1] By using this block diagram, we
investigate the transient behavior of the solution and compare the result
with the foregoing analysis.

CASE 1. $k = 0.2$, $B = 1.5$, AND $B_0 = 0.25$

Figure 9.17 is obtained by the following procedure.   A point $(v(0),\dot{v}(0))$,
i.e., one of the initial conditions, is first prescribed on the $v(\tau)\dot{v}(\tau)$ plane of

[1] Equation (9.36) is identical with (9.18) if 2δ is replaced by k.

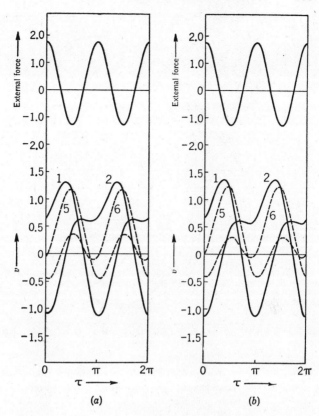

FIGURE 9.18   Waveforms of the $\frac{1}{2}$-harmonic oscillations in the case when $k = 0.2$, $B = 1.5$, and $B_0 = 0.25$. (*a*) Obtained by phase-plane analysis. (*b*) Obtained by analog-computer analysis.

the computer recorder. Then the solution curve, i.e., the trajectory of the point $(v(\tau), \dot{v}(\tau))$ which starts from the initial point $(v(0), \dot{v}(0))$, will ultimately tend to one of the closed curves I and II. By repeating this process for different values of the initial conditions, the whole plane is divided into four regions; the region containing the point m ($= 1, 2, 5,$ or 6) is so determined that the representative point $(v(\tau), \dot{v}(\tau))$, which starts from this region, passes through the point m when $\tau = 2n\pi$, n being a sufficiently large positive integer.[1]

---

[1] A cycle indicator which counts every two cycles of the external force $B \cos 2\tau$ is used for this purpose.

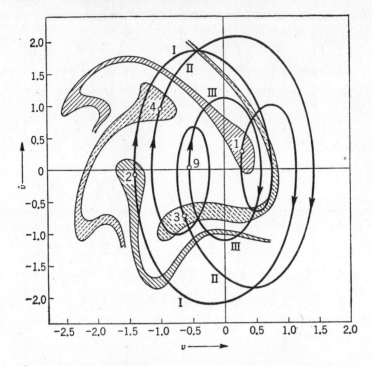

FIGURE 9.19   Regions of initial conditions leading to the ½-harmonic
response and the trajectories of the periodic solutions,
both obtained by analog-computer analysis (see Fig.
9.15).

Figure 9.17 shows a satisfactory agreement with the theoretical
result of Fig. 9.12.   The system parameters are the same for both cases.
Therefore, the assumptions made in the derivation of Eqs. (9.21) may be
accepted.   The time-response curves of the ½-harmonic oscillations are
shown in Fig. 9.18a and b.   The calculated curves in Fig. 9.18a are
obtained by substituting the steady-state values x_0, y_0, and z_0 of Table 9.3
into Eqs. (9.35).   The curves in Fig. 9.18b are obtained by making use of
the analog computer.   As one sees in the figure, there are four ½-harmonics having two different waveforms, and for each of these there are
two oscillations differing in phase by π radians.

CASE 2. $k = 0.1$, $B = 2.0$, AND $B_0 = 0$

By a procedure analogous to that for the first case, we obtain Fig. 9.19,
which again shows an agreement with the theoretical result as given in

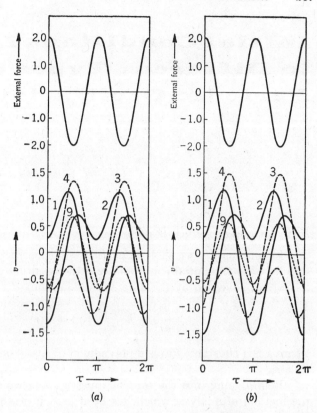

FIGURE 9.20 Waveforms of the harmonic and the ½-harmonic oscillations in the case when $k = 0.1$, $B = 2.0$, and $B_0 = 0$. (a) Obtained by phase-plane analysis. (b) Obtained by analog-computer analysis.

Fig. 9.15. Contrary to the preceding case, an initial condition prescribed in the unshaded region results in the oscillation without ½-harmonic. The time-response curves are illustrated in Fig. 9.20a and b. The curves 1 to 4 show the ½-harmonic oscillations; the curve 9 shows the oscillation without ½-harmonic response.

Chapter 10

Initial Conditions Leading to Different Types of Periodic Oscillations

10.1 Method of Analysis [40]

In this chapter we shall be particularly concerned with the relationship between the initial conditions and the resulting periodic oscillations of systems governed by the differential equation

$$\frac{d^2v}{d\tau^2} + k\frac{dv}{d\tau} + f(v) = g(\tau) \tag{10.1}$$

where k is a constant, $f(v)$ is a polynomial of v, and $g(\tau)$ is periodic in the time τ.

Before going into the present investigation, we shall briefly review the method of analysis which has been used in the preceding chapters. For the sake of brevity, we confine the problem to the analysis of harmonic oscillations under the impression of the external force $g(\tau) = B\cos\tau$. We write the solution of Eq. (10.1) as

$$v(\tau) = x(\tau)\sin\tau + y(\tau)\cos\tau$$

where it is assumed that the amplitudes $x(\tau)$ and $y(\tau)$ are slowly varying functions of the time τ. Under this condition we may derive a set of simultaneous equations of the form

$$\frac{dx}{d\tau} = X(x,y) \quad \text{and} \quad \frac{dy}{d\tau} = Y(x,y) \tag{10.2}$$

where $X(x,y)$ and $Y(x,y)$ are the polynomials of x and y that may readily be found. Upon elimination of τ in Eqs. (10.2), the integral curves, i.e., the trajectories of the representative point $(x(\tau),y(\tau))$, are plotted in the xy plane. A singular point, for which $X(x,y) = 0$ and $Y(x,y) = 0$, corresponds to a periodic solution of Eq. (10.1). For certain values of k and B there exist three singularities, i.e., two stable foci and one saddle point which is intrinsically unstable (Fig. 8.2). The integral curve which

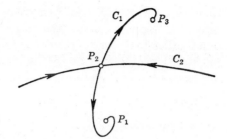

FIGURE 10.1   Fixed points and invariant
curves under the mapping.

tends to the saddle point with increasing τ is the separatrix which divides
the coordinate plane into two regions, such that any initial conditions
prescribed in each of them result in passage to a particular type of harmonic response.

This method of analysis has been used extensively for the study of
harmonic and subharmonic oscillations in the transient state, but it has
the following drawbacks.   First, if the initial conditions are prescribed
at values which are far different from those of the steady state, the
assumption that the amplitude and phase of the oscillation vary slowly
does not hold; therefore, the result obtained by this method may not be
quite accurate.   The second and more serious defect is that, if a number
of steady-state responses are to be expected, this method is practically
inapplicable, since the analysis must be effected by graphical solution in
a higher-dimensional phase space.

This section describes the method of analysis which is applicable in
such situations.[1]   We consider the behavior of a point whose coordinates
are given by $v(\tau)$ and $\dot{v}(\tau)$ in the $v\dot{v}$ plane (dots over v refer to differentiation with respect to τ).   An initial condition is then defined by a point
at $\tau = 0$.   Special attention is directed toward the location of the points
at the instants of $\tau = T, 2T, 3T, \ldots$, where T is the period of the
external force $g(\tau)$.   Mathematically, these points are obtained as the
successive images of the initial point under iterations of mapping from
$\tau = nT$ to $(n + 1)T$, where $n = 0, 1, 2, \ldots$.   A periodic solution of
Eq. (10.1) is then represented by a fixed point of the mapping in the $v\dot{v}$
plane.   To be more specific, let us consider once again the harmonic
oscillation which has been mentioned just now.   Then there are three
fixed points, corresponding to the periodic solutions of (10.1).   They
are shown by P_1, P_2, and P_3 in Fig. 10.1; P_1 and P_3 are stable, and P_2 is
unstable.   Through P_2 there are two curves, C_1 and C_2, which are invariant under the mapping.   Points on C_2 approach P_2 under iterations of

---

[1] A similar method of analysis has also been studied by K. W. Blair and W. S.
Loud [9].   The interested reader is referred to their paper for the mathematical
treatment of the problem.

the mapping, while points on C_1 approach P_2 under iterations of the inverse mapping. Hence the successive images of an initial point will tend either to P_1 or to P_3, depending on which side of C_2 is the initial point. Thus the curve C_2 is the boundary between two regions; in each of them, all initial conditions lead to a particular type of harmonic oscillation. These regions will be called the *domains of attraction*. The behavior of the loci of images is analogous to that of the integral curves in the neighborhood of the saddle point in the xy plane.

The domains of attraction for different types of periodic solutions may be found by the following steps.

1. A periodic solution is expanded into Fourier series, assuming the harmonic or subharmonic frequency as its least frequency. If the periodic solution, either stable or unstable, does exist, the coefficients of the principal terms of the Fourier series may be determined by the method of harmonic balance.

2. A small variation ξ from the periodic solution is governed by the variational equation of (10.1), that is,

$$\frac{d^2\xi}{d\tau^2} + k\frac{d\xi}{d\tau} + \left(\frac{df}{dv}\right)_{v=v_0} \xi = 0 \qquad (10.3)$$

where v_0 is the periodic solution. Equation (10.3) takes the form of Hill's equation and may be solved by an approximate method. Thus we can distinguish between the stable and unstable fixed points and also determine the slope of the invariant curve C_2 at the unstable fixed point P_2 (Fig. 10.1).

3. The boundary between the domains of attraction is the invariant curve C_2, which is the locus of the images that approach the unstable fixed point from both sides. These curves are obtained by starting just on either side of the unstable fixed point and integrating the original equation (10.1) for decreasing time, i.e., by using negative time in (10.1). A digital computer may be used for numerical integration. It is found that, if two initial points are prescribed not exactly on C_2, but on the two sides of C_2 and sufficiently close to each other, the loci of the images which have started from these points nearly coincide with each other after several iterations of the mapping.

In the sections that follow we shall seek the domains of attraction for some representative types of periodic oscillations in nonlinear systems.

10.2 Symmetrical Systems

As an example of Eq. (10.1) we consider the differential equation

$$\frac{d^2v}{d\tau^2} + k\frac{dv}{d\tau} + v^3 = B\cos\tau \qquad (10.4)$$

This equation is unchanged if the sign of v is reversed and τ is shifted by π radians. Therefore the system governed by Eq. (10.4) will be called the symmetrical system. Since the nonlinearity is cubic in v, one may expect a periodic solution with harmonic frequency or subharmonic frequency of order $\frac{1}{3}$ as its least frequency (Sec. 7.2).[1] If the system parameters k and B are appropriately chosen, the periodic solution might be assumed to take the form

$$v_0(\tau) = x_1 \sin \tau + y_1 \cos \tau \tag{10.5}$$

for harmonic response, and

$$v_0(\tau) = x_{\frac{1}{3}} \sin \tfrac{1}{3}\tau + y_{\frac{1}{3}} \cos \tfrac{1}{3}\tau + x_1 \sin \tau + y_1 \cos \tau \tag{10.6}$$

for subharmonic response. Terms of frequency other than those that appear in Eqs. (10.5) and (10.6) are ignored to this order of approximation. It depends on the initial condition as to which response, harmonic or subharmonic, will actually occur. This problem will be studied in Sec. 10.3.

(a) Location of the Fixed Points

The fixed points of the mapping may readily be located in the $v\dot{v}$ plane, once the periodic solutions (10.5) and (10.6) are known. The coefficients of these periodic solutions may be determined by the method of harmonic balance. Substituting Eq. (10.6) into (10.4) and equating the coefficients of the terms containing $\sin \tau$, $\cos \tau$, $\sin \frac{1}{3}\tau$, and $\cos \frac{1}{3}\tau$ separately to zero leads to

$$
\begin{aligned}
A_1 x_1 + k y_1 + \tfrac{1}{4}(x_{\frac{1}{3}}^2 - 3 y_{\frac{1}{3}}^2) x_{\frac{1}{3}} &= 0 \\
k x_1 - A_1 y_1 - \tfrac{1}{4}(3 x_{\frac{1}{3}}^2 - y_{\frac{1}{3}}^2) y_{\frac{1}{3}} - B &= 0 \\
\tfrac{3}{4}(x_{\frac{1}{3}}^2 - y_{\frac{1}{3}}^2) x_1 + \tfrac{3}{2} x_{\frac{1}{3}} y_{\frac{1}{3}} y_1 + A_{\frac{1}{3}} x_{\frac{1}{3}} + k y_{\frac{1}{3}} &= 0 \\
\tfrac{3}{2} x_{\frac{1}{3}} y_{\frac{1}{3}} x_1 - \tfrac{3}{4}(x_{\frac{1}{3}}^2 - y_{\frac{1}{3}}^2) y_1 + k x_{\frac{1}{3}} - A_{\frac{1}{3}} y_{\frac{1}{3}} &= 0
\end{aligned}
\tag{10.7}
$$

$$\text{where} \quad A_1 = 1 - \tfrac{3}{4}(R_1 + 2R_{\frac{1}{3}}) \qquad A_{\frac{1}{3}} = \tfrac{1}{3} - \tfrac{3}{4}(2R_1 + R_{\frac{1}{3}})$$
$$R_1 = r_1^2 = x_1^2 + y_1^2 \qquad R_{\frac{1}{3}} = r_{\frac{1}{3}}^2 = x_{\frac{1}{3}}^2 + y_{\frac{1}{3}}^2$$

from which one may derive the relations to determine R_1 and $R_{\frac{1}{3}}$, namely,

$$
\begin{aligned}
(9A_1 R_1 - A_{\frac{1}{3}} R_{\frac{1}{3}})^2 + k^2(9R_1 + R_{\frac{1}{3}})^2 &= 81 B^2 R_1 \\
(A_{\frac{1}{3}}^2 + k^2 - \tfrac{81}{16} R_1 R_{\frac{1}{3}}) R_{\frac{1}{3}} &= 0
\end{aligned}
\tag{10.8}
$$

Through use of Eqs. (10.7) and (10.8), the coefficients of the periodic

---

[1] A subharmonic oscillation of order $\frac{1}{2}$ may exist over a narrow range of the system parameters. However, since this type of oscillation is likely to occur when the system is unsymmetrical, the case will be discussed later, in Sec. 10.4.

solutions are found to be

$$x_1 = \frac{k(9R_1 + R_{\frac{1}{3}})}{9B} \qquad y_1 = \frac{-(9A_1R_1 - A_{\frac{1}{3}}R_{\frac{1}{3}})}{9B} \tag{10.9}$$

and
$$
\begin{aligned}
x_{\frac{1}{3}} &= r_{\frac{1}{3}}\cos\theta_{\frac{1}{3}} & r_{\frac{1}{3}}\cos(\theta_{\frac{1}{3}} + 120°) & \quad r_{\frac{1}{3}}\cos(\theta_{\frac{1}{3}} + 240°) \\
y_{\frac{1}{3}} &= r_{\frac{1}{3}}\sin\theta_{\frac{1}{3}} & r_{\frac{1}{3}}\sin(\theta_{\frac{1}{3}} + 120°) & \quad r_{\frac{1}{3}}\sin(\theta_{\frac{1}{3}} + 240°)
\end{aligned}
$$

where
$$\tag{10.10}$$

$$\cos 3\theta_{\frac{1}{3}} = -\frac{-4(A_1x_1 - ky_1)}{9R_1r_{\frac{1}{3}}} \qquad \sin 3\theta_{\frac{1}{3}} = \frac{-4(kx_1 + A_{\frac{1}{3}}y_1)}{9R_1r_{\frac{1}{3}}}$$

From the second equation of (10.8) one sees that either

$$A_{\frac{1}{3}}^2 + k^2 - {}^{81}\!/_{16}R_1R_{\frac{1}{3}} = 0 \qquad \text{or} \qquad R_{\frac{1}{3}} = 0$$

When $R_{\frac{1}{3}} = 0$, there will be no subharmonic response, and Eqs. (10.9) with $R_{\frac{1}{3}} = 0$ give the coefficients of the harmonic solution (10.5).[1]

(b) Stability Investigation

We are now to determine the stability of the periodic solutions (thus, of the fixed points) as given by Eqs. (10.9) and (10.10). By proceeding as in Sec. 7.3b, we consider a small variation $\xi(\tau)$ defined by

$$v(\tau) = v_0(\tau) + \xi(\tau) \tag{10.11}$$

By substituting this into (10.4) and introducing a new variable $\eta(\tau)$ defined by

$$\xi(\tau) = e^{-\frac{1}{2}k\tau}\eta(\tau) \tag{10.12}$$

we obtain the variational equation

$$\frac{d^2\eta}{d\tau^2} + \left(-\frac{1}{4}k^2 + 3v_0{}^2\right)\eta = 0 \tag{10.13}$$

Inserting $v_0(\tau)$ as given by Eq. (10.6) into (10.13) leads to a Hill's equation of the form

$$\frac{d^2\eta}{d\tau^2} + \left[\theta_0 + 2\sum_{n=1}^{3}\theta_n\cos\left(2n\frac{\tau}{3} - \epsilon_n\right)\right]\eta = 0$$

where

$$
\begin{aligned}
\theta_0 &= -\tfrac{1}{4}k^2 + \tfrac{3}{2}(R_1 + R_{\frac{1}{3}}) \\
\theta_n{}^2 &= \theta_{ns}{}^2 + \theta_{nc}{}^2 & \epsilon_n &= \tan^{-1}\frac{\theta_{ns}}{\theta_{nc}} \\
\theta_{1s} &= \tfrac{3}{2}(x_1y_{\frac{1}{3}} - y_1x_{\frac{1}{3}} + x_{\frac{1}{3}}y_{\frac{1}{3}}) & \theta_{1c} &= \tfrac{3}{2}(x_1x_{\frac{1}{3}} + y_1y_{\frac{1}{3}} - \tfrac{1}{2}x_{\frac{1}{3}}{}^2 \\
& & & \qquad + \tfrac{1}{2}y_{\frac{1}{3}}{}^2) \\
\theta_{2s} &= \tfrac{3}{2}(x_1y_{\frac{1}{3}} + y_1x_{\frac{1}{3}}) & \theta_{2c} &= \tfrac{3}{2}(-x_1x_{\frac{1}{3}} + y_1y_{\frac{1}{3}}) \\
\theta_{3s} &= \tfrac{3}{2}x_1y_1 & \theta_{3c} &= \tfrac{3}{2}(-\tfrac{1}{2}x_1{}^2 + \tfrac{1}{2}y_1{}^2)
\end{aligned}
\tag{10.14}
$$

[1] By following the procedure mentioned in Sec. 1.5, we may obtain a higher approximation of the periodic solution in which higher harmonics are taken into account.

By Floquet's theory, the solution of Eqs. (10.14) may be written in the form

$$\eta(\tau) = c_1 e^{\mu\tau}\phi(\tau) + c_2 e^{-\mu\tau}\psi(\tau) \tag{10.15}$$

where μ is the characteristic exponent and is dependent upon the parameters θ, $\phi(\tau)$ and $\psi(\tau)$ are periodic in τ, and c_1 and c_2 are arbitrary constants. From Eqs. (10.12) and (10.15) one sees that the variation $\xi(\tau)$ tends to zero with increasing τ, provided that the damping $k/2$ is greater than μ (>0). Hence the stability condition for the periodic solution $v_0(\tau)$ is given by $\frac{1}{2}k - \mu > 0$. When μ is computed to a first approximation, this condition leads to (Sec. 4.2)

$$\left[\theta_0 - \left(\frac{n}{3}\right)^2\right]^2 + 2\left[\theta_0 + \left(\frac{n}{3}\right)^2\right]\left(\frac{k}{2}\right)^2 + \left(\frac{k}{2}\right)^4 > \theta_n^2 \tag{10.16}$$

$$n = 1, 2, 3$$

Substituting the parameters θ as given by Eqs. (10.14) into (10.16) leads to

$$(R_1 + R_{\frac{1}{3}} - \tfrac{2}{27})^2 - (R_1 + \tfrac{1}{4}R_{\frac{1}{3}} + \tfrac{2}{9}A_{\frac{1}{3}})R_{\frac{1}{3}} + \tfrac{4}{81}k^2 > 0$$
$$\text{for } n = 1$$
$$(R_1 + R_{\frac{1}{3}} - \tfrac{8}{27})^2 - R_1 R_{\frac{1}{3}} + \tfrac{16}{81}k^2 > 0 \tag{10.17}$$
$$\text{for } n = 2$$
$$(R_1 + R_{\frac{1}{3}} - \tfrac{2}{3})^2 - \tfrac{1}{4}R_1^2 + \tfrac{4}{9}k^2 > 0$$
$$\text{for } n = 3$$

If the condition for $n = m$ is not satisfied, the periodic solution, (10.5) or (10.6), becomes unstable owing to the buildup of a self-excited oscillation having the frequency $m/3$.

CASE 1. HARMONIC RESPONSE

Since $R_{\frac{1}{3}} = 0$ in this case, the first and second conditions of (10.17) are satisfied. The third condition is reduced to

$$2\tfrac{7}{16}R_1^2 - 3R_1 + k^2 + 1 > 0 \tag{10.18}$$

This is the stability condition for the periodic solution (10.5) (Sec. 5.1c).

CASE 2. SUBHARMONIC RESPONSE ($\frac{1}{3}$-HARMONIC)

For any combinations of R_1 and $R_{\frac{1}{3}}$ calculated from Eqs. (10.8), one can verify that the second and third conditions of (10.17) are satisfied. By virtue of Eqs. (10.8) the first condition leads to

$$\tfrac{3}{2}R_1 + R_{\frac{1}{3}} - \tfrac{4}{3} > 0 \tag{10.19}$$

This is the stability condition for the periodic solution (10.6) (Sec. 7.3b).

10.3 Domains of Attraction for Harmonic and ⅓-harmonic Responses

As mentioned in Sec. 10.1, the boundary between the two domains of attraction for harmonic response is the locus of the images $(v_0(2n\pi), \dot{v}_0(2n\pi))$ that approach the unstable fixed point with increasing time. This locus may be obtained by integrating Eq. (10.4) for decreasing time, i.e., by using negative time in the equation. The initial conditions, i.e., the initial points of integration, should be on the invariant curve C_2 and may preferably be close to the unstable fixed point P_2 (Fig. 10.1). The location of the fixed points may readily be determined from the periodic solutions, Eqs. (10.5) and (10.6), in which the coefficients are to be found by using Eqs. (10.8) to (10.10). The stability of the fixed points will be studied by the conditions (10.18) and (10.19). We are particularly interested in the fixed points that are unstable. The slope of the invariant curve C_2 at the unstable fixed point may be determined by the following procedure: From Eqs. (10.12) and (10.15) the variation $\xi(\tau)$ from the periodic solution $v_0(\tau)$ is given by

$$\xi(\tau) = c_1 e^{(-\frac{1}{2}k+\mu)\tau}\phi(\tau) + c_2 e^{(-\frac{1}{2}k-\mu)\tau}\psi(\tau) \tag{10.20}$$

In the neighborhood of the unstable fixed point, the images on the invariant curves C_1 and C_2 satisfy the condition that

$$\frac{\xi(0)}{\xi(0)} = \frac{\xi(T)}{\xi(T)} = \frac{\xi(2T)}{\xi(2T)} = \cdots \quad (= \text{slope of the invariant curve})$$

where $T = 2\pi$ for harmonic response and $T = 6\pi$ for ⅓-harmonic response. Hence it follows that either c_1 or c_2 must be zero. On the invariant curve C_2 the successive images approach the unstable fixed point with increasing time. Therefore these images are represented by the points $(\xi(nT), \xi(nT))$, where $\xi(\tau)$ is given by

$$\xi(\tau) = c_2 e^{(-\frac{1}{2}k-\mu)\tau}\psi(\tau)$$

Hence the slope of the invariant curve C_2, that is, the direction of the boundary at the unstable fixed point, is given by[1]

$$\alpha = \frac{\xi(0)}{\xi(0)} = -\left(\frac{1}{2}k + \mu\right) + \frac{\psi(0)}{\psi(0)} \tag{10.21}$$

Thus the initial point of integration may be located on the line segment which passes through the unstable fixed point with slope α.

NUMERICAL ANALYSIS

As mentioned in Sec. 10.2, the harmonic and ⅓-harmonic solutions are to be expected for Eq. (10.4). From the analyses in Secs. 5.1 and 7.3b,

---

[1] The reader is referred to Appendix III for the calculation of the characteristic exponent μ and the periodic function $\psi(\tau)$ in the solution given by Eq. (10.15).

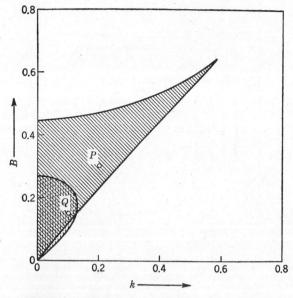

    : Harmonic (resonant and nonresonant)

    : 1/3-harmonic

FIGURE 10.2   Regions in which different types of oscillation are sustained.

one may obtain the regions of the system parameters k and B in which
harmonic and $\frac{1}{3}$-harmonic oscillations occur. The result is shown in
Fig. 10.2. In the area hatched by full lines, one obtains two different
types of harmonic response, i.e., resonant and nonresonant oscillations.
Which one occurs depends on the initial conditions. Outside this region
the harmonic oscillation is uniquely obtained. The dashed area is the
region of $\frac{1}{3}$-harmonic response.

(a) Harmonic Response

We first consider the case in which the system parameters are given
by $k = 0.2$ and $B = 0.3$ in Eq. (10.4), and write

$$\frac{d^2v}{d\tau^2} + 0.2\frac{dv}{d\tau} + v^3 = 0.3 \cos \tau \qquad (10.22)$$

The location of these parameters is indicated by point P in Fig. 10.2.
Thus we have only harmonic response, i.e., the resonant and nonresonant
oscillations, but no subharmonic oscillation.

For these particular values of the parameters, the periodic solutions

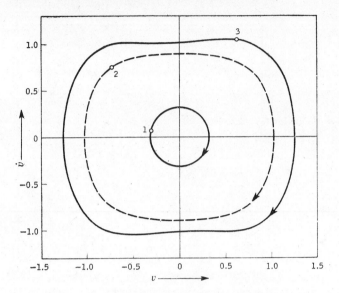

FIGURE 10.3   Trajectories of the solutions for Eq. (10.22).

may be determined from Eqs. (10.8) and (10.9). Their stability may be discussed by using condition (10.18). The result is as follows.

$$v_{01} = 0.067 \sin \tau - 0.310 \cos \tau + 0.001 \sin 3\tau - 0.001 \cos 3\tau$$
$$v_{02} = 0.671 \sin \tau - 0.744 \cos \tau + 0.026 \sin 3\tau + 0.022 \cos 3\tau \quad (10.23)$$
$$v_{03} = 0.988 \sin \tau + 0.684 \cos \tau + 0.021 \sin 3\tau - 0.061 \cos 3\tau$$

v_{01}, v_{03} being stable and v_{02} being unstable.

By using Eqs. (10.23) one may readily locate the fixed points in the $v\dot{v}$ plane. The fixed points are invariant under iterations of the mapping from $\tau = 2n\pi$ to $2(n + 1)\pi$. We are particularly interested in the fixed point that is unstable, since the boundary between domains of attraction contains such a point. The direction of the boundary curve at the unstable fixed point may be determined as follows: Substitution of the unstable solution v_{02} into Eq. (10.13) leads to a Hill's equation of the form

$$\frac{d^2\eta}{d\tau^2} + \left[\theta_0 + 2 \sum_{n=1}^{3} \theta_n \cos (2n\tau - \epsilon_n) \right] \eta = 0$$

where[1] $\theta_0 = 1.497$ (10.24)

$\theta_1 = 0.803 \qquad \epsilon_1 = -1.473$
$\theta_2 = -0.051 \qquad \epsilon_2 = 0.129$
$\theta_3 = -0.001 \qquad \epsilon_3 = -1.417$

[1] The arguments ϵ are measured in radians.

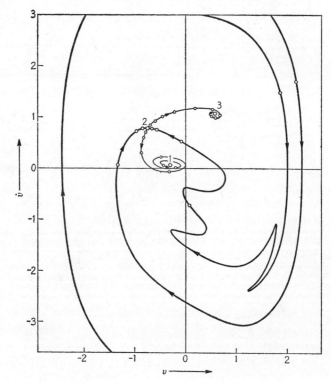

FIGURE 10.4   The loci of image points under iterations of the
mapping (harmonic response).

A particular solution of Eq. (10.24) which tends to zero as τ increases is
given by
$$\eta(\tau) = e^{-\mu\tau}\psi(\tau)$$
where

$$\mu = 0.251 \tag{10.25}$$
$$\psi(\tau) = \sin(\tau + 1.070) + 0.104\sin(3\tau - 0.737)$$
$$- 0.005\sin(5\tau + 0.898)$$

Substituting μ, $\psi(0)$, and $\dot{\psi}(0)$ as given by Eqs. (10.25) into (10.21) gives
us the direction α of the boundary curve at the unstable fixed point; thus
we obtain

$$\alpha = -0.106$$

The fixed points and the related properties are listed in Table 10.1.

The trajectories of the periodic solutions, i.e., the loci of the point
$(v_0(\tau), \dot{v}_0(\tau))$ on the $v\dot{v}$ plane, are shown in Fig. 10.3.   The outer and inner

Table 10.1   Fixed Points and Related Properties Correlated with the Periodic Solutions of Eq. (10.22)

Fixed point	Response	v	$\dot{v}$	α	Stability
1	Harmonic	-0.311	0.069		Stable
2	Harmonic	-0.722	0.748	-0.106	Unstable
3	Harmonic	0.624	1.052		Stable

trajectories represent the resonant and nonresonant oscillations, v_{03} and v_{01}, respectively, while the intermediate one (shown by broken line) represents the unstable oscillation v_{02}. The location of the fixed points is also shown in the figure. Following the procedure described in Sec. 10.1, successive images of the mapping for harmonic response are shown in Fig. 10.4. The boundary between the two domains of attraction is shown by the thick line, on which the image points approach the unstable fixed point 2 with increasing time. This boundary was obtained by starting just on both sides (in the direction α) of the unstable fixed point and integrating Eq. (10.22) for decreasing time. Both analog and digital computers were used for this purpose. These domains of attraction agree with the result obtained by the phase-plane analysis in Sec. 8.4 (see Fig. 8.5).

(b) Harmonic and ⅓-harmonic Responses

Secondly, we consider the case in which the system parameters are given by $k = 0.1$ and $B = 0.15$ in Eq. (10.4), and write

$$\frac{d^2v}{d\tau^2} + 0.1\,\frac{dv}{d\tau} + v^3 = 0.15 \cos \tau \qquad (10.26)$$

Since these parameters are located at point Q in Fig. 10.2, one may expect the harmonic oscillations (resonant and nonresonant) and the subharmonic oscillation of order ⅓ in the steady state.

For these particular values of the parameters, the periodic solutions, Eqs. (10.5) and (10.6), may be determined from Eqs. (10.8) to (10.10). Their stability is studied by conditions (10.18) and (10.19). The result is shown in what follows. For harmonic response,

$$
\begin{aligned}
v_{01} &= 0.011 \sin \tau - 0.153 \cos \tau \\
v_{02} &= 0.806 \sin \tau - 0.716 \cos \tau + 0.023 \sin 3\tau + 0.037 \cos 3\tau \quad (10.27) \\
v_{03} &= 0.960 \sin \tau + 0.686 \cos \tau + 0.019 \sin 3\tau - 0.040 \cos 3\tau
\end{aligned}
$$

v_{01}, v_{03} being stable and v_{02} being unstable.   For subharmonic response,

$$v_{04} = 0.063 \sin \tfrac{1}{3}\tau + 0.358 \cos \tfrac{1}{3}\tau + 0.032 \sin \tau - 0.180 \cos \tau$$
$$v_{05} = -0.342 \sin \tfrac{1}{3}\tau - 0.124 \cos \tfrac{1}{3}\tau + 0.032 \sin \tau - 0.180 \cos \tau$$
$$v_{06} = 0.278 \sin \tfrac{1}{3}\tau - 0.234 \cos \tfrac{1}{3}\tau + 0.032 \sin \tau - 0.180 \cos \tau$$
$$v_{07} = 0.149 \sin \tfrac{1}{3}\tau + 0.226 \cos \tfrac{1}{3}\tau + 0.025 \sin \tau - 0.171 \cos \tau \qquad (10.28)$$
$$v_{08} = -0.271 \sin \tfrac{1}{3}\tau + 0.016 \cos \tfrac{1}{3}\tau + 0.025 \sin \tau - 0.171 \cos \tau$$
$$v_{09} = 0.122 \sin \tfrac{1}{3}\tau - 0.242 \cos \tfrac{1}{3}\tau + 0.025 \sin \tau - 0.171 \cos \tau$$

v_{04}, v_{05}, v_{06} being stable and v_{07}, v_{08}, v_{09} being unstable.

By using Eqs. (10.27) and (10.28), one may readily locate the fixed points in the $v\dot{v}$ plane.   For harmonic response the fixed points are invariant under iterations of the mapping from $\tau = 2n\pi$ to $2(n+1)\pi$; for subharmonic response the fixed points are invariant under every third iterate of the mapping.   By a procedure analogous to that for the foregoing case, we may calculate the direction α of the boundary curve at the unstable fixed point through use of Eq. (10.21).   The fixed points and the related properties thus obtained are listed in Table 10.2.

The trajectories of the stable solutions are shown in Fig. 10.5.   The location of the stable fixed points is also shown in the figure.   It is noted that the fixed points 4 to 6, corresponding to the subharmonic oscillation, lie on the same trajectory and that, under iterations of the mapping,

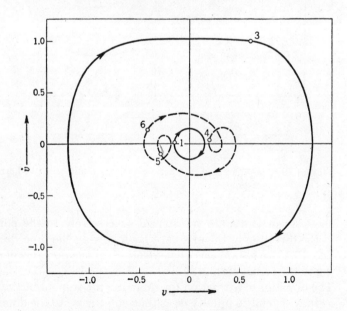

FIGURE 10.5   Trajectories of the stable solutions for Eq. (10.26).

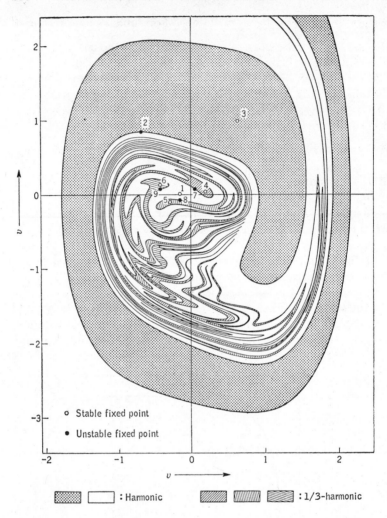

FIGURE 10.6   Domains of attraction for harmonic and ⅓-harmonic responses.

these fixed points are transferred successively to the points that follow in the direction of the arrows.   By following the procedure mentioned in Sec. 10.1, we obtain the whole diagram of the domains of attraction leading to the harmonic and subharmonic responses, as illustrated in Fig. 10.6. The domains of attraction for subharmonic response have narrowing tails as they extend to infinity or as they come close to the domain of harmonic response containing the fixed point 3.   These extremely narrow tails are omitted in the figure, because the computation becomes too laborious.

Table 10.2   Fixed Points and Related Properties Correlated with the Periodic Solutions of Eq. (10.26)

Fixed point	Response	v	$\dot{v}$	α	Stability
1	Harmonic	-0.153	0.011		Stable
2	Harmonic	-0.679	0.876	-0.020	Unstable
3	Harmonic	0.646	1.016		Stable
4	$\frac{1}{3}$-harmonic	0.178	0.053		Stable
5	$\frac{1}{3}$-harmonic	-0.304	-0.082		Stable
6	$\frac{1}{3}$-harmonic	-0.413	0.124		Stable
7	$\frac{1}{3}$-harmonic	0.056	0.075	-0.644	Unstable
8	$\frac{1}{3}$-harmonic	-0.155	-0.065	-0.164	Unstable
9	$\frac{1}{3}$-harmonic	-0.413	0.066	0.263	Unstable

10.4 Unsymmetrical Systems

As an example of Eq. (10.1) we consider the differential equation

$$\frac{d^2v}{d\tau^2} + k\frac{dv}{d\tau} + v^3 = B\cos\tau + B_0 \tag{10.29}$$

where the asymmetry appears as a unidirectional component of the external force.[1]   In addition to the responses mentioned in Sec. 10.2, the subharmonic oscillation of order $\frac{1}{2}$ may be expected in this case.   The periodic solution for Eq. (10.29) might be assumed by

$$v_0(\tau) = x_1\sin\tau + y_1\cos\tau + z \tag{10.30}$$

for harmonic response, and

$$v_0(\tau) = x_{\frac{1}{2}}\sin\tfrac{1}{2}\tau + y_{\frac{1}{2}}\cos\tfrac{1}{2}\tau + x_1\sin\tau + y_1\cos\tau + z \tag{10.31}$$

or $\quad v_0(\tau) = x_{\frac{1}{3}}\sin\tfrac{1}{3}\tau + y_{\frac{1}{3}}\cos\tfrac{1}{3}\tau + x_1\sin\tau + y_1\cos\tau + z \tag{10.32}$

for subharmonic response.   Since the system is unsymmetrical, the constant term z of zero frequency is added to the solution.   If the system parameters k, B, and B_0 are appropriately chosen, the resulting response will be one of the types given by Eqs. (10.30) to (10.32), depending on different values of the initial condition.

By a procedure analogous to that of Sec. 10.2, the coefficients of the periodic solutions are determined.   The conditions for stability of the periodic solutions are also derived by solving the variational equations of the Hill type.

[1] Equation (10.1) with unsymmetrical nonlinearity, $f(v) = c_1 v + c_2 v^2 + c_3 v^3$, may readily be transformed to one with symmetrical nonlinearity but with unsymmetrical external force (see Sec. 5.2a).

(a) Harmonic Response

The coefficients of the periodic solution (10.30) are found to be

$$x_1 = \frac{kR_1}{B} \qquad y_1 = \frac{-A_1R_1}{B}$$

where   (10.33)

$$A_1 = 1 - \tfrac{3}{4}(R_1 + 4Z) \qquad R_1 = r_1{}^2 = x_1{}^2 + y_1{}^2 \qquad Z = z^2$$

in which the unknown quantities R_1 and Z may be determined by solving the simultaneous equations

$$(A_1{}^2 + k^2)R_1 = B^2 \qquad (\tfrac{3}{2}R_1 + Z)z = B_0 \qquad (10.34)$$

By making use of inequality (4.9) in Sec. 4.2, we obtain the stability conditions

$$
\begin{array}{ll}
(R_1 + 2Z - \tfrac{1}{6})^2 - 4R_1Z + \tfrac{1}{9}k^2 > 0 & \text{for } n = 1 \\
(R_1 + 2Z - \tfrac{2}{3})^2 - \tfrac{1}{4}R_1{}^2 + \tfrac{4}{9}k^2 > 0 & \text{for } n = 2
\end{array}
\qquad (10.35)
$$

If the condition for $n = m$ is not satisfied, the periodic solution, Eq. (10.30), becomes unstable, owing to the buildup of a self-excited oscillation having the frequency $m/2$.

(b) ½-harmonic Response

The coefficients of the periodic solution, Eq. (10.31), are found to be

$$x_1 = \frac{k(4R_1 + R_{\frac{1}{2}})}{4B} \qquad y_1 = \frac{-(4A_1R_1 - A_{\frac{1}{2}}R_{\frac{1}{2}})}{4B}$$

$$
\begin{array}{ll}
x_{\frac{1}{2}} = r_{\frac{1}{2}} \cos \theta_{\frac{1}{2}} & r_{\frac{1}{2}} \cos (\theta_{\frac{1}{2}} + 180°) \\
y_{\frac{1}{2}} = r_{\frac{1}{2}} \sin \theta_{\frac{1}{2}} & r_{\frac{1}{2}} \sin (\theta_{\frac{1}{2}} + 180°)
\end{array}
$$

where   (10.36)

$$A_1 = 1 - \tfrac{3}{4}(R_1 + 2R_{\frac{1}{2}} + 4Z)$$
$$A_{\frac{1}{2}} = \tfrac{1}{2} - \tfrac{3}{2}(2R_1 + R_{\frac{1}{2}} + 4Z)$$
$$R_1 = r_1{}^2 = x_1{}^2 + y_1{}^2 \qquad R_{\frac{1}{2}} = r_{\frac{1}{2}}{}^2 = x_{\frac{1}{2}}{}^2 + y_{\frac{1}{2}}{}^2 \qquad Z = z^2$$
$$\cos 2\theta_{\frac{1}{2}} = \frac{-(kx_1 + A_{\frac{1}{2}}y_1)}{6R_1z} \qquad \sin 2\theta_{\frac{1}{2}} = \frac{A_{\frac{1}{2}}x_1 - ky_1}{6R_1z}$$

in which the unknown quantities R_1, $R_{\frac{1}{2}}$, and Z may be determined by solving the simultaneous equations

$$(4A_1R_1 - A_{\frac{1}{2}}R_{\frac{1}{2}})^2 + k^2(4R_1 + R_{\frac{1}{2}})^2 - 16B^2R_1 = 0$$
$$A_{\frac{1}{2}}{}^2 + k^2 - 36R_1Z = 0$$
$$\tfrac{3}{2}\left(R_1 + R_{\frac{1}{2}} + \tfrac{2}{3}Z\right)z + \frac{A_{\frac{1}{2}}R_{\frac{1}{2}}}{8z} - B_0 = 0$$

(10.37)

The stability conditions may also be written as

$$
\begin{aligned}
(R_1 + R_{\frac{1}{2}} + 2Z - \tfrac{1}{24})^2 & \\
+ (R_1 + R_{\frac{1}{2}} - \tfrac{1}{3})R_{\frac{1}{2}} + \tfrac{1}{36}k^2 &> 0 && \text{for } n = 1 \\
(R_1 + R_{\frac{1}{2}} + 2Z - \tfrac{1}{6})^2 & \\
- (4R_1Z + \tfrac{1}{4}R_{\frac{1}{2}}^2 + \tfrac{1}{3}A_{\frac{1}{2}}R_{\frac{1}{2}}) + \tfrac{1}{9}k^2 &> 0 && \text{for } n = 2 \\
(R_1 + R_{\frac{1}{2}} + 2Z - \tfrac{3}{8})^2 - R_1R_{\frac{1}{2}} + \tfrac{1}{4}k^2 & > 0 && \text{for } n = 3 \\
(R_1 + R_{\frac{1}{2}} + 2Z - \tfrac{2}{3})^2 - \tfrac{1}{4}R_1^2 + \tfrac{4}{9}k^2 & > 0 && \text{for } n = 4
\end{aligned}
\tag{10.38}
$$

If the condition for $n = m$ is not satisfied, the periodic solution, Eq. (10.31), becomes unstable owing to the buildup of a self-excited oscillation having the frequency $m/4$. The condition for $n = 4$ is superfluous in this particular case, since it is always satisfied by the quantities obtained from Eqs. (10.37). Therefore the conditions for $n = 1, 2,$ and 3 must be ascertained for stability of the periodic solution.

(c) ⅓-harmonic Response

The coefficients of the periodic solution, Eq. (10.32), are found to be

$$
x_1 = \frac{k(9R_1 + R_{\frac{1}{2}})}{9B} \qquad y_1 = \frac{-(9A_1R_1 - A_{\frac{1}{2}}R_{\frac{1}{2}})}{9B}
$$

$$
\begin{aligned}
x_{\frac{1}{2}} &= r_{\frac{1}{2}}\cos\theta_{\frac{1}{2}} & r_{\frac{1}{2}}\cos(\theta_{\frac{1}{2}} + 120°) & \quad r_{\frac{1}{2}}\cos(\theta_{\frac{1}{2}} + 240°) \\
y_{\frac{1}{2}} &= r_{\frac{1}{2}}\sin\theta_{\frac{1}{2}} & r_{\frac{1}{2}}\sin(\theta_{\frac{1}{2}} + 120°) & \quad r_{\frac{1}{2}}\sin(\theta_{\frac{1}{2}} + 240°)
\end{aligned}
$$

where

$$
\begin{aligned}
A_1 &= 1 - \tfrac{3}{4}(R_1 + 2R_{\frac{1}{2}} + 4Z) \\
A_{\frac{1}{2}} &= \tfrac{1}{3} - \tfrac{3}{4}(2R_1 + R_{\frac{1}{2}} + 4Z)
\end{aligned}
\tag{10.39}
$$

$$
R_1 = r_1^2 = x_1^2 + y_1^2 \qquad R_{\frac{1}{2}} = r_{\frac{1}{2}}^2 = x_{\frac{1}{2}}^2 + y_{\frac{1}{2}}^2 \qquad Z = z^2
$$

$$
\cos 3\theta_{\frac{1}{2}} = \frac{-4(A_{\frac{1}{2}}x_1 - ky_1)}{9R_1r_{\frac{1}{2}}} \qquad \sin 3\theta_{\frac{1}{2}} = \frac{-4(kx_1 + A_{\frac{1}{2}}y_1)}{9R_1r_{\frac{1}{2}}}
$$

in which the unknown quantities R_1, $R_{\frac{1}{2}}$, and Z may be determined by solving the simultaneous equations

$$
\begin{aligned}
(9A_1R_1 - A_{\frac{1}{2}}R_{\frac{1}{2}})^2 + k^2(9R_1 + R_{\frac{1}{2}})^2 - 81B^2R_1 &= 0 \\
A_{\frac{1}{2}}^2 + k^2 - \tfrac{81}{16}R_1R_{\frac{1}{2}} &= 0 \\
\tfrac{3}{2}(R_1 + R_{\frac{1}{2}} + \tfrac{2}{3}Z)z - B_0 &= 0
\end{aligned}
\tag{10.40}
$$

The stability conditions are

$$
\begin{aligned}
(R_1 + R_{\frac{1}{2}} + 2Z - \tfrac{1}{54})^2 - 4R_{\frac{1}{2}}Z + \tfrac{1}{81}k^2 &> 0 && \text{for } n = 1 \\
(R_1 + R_{\frac{1}{2}} + 2Z - \tfrac{2}{27})^2 & \\
- (R_1 + \tfrac{1}{4}R_{\frac{1}{2}} + \tfrac{2}{9}A_{\frac{1}{2}})R_{\frac{1}{2}} + \tfrac{4}{81}k^2 &> 0 && \text{for } n = 2 \\
(R_1 + R_{\frac{1}{2}} + 2Z - \tfrac{1}{6})^2 - 4R_1Z + \tfrac{1}{9}k^2 &> 0 && \text{for } n = 3 \\
(R_1 + R_{\frac{1}{2}} + 2Z - \tfrac{8}{27})^2 - R_1R_{\frac{1}{2}} + \tfrac{16}{81}k^2 &> 0 && \text{for } n = 4 \\
(R_1 + R_{\frac{1}{2}} + 2Z - \tfrac{25}{54})^2 + \tfrac{25}{81}k^2 &> 0 && \text{for } n = 5 \\
(R_1 + R_{\frac{1}{2}} + 2Z - \tfrac{2}{3})^2 - \tfrac{1}{4}R_1^2 + \tfrac{4}{9}k^2 &> 0 && \text{for } n = 6
\end{aligned}
\tag{10.41}
$$

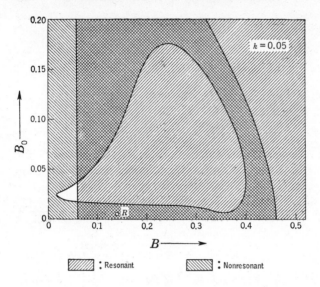

FIGURE 10.7   Regions in which the harmonic oscillations
(resonant and nonresonant) are sustained.

If the condition for $n = m$ is not satisfied, the periodic solution, Eq. (10.32), becomes unstable owing to the buildup of a self-excited oscillation having the frequency $m/6$. The conditions for $n = 4, 5$, and 6 are superfluous in this particular case, since they are always satisfied by the quantities obtained from Eqs. (10.40). Therefore the conditions for $n = 1, 2$, and 3 must be ascertained for stability of the periodic solution.

10.5 Domains of Attraction for Harmonic, ½-harmonic, and ⅓-harmonic Responses

By a procedure analogous to that of Sec. 10.3, we may determine, in the $v\dot{v}$ plane, the domains of attraction for the respective types of response for Eq. (10.29).

NUMERICAL ANALYSIS

Figures 10.7 and 10.8 show the regions of harmonic and subharmonic oscillations, respectively. The variable parameters are B and B_0, and k is kept constant. It is obvious from the form of Eq. (10.29) that the regions of those periodic oscillations also appear, for negative values of B_0, symmetrically about the B axis. The appearance of the unshaded area inside the region of nonresonant oscillations in Fig. 10.7 is due to the buildup of unstable oscillations having the ½-harmonic frequency   The

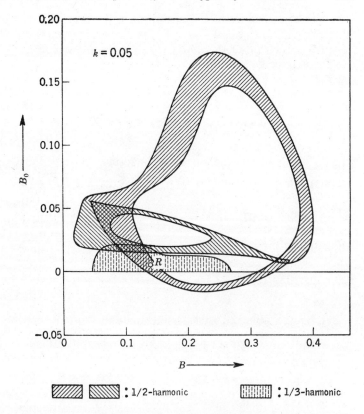

FIGURE 10.8   Regions in which the subharmonic oscillations ($\frac{1}{2}$-harmonic and $\frac{1}{3}$-harmonic) are sustained.

first condition of (10.35) is not satisfied in this case.   Similarly, in the unshaded areas surrounded by the region of $\frac{1}{2}$-harmonic oscillations in Fig. 10.8, the buildup of unstable oscillations having the $\frac{1}{4}$- or $\frac{3}{4}$-harmonic frequency disturbs the original $\frac{1}{2}$-harmonic oscillation.   The first and third conditions of (10.38) are not satisfied in this case (see Secs. 7.6b and 7.8).

We consider the case in which the system parameters are given by $k = 0.05$, $B = 0.14$, and $B_0 = 0.005$ in Eq. (10.29), and write

$$\frac{d^2v}{d\tau^2} + 0.05 \frac{dv}{d\tau} + v^3 = 0.14 \cos \tau + 0.005 \qquad (10.42)$$

Since these parameters are located at point R in Figs. 10.7 and 10.8, one may expect the harmonic oscillations (resonant and nonresonant) and the subharmonic oscillations of order $\frac{1}{2}$ and $\frac{1}{3}$ in the steady state.

For these values of the parameters, the periodic solutions and their stability are investigated by the use of Eqs. (10.33) to (10.41). The result is shown in what follows. For harmonic response,

$v_{01} = 0.111 + 0.008 \sin \tau - 0.147 \cos \tau$.

$v_{02} = 0.003 + 0.399 \sin \tau - 0.983 \cos \tau + 0.038 \sin 3\tau$
$$- 0.016 \cos 3\tau \quad (10.43)$$

$v_{03} = 0.002 + 0.488 \sin \tau + 1.088 \cos \tau + 0.059 \sin 3\tau$
$$+ 0.020 \cos 3\tau$$

v_{01}, v_{03} being stable and v_{02} being unstable. For $\frac{1}{2}$-harmonic response,

$v_{04} = 0.069 + 0.079 \sin \frac{1}{2}\tau + 0.526 \cos \frac{1}{2}\tau + 0.040 \sin \tau$
$$- 0.180 \cos \tau$$

$v_{05} = 0.069 - 0.079 \sin \frac{1}{2}\tau - 0.526 \cos \frac{1}{2}\tau + 0.040 \sin \tau$
$$- 0.180 \cos \tau$$

$v_{06} = 0.039 + 0.255 \sin \frac{1}{2}\tau + 0.416 \cos \frac{1}{2}\tau + 0.039 \sin \tau \qquad (10.44)$
$$- 0.220 \cos \tau$$

$v_{07} = 0.039 - 0.255 \sin \frac{1}{2}\tau - 0.416 \cos \frac{1}{2}\tau + 0.039 \sin \tau$
$$- 0.220 \cos \tau$$

v_{04}, v_{05} being stable and v_{06}, v_{07} being unstable. For $\frac{1}{3}$-harmonic response,

$v_{08} = 0.019 + 0.033 \sin \frac{1}{3}\tau + 0.388 \cos \frac{1}{3}\tau + 0.016 \sin \tau$
$$- 0.166 \cos \tau$$

$v_{09} = 0.019 - 0.353 \sin \frac{1}{3}\tau - 0.165 \cos \frac{1}{3}\tau + 0.016 \sin \tau$
$$- 0.166 \cos \tau$$

$v_{010} = 0.019 + 0.320 \sin \frac{1}{3}\tau - 0.223 \cos \frac{1}{3}\tau + 0.016 \sin \tau$
$$- 0.166 \cos \tau$$

$v_{011} = 0.039 + 0.177 \sin \frac{1}{3}\tau + 0.165 \cos \frac{1}{3}\tau + 0.012 \sin \tau \qquad (10.45)$
$$- 0.160 \cos \tau$$

$v_{012} = 0.039 - 0.233 \sin \frac{1}{3}\tau - 0.067 \cos \frac{1}{3}\tau + 0.012 \sin \tau$
$$- 0.160 \cos \tau$$

$v_{013} = 0.039 + 0.055 \sin \frac{1}{3}\tau - 0.236 \cos \frac{1}{3}\tau + 0.012 \sin \tau$
$$- 0.160 \cos \tau$$

v_{08}, v_{09}, v_{010} being stable and $v_{011}, v_{012}, v_{013}$ being unstable.

By using Eqs. (10.43), (10.44), and (10.45), one may readily locate the fixed points in the $v\dot{v}$ plane. By proceeding as before, we calculate the direction α of the boundary curve at the unstable fixed point through use of Eq. (10.21). The fixed points and the related properties thus obtained are listed in Table 10.3.

The trajectories of the stable solutions are shown in Fig. 10.9. The location of the stable fixed points is also shown in the figure. The domains of attraction leading to the harmonic, $\frac{1}{2}$-harmonic, and $\frac{1}{3}$-harmonic responses are illustrated in Fig. 10.10. The boundaries between

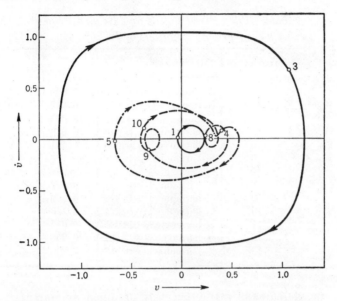

FIGURE 10.9   Trajectories of the stable solutions for Eq. (10.42).

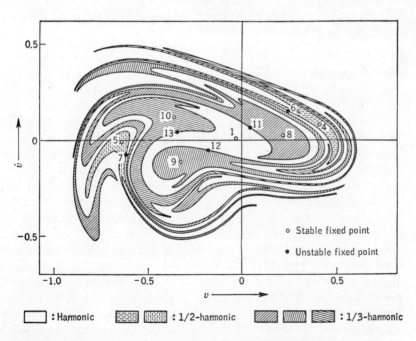

FIGURE 10.10   Domains of attraction for harmonic, ½-harmonic, and ⅓-harmonic responses.

Table 10.3   *Fixed Points and Related Properties Correlated with the Periodic Solutions of Eq. (10.42)*

Fixed point	Response	v	$\dot{v}$	α	Stability
1	Harmonic	-0.036	0.008		Stable
2	Harmonic	-0.996	0.513	1.054	Unstable
3	Harmonic	1.111	0.665		Stable
4	$\frac{1}{2}$-harmonic	0.415	0.080		Stable
5	$\frac{1}{2}$-harmonic	-0.638	-0.001		Stable
6	$\frac{1}{2}$-harmonic	0.235	0.166	-0.601	Unstable
7	$\frac{1}{2}$-harmonic	-0.597	-0.088	2.994	Unstable
8	$\frac{1}{3}$-harmonic	0.241	0.027		Stable
9	$\frac{1}{3}$-harmonic	-0.313	-0.102		Stable
10	$\frac{1}{3}$-harmonic	-0.371	0.123		Stable
11	$\frac{1}{3}$-harmonic	0.045	0.071	-0.674	Unstable
12	$\frac{1}{3}$-harmonic	-0.187	-0.066	-0.194	Unstable
13	$\frac{1}{3}$-harmonic	-0.357	0.030	0.199	Unstable

the domains of attraction were obtained by starting just on both sides (in the direction α) of the unstable fixed points and integrating Eq. (10.42) for decreasing time. Similarly to the case of Fig. 10.6, the domain of attraction leading to the fixed point 3 exists outside those domains, but it is omitted in the figure.

10.6 *Experimental Investigation*

(a) *Symmetrical System* [33]

In Sec. 8.7 we considered the case in which the transient oscillation initiated in the electric circuit of Fig. 8.11 results in the harmonic response. In this section the same symmetrical system is considered, but the steady-state response will be either harmonic or $\frac{1}{3}$-harmonic oscillations, depending on different values of the initial condition.

As is apparent from Fig. 9.2, three types of response, i.e., the resonant, nonresonant, and $\frac{1}{3}$-harmonic oscillations, are obtained if the capacitance C and the applied voltage V are chosen within the area common to the two regions, one by full lines and the other by dashed lines. Figure 10.11 shows the result of an experiment performed under such a condition.[1] By comparison of this figure with the results of the preceding experiments as illustrated in Figs. 8.12 and 9.6, it may be seen that the regions of the $\frac{1}{3}$-harmonic oscillation are contracted by the appearance of the regions of the resonant oscillation and are distributed in narrower bands which are very close to one another.

[1] Typical waveforms of the oscillations were already shown in Fig. 4.1 in Sec. 4.1.

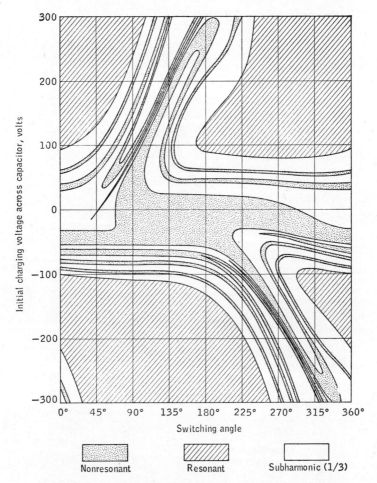

Nonresonant   Resonant   Subharmonic (1/3)

FIGURE 10.11   Regions of initial conditions leading to the harmonic (resonant and nonresonant) and ⅓-harmonic responses.

(b) Unsymmetrical System [36]

As mentioned in Sec. 7.8, a subharmonic oscillation of order ½ may occur in the electric circuit of Fig. 7.24 owing to the asymmetry of the saturable core. The differential equation which governs the system is given by Eq. (7.36) or (10.29). Since these equations contain a cubic term in v, one may expect the occurrence of the subharmonic oscillation of order ⅓ also.

By making use of this circuit, we perform an experiment. The

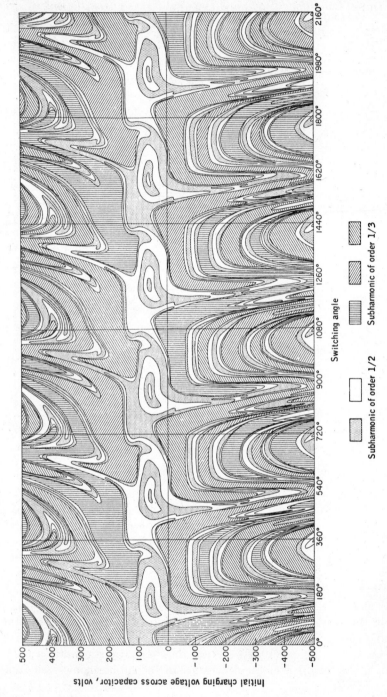

Initial charging voltage across capacitor, volts

Switching angle

Subharmonic of order 1/2   Subharmonic of order 1/3

Subharmonic of order 1/2   Subharmonic of order 1/3

FIGURE 10.12 Regions of initial conditions leading to the ½-harmonic and ⅓-harmonic responses.

260

system parameters are so chosen that either the subharmonic oscillation of order ½ or that of order ⅓ occurs, as obtained by use of suitable different values of the initial condition.   An experimental result is shown in Fig. 10.12.   The regions of initial conditions which give rise to two kinds of ½-harmonic oscillations [Eqs. (10.36)] are identified by dotted regions and unshaded regions.[1]   Each region appears alternately at every two cycles of the applied voltage.   Similarly, the regions of initial conditions resulting in three kinds of ⅓-harmonic oscillations [Eqs. (10.39)] are identified by different directions of the shaded line.   Each region appears at every three cycles of the applied voltage.   Hence the regions in toto result in a repeating of the same configuration at every six cycles of the applied voltage.[2]

[1] In order to discriminate between the two kinds of oscillations, their phase angles with respect to the applied voltage were observed by making use of an oscilloscope which has a time sweep synchronized with the electronic switch of Fig. VI.1.

[2] Since various patterns of the initial condition were obtained for the ½-harmonic response (Fig. 9.16), we may have a wide variety of such configurations when the harmonic, ½-harmonic, and ⅓-harmonic responses are expected.

Chapter 11

Almost Periodic Oscillations

11.1 Introduction

When a periodic force is applied to a nonlinear system, the resulting oscillation is usually, but not necessarily, periodic. When it is periodic, the fundamental period of the oscillation is the same as, or equal to an integral multiple of, the period of the external force. To these phenomena the terms harmonic oscillation and subharmonic oscillation have been respectively applied. There are also certain special cases in which the response of a nonlinear system is not periodic even when subjected to a periodic excitation. This chapter is concerned with a type of oscillation such that the amplitude and phase of the oscillation vary slowly but periodically after some transients have died out. Furthermore, the ratio between the period for amplitude variation and the period of the external force is in general incommensurable; thus, there is no periodicity in those oscillations. Since such oscillations may be represented by Fourier polynomials with incommensurable frequencies, they will be referred to as *almost periodic oscillations.*[1]

It was mentioned in Chap. 8 that a transient oscillation is described by the behavior of the representative point which moves along the integral curve of Eq. (8.1). A periodic oscillation is then correlated with a singular point of this equation, which in turn fixes the amplitude and phase of the oscillation. One may infer that an almost periodic oscillation, if it exists, is correlated with a limit cycle. Since an almost periodic oscillation is affected by amplitude and phase modulation, the representa-

---

[1] A function $f(x)$ is called *almost periodic* if the following two conditions are satisfied:

1. For any given $\epsilon > 0$ as small as we please, there exists a set of τ such that

$$|f(x + \tau) - f(x)| < \epsilon \qquad -\infty < x < +\infty$$

2. A length $L = L(\epsilon)$ exists such that any interval $\alpha < x < \alpha + L$ contains at least one number τ of the set.

The number τ is called the *translation number*, and the length L is called the *inclusion interval* of the set. For the theory of almost periodic functions see Refs. 7 and 11.

tive point keeps on moving along a limit cycle toward which neighboring integral curves tend as time increases.

Two representative cases of the almost periodic oscillation will be studied in the present chapter [37]. The first is the case in which a harmonic oscillation in a resonant circuit becomes unstable and changes into an almost periodic oscillation. The second case deals with the almost periodic oscillation which develops from a subharmonic oscillation of order $\frac{1}{2}$ in a parametric excitation circuit.[1]

11.2 Almost Periodic Oscillations in a Resonant Circuit with Direct Current Superposed

It was mentioned in Sec. 8.3 that Eq. (8.12) has no limit cycle when $X(x,y)$ and $Y(x,y)$ are given by Eqs. (8.4). The system considered there is symmetrical; an example of the system was given in Sec. 5.1 (see Fig. 5.1). In Secs. 5.2 and 5.3 we investigated the harmonic oscillation in unsymmetrical systems and obtained three types of steady-state response (i.e., the resonant, subresonant, and nonresonant oscillations). The experiment was performed on the circuit of Fig. 5.7. In this experiment we also found a steady-state response such that it does not settle to one of those types of harmonic oscillation, but alternately repeats two different types of harmonic oscillation which roughly correspond to the subresonant and nonresonant oscillations.[2] The occurrence of this sort of oscillation—an almost periodic oscillation—is due to the interaction between the a-c and d-c circuits of Fig. 5.7.[3]

(a) Fundamental Equations

We now consider an electric circuit as shown in Fig. 11.1. The inductor L of Fig. 5.7 is removed in the present connection, so that the current in the d-c circuit may not be constant. Following the notation in the figure, the equations for the circuit may be written as follows:

$$n \frac{d}{dt} (\phi_1 + \phi_2) + R_1 i_R = E_1 \sin \omega t$$

$$n \frac{d}{dt} (\phi_1 - \phi_2) + R_2 i_2 = E_0 \tag{11.1}$$

$$R_1 i_R = \frac{1}{C} \int i_C \, dt \qquad i_1 = i_R + i_C$$

---

[1] An almost periodic oscillation may also occur in a self-oscillatory system under the action of an external force. Cases of this type will be studied in Chap. 13.

[2] A similar phenomenon was also observed by Thomson [105], Hámos [30], and Bessonov [8].

[3] It was found that an almost periodic oscillation occurs only when the impedance of the d-c circuit is sufficiently low.

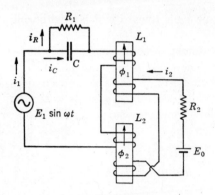

FIGURE 11.1   Resonant circuit containing saturable reactors with secondary d-c windings.

where ϕ_1 and ϕ_2 are the magnetic fluxes in the cores L_1 and L_2, respectively, and n is the number of turns of the coils wound around the cores (the same number of turns is assumed for each coil). The nonlinear characteristics of the cores are assumed to be

$$c_3\phi_1{}^3 = ni_1 + ni_2 \qquad c_3\phi_2{}^3 = ni_1 - ni_2 \tag{11.2}$$

where c_3 is a constant dependent on the nature of the cores. Proceeding as in Sec. 5.1a, we introduce the nondimensional variables defined by

$$i_1 = I_n u_1 \qquad i_2 = I_n u_2 \qquad \phi_1 = \Phi_n v_1 \qquad \phi_2 = \Phi_n v_2 \tag{11.3}$$

and fix the base quantities I_n and Φ_n by

$$n\omega^2 C\Phi_n = I_n \qquad c_3\Phi_n{}^3 = nI_n \tag{11.4}$$

Then Eqs. (11.1) may be transformed to

$$\frac{d^2}{d\tau^2}(v_1 + v_2) + k_1\frac{d}{d\tau}(v_1 + v_2) + u_1 = B\cos\tau$$

$$\frac{d}{d\tau}(v_1 - v_2) + k_2 u_2 = B_0 \tag{11.5}$$

where
$$k_1 = \frac{1}{\omega C R_1} \qquad k_2 = \omega C R_2 \qquad \tau = \omega t - \tan^{-1} k_1$$

$$B = \frac{E_1}{n\omega\Phi_n}\sqrt{1 + k_1{}^2} \qquad B_0 = \frac{E_0}{n\omega\Phi_n}$$

The nonlinear characteristics (11.2) may also be expressed by

$$v_1{}^3 = u_1 + u_2 \qquad v_2{}^3 = u_1 - u_2 \tag{11.6}$$

Introducing still another set of new variables

$$a = v_1 + v_2 \qquad b = v_1 - v_2 \tag{11.7}$$

yields, from Eqs. (11.5) and (11.6), the following simultaneous equations:

$$\frac{d^2a}{d\tau^2} + k_1 \frac{da}{d\tau} + \frac{1}{8}(a^2 + 3b^2)a = B \cos \tau$$

$$\frac{db}{d\tau} + \frac{1}{8}k_2(3a^2 + b^2)b = B_0 \tag{11.8}$$

Since we are concerned with the harmonic oscillation which has the same frequency as the impressed voltage, the variables a and b may be assumed by the form

$$a = x(\tau) \sin \tau + y(\tau) \cos \tau \qquad b = z(\tau) \tag{11.9}$$

where $x(\tau)$, $y(\tau)$, and $z(\tau)$ are slowly varying functions of the time τ.

Substituting Eqs. (11.9) into (11.8) and equating the coefficients of the terms containing $\cos \tau$ and $\sin \tau$ and the nonoscillatory terms separately to zero leads to

$$\frac{dx}{d\tau} = \frac{1}{2}(-k_1 x + Ay + B) \equiv X(x,y,z)$$

$$\frac{dy}{d\tau} = \frac{1}{2}(-Ax - k_1 y) \equiv Y(x,y,z) \tag{11.10}$$

$$\frac{dz}{d\tau} = B_0 - \frac{1}{16}k_2(3r^2 + 2z^2)z \equiv Z(x,y,z)$$

with
$$A = 1 - \tfrac{3}{32}(r^2 + 4z^2) \qquad r^2 = x^2 + y^2$$

under the assumptions that $x(\tau)$, $y(\tau)$, and $z(\tau)$ are slowly varying functions of the time τ, so that $d^2x/d\tau^2$, $d^2y/d\tau^2$, and $d^2z/d\tau^2$ may be neglected and that k_1 is a sufficiently small quantity and hence $k_1\, dx/d\tau$, $k_1\, dy/d\tau$, and $k_1\, dz/d\tau$ are also discarded. The result which is obtained from Eqs. (11.10) may be considered to be legitimate as far as we deal with the harmonic oscillation.

(b) *Periodic Solutions and Their Stability*

We consider the periodic state in which $x(\tau)$, $y(\tau)$, and $z(\tau)$ are constant, that is,

$$\frac{dx}{d\tau} = 0 \qquad \frac{dy}{d\tau} = 0 \qquad \text{and} \qquad \frac{dz}{d\tau} = 0 \tag{11.11}$$

Substituting these conditions in Eqs. (11.10) gives the steady-state amplitude $r_0\ (= \sqrt{x_0^2 + y_0^2})$ and the unidirectional component z_0 of the periodic solution as follows:

$$(A^2 + k_1^2)r_0^2 = B^2 \qquad \tfrac{1}{16}k_2(3r_0^2 + 2z_0^2)z_0 = B_0 \tag{11.12}$$

The components x_0, y_0 of the amplitude r_0 are found to be

$$x_0^2 = \frac{r_0^2}{1 + (A/k_1)^2} \qquad y_0^2 = \frac{r_0^2}{1 + (k_1/A)^2} \qquad (11.13)$$

The periodic solution, i.e., the equilibrium state of the system (11.10) is correlated with the singular point (x_0, y_0, z_0) in the xyz space. If the singular point is stable, the corresponding periodic solution is also stable; if not, it is unstable. The stability of the singular point is studied by considering the behavior of integral curves in the neighborhood of that singular point. To this end we consider small variations ξ, η, and ζ, respectively, from the steady-state values x_0, y_0, and z_0 and determine whether these variations approach zero or not with increase of the time τ. From Eqs. (11.10) we obtain

$$\frac{d\xi}{d\tau} = a_{11}\xi + a_{12}\eta + a_{13}\zeta$$

$$\frac{d\eta}{d\tau} = a_{21}\xi + a_{22}\eta + a_{23}\zeta \qquad (11.14)$$

$$\frac{d\zeta}{d\tau} = a_{31}\xi + a_{32}\eta + a_{33}\zeta$$

with

$$a_{11} = \left(\frac{\partial X}{\partial x}\right)_0 = \frac{1}{2}\left(-\frac{3}{16}x_0 y_0 - k_1\right)$$

$$a_{12} = \left(\frac{\partial X}{\partial y}\right)_0 = \frac{1}{2}\left(A - \frac{3}{16}y_0^2\right)$$

$$a_{13} = \left(\frac{\partial X}{\partial z}\right)_0 = -\frac{3}{8}y_0 z_0$$

$$a_{21} = \left(\frac{\partial Y}{\partial x}\right)_0 = \frac{1}{2}\left(-A + \frac{3}{16}x_0^2\right)$$

$$a_{22} = \left(\frac{\partial Y}{\partial y}\right)_0 = \frac{1}{2}\left(\frac{3}{16}x_0 y_0 - k_1\right) \qquad (11.14a)$$

$$a_{23} = \left(\frac{\partial Y}{\partial z}\right)_0 = \frac{3}{8}x_0 z_0$$

$$a_{31} = \left(\frac{\partial Z}{\partial x}\right)_0 = -\frac{3}{8}k_2 x_0 z_0$$

$$a_{32} = \left(\frac{\partial Z}{\partial y}\right)_0 = -\frac{3}{8}k_2 y_0 z_0$$

$$a_{33} = \left(\frac{\partial Z}{\partial z}\right)_0 = -\frac{3}{16}k_2(r_0^2 + 2z_0^2)$$

where $\left(\dfrac{\partial X}{\partial x}\right)_0$, $\cdots$, $\left(\dfrac{\partial Z}{\partial z}\right)_0$ denote the values of $\dfrac{\partial X}{\partial x}$, $\cdots$, $\dfrac{\partial Z}{\partial z}$ at $x = x_0$, $y = y_0$, and $z = z_0$, respectively.

The characteristic equation of the system (11.14) is given by

$$\begin{vmatrix} a_{11} - \lambda & a_{12} & a_{13} \\ a_{21} & a_{22} - \lambda & a_{23} \\ a_{31} & a_{32} & a_{33} - \lambda \end{vmatrix} = 0 \qquad (11.15)$$

or

$$\lambda^3 + b_1\lambda^2 + b_2\lambda + b_3 = 0$$

where

$$b_1 = -(a_{11} + a_{22} + a_{33})$$

$$b_2 = \begin{vmatrix} a_{11} & a_{12} \\ a_{21} & a_{22} \end{vmatrix} + \begin{vmatrix} a_{11} & a_{13} \\ a_{31} & a_{33} \end{vmatrix} + \begin{vmatrix} a_{22} & a_{23} \\ a_{32} & a_{33} \end{vmatrix} \qquad (11.16)$$

$$b_3 = -\begin{vmatrix} a_{11} & a_{12} & a_{13} \\ a_{21} & a_{22} & a_{23} \\ a_{31} & a_{32} & a_{33} \end{vmatrix} \equiv \Delta$$

The variations ξ, η, and ζ approach zero with the time τ, provided that the real part of λ is negative. In this case the corresponding periodic solution is stable. The stability condition is given by the Routh-Hurwitz criterion (see Sec. 3.3), that is,

$$b_1 > 0 \qquad b_1b_2 - b_3 > 0 \qquad b_3 > 0 \qquad (11.17)$$

The first condition of (11.17) is always fulfilled, because, by virtue of Eqs. (11.14a),

$$b_1 = -(a_{11} + a_{22} + a_{33}) = k_1 + \tfrac{3}{16}k_2(r_0^2 + 2z_0^2) > 0$$

Substituting Eqs. (11.14a) and (11.16) into the second and third conditions of (11.17) gives us

$$k_1\left(\frac{B^2}{r_0^2} - \frac{3}{16}Ar_0^2\right) + \frac{3}{4}k_2\left\{k_1\left[k_1 + \frac{3}{16}k_2(r_0^2 + 2z_0^2)\right]\right.$$

$$\left. \times (r_0^2 + 2z_0^2) - \frac{3}{8}Ar_0^2z_0^2\right\} > 0 \quad (11.18)$$

and

$$\frac{B^2}{r_0^2}(r_0^2 + 2z_0^2) - \frac{3}{16}Ar_0^2(r_0^2 - 6z_0^2) > 0$$

These are the stability conditions of the periodic solution. By making use of Eqs. (11.12), the last condition may also be written as

$$\frac{3}{4}Ar_0^2z_0\frac{dB_0}{dr_0^2} = \left(\frac{B^2}{r_0^2} - \frac{3}{16}Ar_0^2\right)z_0\frac{dB_0}{dz_0^2} > 0 \qquad (11.19)$$

Hence the vertical tangency of the characteristic curve ($B_0r_0^2$ or $B_0z_0^2$ relation) results at the stability limit of (11.19).

(c) Almost Periodic Oscillations

When the system (11.10) has a stable limit cycle in the xyz space, the representative point, whose coordinates are $x(\tau)$, $y(\tau)$, and $z(\tau)$, keeps on

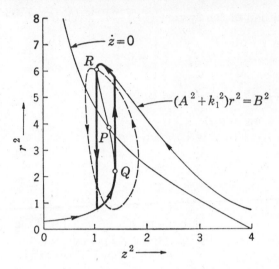

FIGURE 11.2 Limit cycle with discontinuities for
$k_2 \rightarrow 0$.

moving along the limit cycle as τ increases; in other words, an almost periodic oscillation occurs. In order to explain the occurrence of this kind of oscillation, we consider a special case in which k_2 and B_0 are much less than k_1. Then, from Eqs. (11.10), it is evident that $dz/d\tau \ll dx/d\tau$ and $dz/d\tau \ll dy/d\tau$, so that the representative point is first governed by

$$\frac{dx}{d\tau} = \frac{1}{2}\left(-k_1 x + Ay + B\right) \qquad \frac{dy}{d\tau} = \frac{1}{2}\left(-Ax - k_1 y\right)$$

and approaches the characteristic curve defined by $dx/d\tau = 0$ and $dy/d\tau = 0$, or

$$(A^2 + k_1^2)r^2 = B^2 \tag{11.20}$$

During this transient $z(\tau)$ is held nearly constant. After this period $dx/d\tau$, $dy/d\tau$, and $dz/d\tau$ will all be of the same order in magnitude. In Fig. 11.2 is shown the characteristic curve (11.20) for which $k_1 = 0.2$ and $B = 0.5$. Also plotted in the figure is the curve represented by

$$\frac{dz}{d\tau} \left(= \dot{z}\right) = B_0 - \frac{1}{16}k_2(3r^2 + 2z^2)z = 0 \tag{11.21}$$

for a particular case of $B_0 = k_2$. The intersection P of these curves represents an equilibrium state, since the point P is satisfied by Eqs. (11.11). However, it is readily verified by the conditions (11.18) that this equilibrium state is unstable. Since $\dot{z}$ is negative in the region above the curve (11.21) and positive below the curve, the representative point

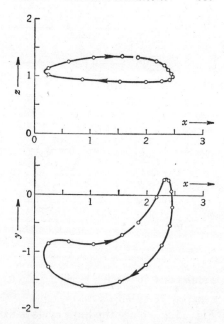

FIGURE 11.3 Limit cycle correlated with an almost periodic oscillation in the resonant circuit of Fig. 11.1.

will gradually move in the direction of the arrows with increasing τ. Hence, discontinuous jumps occur at the limiting points Q and R, and the representative point keeps on moving near the limit cycle represented by the thick line in the figure.

The description so far explains the occurrence of the limit cycle for the case in which the system parameters k_2 and B_0 are very small. As expected from the third equation of (11.10), the period required for the representative point to complete one revolution along the limit cycle decreases with the increase in k_2 and B_0 even if their ratio is kept constant. A more concrete example of the limit cycle will be given in what follows.

NUMERICAL ANALYSIS

A numerical analysis of the system (11.10) was carried out for the parameters as given by

$$k_1 = 0.2 \qquad k_2 = 0.03 \qquad B = 0.5 \qquad \text{and} \qquad B_0 = 0.03$$

After a sufficiently long period of the time τ, the representative point moves along the limit cycle as illustrated in Fig. 11.3. The time intervals between two successive points on the limit cycle are 2π, or equal to one cycle of the impressed voltage. The period required for the representative point to complete one revolution along the limit cycle is $2\pi \times 15.5 \cdots$; this period decreases continuously as k_2 and B_0 increase. The projection of the limit cycle on the r^2z^2 plane is shown dashed in

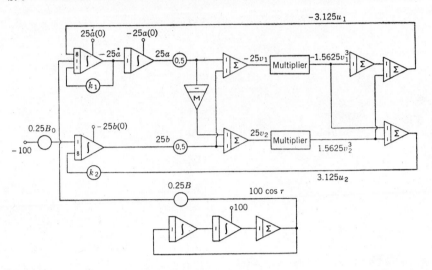

FIGURE 11.4   Computer block diagram for Eqs. (11.22).

Fig. 11.2.   Once the limit cycle is fixed in the xyz space, the waveform of the corresponding almost periodic oscillation in terms of a and b may readily be obtained by using Eqs. (11.9).

(d) Analog-computer Analysis

The numerical analysis described in the preceding part is compared with the solution obtained by using an analog computer. Figure 11.4 shows the block diagram of a computer setup for the solution of Eqs. (11.5) with the nonlinearities (11.6); namely,

$$\frac{d^2a}{d\tau^2} + k_1 \frac{da}{d\tau} + u_1 = B \cos \tau \qquad \frac{db}{d\tau} + k_2 u_2 = B_0$$

where
$$a = v_1 + v_2 \qquad\qquad b = v_1 - v_2$$
$$u_1 = \tfrac{1}{2}(v_1{}^3 + v_2{}^3) \qquad u_2 = \tfrac{1}{2}(v_1{}^3 - v_2{}^3)$$

(11.22)

The symbols in the figure follow the conventional notation.[1]

By using this setup, we obtain the waveforms of a and b as illustrated in Fig. 11.5. The successive points on the curves show the instants when $\tau = 2n\pi$, n being 1, 2, 3, ... In order to compare this result with the preceding analysis, we make use of Eqs. (11.9) and seek the limit cycle correlated with the almost periodic oscillation of Fig. 11.5. Thus we obtain Fig. 11.6, which shows a satisfactory agreement with the theoretical result of Fig. 11.3. Since the computer solution is obtained directly by solving Eqs. (11.22), the assumptions made in the derivation of Eqs. (11.10) may be accepted.

[1] See Fig. 7.21.

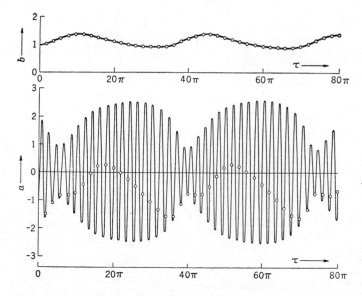

FIGURE 11.5   Waveforms of an almost periodic oscillation obtained by analog-computer analysis. The system parameters are the same as in Fig. 11.3.

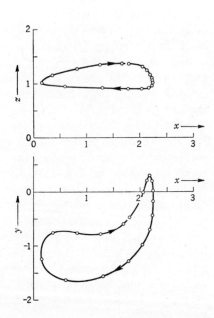

FIGURE 11.6   Limit cycle correlated with the almost periodic oscillation of Fig. 11.5.

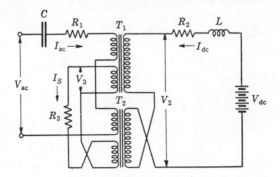

FIGURE 11.7   Resonant circuit containing saturable reactors with secondary and
tertiary windings.

11.3 Experimental Investigation

As we have already remarked, an almost periodic oscillation occurs
in the circuit of Fig. 11.1 provided that the resistor R_2 and consequently
the voltage E_0—in order to keep the d-c component of the secondary
current i_2 to a moderate value—are sufficiently low.   Under this condition the secondary current i_2 is not constant, but oscillates with a low
frequency resulting in the amplitude and phase modulation of the primary current i_1.   The secondary current i_2 also contains a considerable
amount of the second-harmonic component when the cores are saturated.
Such an anomalous oscillation may also occur in the alternative circuit
of Fig. 11.7.   In this circuit the impedance of the secondary circuit is
high owing to the presence of the inductance L, but the tertiary windings
are provided on the cores and connected in series with opposite polarity
through a low resistor R_3.   The function of the tertiary circuit is to
annul the effect of the inductance L.

Proceeding as in Sec. 5.3, we seek the amplitude characteristic of the
oscillating current I_{ac} against the applied voltage V_{ac} of Fig. 11.7.   The
result is shown in Fig. 11.8.   When $I_{dc} = 1.0$ amp, almost periodic
oscillations occur in the range of V_{ac} as indicated in the figure; the
amplitude of the oscillating current I_{ac} varies slowly but periodically to
the extent of the hatched area.   The subresonant oscillation which was
obtained in Fig. 5.8 is scarcely observed in the present experiment.
Therefore, the almost periodic oscillation may be considered to occur in
place of the subresonant oscillation.   Figure 11.9 shows the regions in
which different types of oscillation are sustained.   The almost periodic
oscillation occurs in the hatched area, most remarkably in the range of

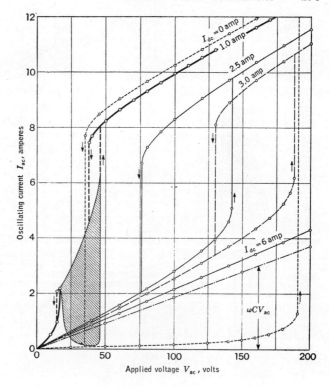

FIGURE 11.8 Amplitude characteristic of harmonic oscillations; almost periodic oscillations occur in the
hatched area when $I_{dc} = 1.0$ amp.

$I_{dc} = 0.8$ to 2.1 amp. Further details are given in the legend of the figure.

The waveforms of almost periodic oscillations are shown in Fig. 11.10. Oscillogram (a) is taken in the circuit of Fig. 5.7; oscillograms (b) to (d) in the circuit of Fig. 11.7. We see in these oscillograms that the amplitude of the oscillation varies periodically, thus presenting a type of beat oscillation. The frequency of the amplitude variation becomes higher as R_3 increases. In the preceding analysis we set up the circuit equations with respect to the magnetic flux in the core. It is the voltage V_3 (Figs. 5.7 and 11.7) that is directly related to the flux, since the time derivative of the flux is proportional to V_3. Thus, in oscillogram (c), the waveform of V_3 is closely observed against V_{ac}; its vectorial representation is shown in Fig. 11.11. Taking V_{ac} as the reference vector, the end point of the vector V_3 moves along the closed trajectory in the direction

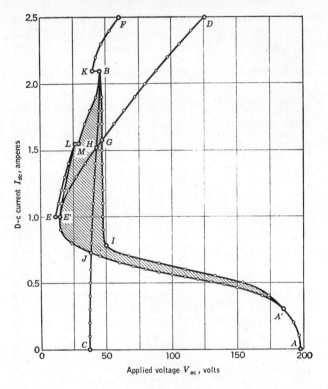

FIGURE 11.9   Regions of different types of oscillation. *AA'IGD*: boundary line on which resonant oscillation starts when V_{ac} is increased. *CJHBKF*: boundary line on which resonant oscillation stops when V_{ac} is decreased. *A'JE'HG*: boundary line on which almost periodic oscillation starts when V_{ac} is increased. *A'IGB*: boundary line on which almost periodic oscillation stops when V_{ac} is increased. *JHB*: boundary line on which almost periodic oscillation starts when V_{ac} is decreased. *A'JE'LMB*: boundary line on which almost periodic oscillation stops when V_{ac} is decreased. *EL*: transition line from subresonant to nonresonant oscillation for decreasing V_{ac}.

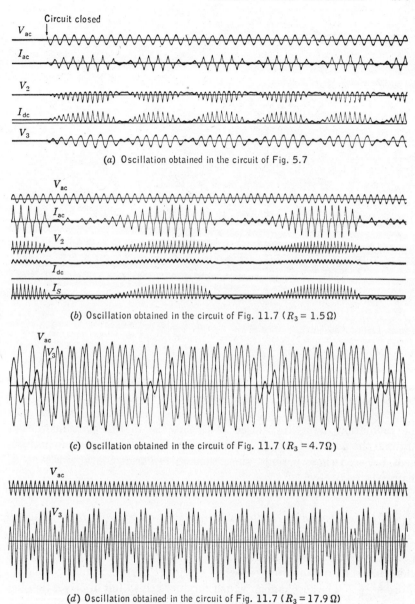

(a) Oscillation obtained in the circuit of Fig. 5.7

(b) Oscillation obtained in the circuit of Fig. 11.7 ($R_3 = 1.5\,\Omega$)

(c) Oscillation obtained in the circuit of Fig. 11.7 ($R_3 = 4.7\,\Omega$)

(d) Oscillation obtained in the circuit of Fig. 11.7 ($R_3 = 17.9\,\Omega$)

FIGURE 11.10   Waveforms of almost periodic oscillations.

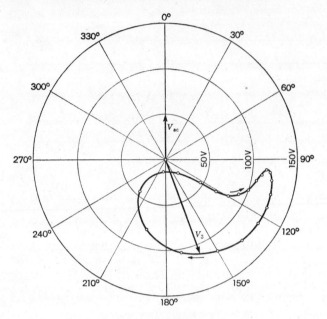

FIGURE 11.11   Closed trajectory of the voltage vector V_3.

of arrows.[1]   The time interval between two successive points on the trajectory corresponds to a half cycle of the applied voltage.   Hence the period of one beat, i.e., the period required for the vector V_3 to complete one revolution along the closed trajectory, is, in this particular case, 10 times the period of the applied voltage.   However, there is no entrainment between these periods, since their ratio varies continuously as R_3 changes.

11.4 *Almost Periodic Oscillations in a Parametric Excitation Circuit*

A schematic circuit is shown in Fig. 11.12.   Under the impression of a sinusoidal voltage $E_1 \sin 2\omega t$, this circuit may produce an oscillation which has the fundamental frequency ω, that is, a subharmonic oscillation of order $\frac{1}{2}$.   The mechanism which produces this kind of oscillation has been known as *parametric excitation*.[2]   However, under certain conditions, the subharmonic oscillation is not maintained in a stable state, but is affected by amplitude and phase modulation; in other words, an almost

---

[1] One sees qualitative agreement between this trajectory and the theoretical result of Fig. 11.3.

[2] This principle is applied to the design of logical circuits in digital computers [27].

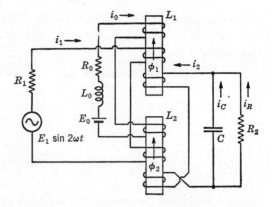

FIGURE 11.12 Parametric excitation circuit which produces a subharmonic oscillation of order ½. An almost periodic oscillation also occurs under certain conditions.

periodic oscillation occurs. We shall be particularly interested in such a case.

(a) Fundamental Equations

Following the notation in Fig. 11.12, the equations for the circuit may be written as follows:

$$n \frac{d}{dt} (\phi_1 + \phi_2) + R_1 i_1 = E_1 \sin 2\omega t$$

$$n \frac{d}{dt} (\phi_1 - \phi_2) = - \frac{1}{C} \int i_C \, dt = - R_2 i_R \qquad (11.23)$$

$$i_2 = i_R + i_C$$

It is assumed that the current i_0 is kept constant owing to the high inductance of L_0. By a procedure analogous to that of Sec. 11.2a, Eqs. (11.23) are transformed to

$$\frac{d}{d\tau} (v_1 + v_2) + k_1 u_1 = B \sin 2\tau$$

$$\frac{d^2}{d\tau^2} (v_1 - v_2) + k_2 \frac{d}{d\tau} (v_1 - v_2) + u_2 = 0 \qquad (11.24)$$

where $\quad i_1 = I_n u_1 \qquad i_2 = I_n u_2 \qquad \phi_1 = \Phi_n v_1 \qquad \phi_2 = \Phi_n v_2$

$$k_1 = \omega C R_1 \qquad k_2 = \frac{1}{\omega C R_2} \qquad B = \frac{E_1}{n \omega \Phi_n} \qquad \tau = \omega t$$

The base quantities I_n and Φ_n may be fixed by Eqs. (11.4). The non-linear characteristics of the cores L_1 and L_2 are expressed, after normalization, by

$$v_1{}^3 = u_0 + u_1 + u_2 \qquad v_2{}^3 = u_0 + u_1 - u_2 \qquad (11.25)$$

where $i_0 = I_n u_0$. By virtue of Eqs. (11.25), Eqs. (11.24) lead to

$$\frac{da}{d\tau} + \frac{1}{8} k_1[(a^2 + 3b^2)a - 8u_0] = B \sin 2\tau$$

$$\frac{d^2b}{d\tau^2} + k_2 \frac{db}{d\tau} + \frac{1}{8}(3a^2 + b^2)b = 0 \qquad (11.26)$$

where a and b are defined, as before, by Eqs. (11.7).

We assume that k_1 is small. The first equation of (11.26) then has an approximate solution

$$a = -\frac{B}{2} \cos 2\tau + a_0 \qquad (11.27)$$

where a_0 is an integration constant. The second equation of (11.26), upon substitution of Eqs. (11.27), leads to a form of Hill's equation with terms for damping and nonlinearity. The solution may have the fundamental period 2π, that is, twice the period of the applied voltage. Hence, an approximate solution for Eqs. (11.26) may be expected to have the form

$$a = -w \cos 2\tau + z(\tau) \qquad w = \frac{B}{2}$$

$$b = x(\tau) \sin \tau + y(\tau) \cos \tau \qquad (11.28)$$

where $x(\tau)$, $y(\tau)$, and $z(\tau)$ are slowly varying functions of the time τ. Substituting Eqs. (11.28) into (11.26) and equating the coefficients of the terms containing $\cos \tau$ and $\sin \tau$ and the nonoscillatory term separately to zero leads to

$$\frac{dx}{d\tau} = \frac{1}{2}\left(-k_2 x + Ay + \frac{3}{8} wyz\right) \equiv X(x,y,z)$$

$$\frac{dy}{d\tau} = \frac{1}{2}\left(-Ax - k_2 y + \frac{3}{8} wxz\right) \equiv Y(x,y,z)$$

$$\frac{dz}{d\tau} = -\frac{1}{8} k_1\left[\left(\frac{3}{2} r^2 + z^2 + \frac{3}{2} w^2\right)z + \frac{3}{4} w(x^2 - y^2) - 8u_0\right] \qquad (11.29)$$

$$\equiv Z(x,y,z)$$

where $\qquad A = 1 - \frac{3}{32}(r^2 + 4z^2 + 2w^2) \qquad r^2 = x^2 + y^2$

It should, however, be remembered that the assumptions mentioned in Sec. 11.2a must be made for the derivation of Eqs. (11.29).

(b) Periodic Solutions and Their Stability

The periodic solution for which the components $x(\tau)$, $y(\tau)$, and $z(\tau)$ are constant is determined by

$$X(x_0,y_0,z_0) = 0 \qquad Y(x_0,y_0,z_0) = 0 \qquad \text{and} \qquad Z(x_0,y_0,z_0) = 0 \quad (11.30)$$

where the subscript 0 is used to designate the values of x, y, and z for the periodic solution. Eliminating x_0 and y_0 in Eqs. (11.30) gives the following relations to determine r_0 ($= \sqrt{x_0{}^2 + y_0{}^2}$) and z_0, that is,

$$A^2 + k_2{}^2 = \left(\frac{3}{8} wz_0\right)^2 \qquad \left(\frac{3}{2} r_0{}^2 + z_0{}^2 + \frac{3}{2} w^2\right)z_0 + \frac{2Ar_0{}^2}{z_0} = 8u_0 \quad (11.31)$$

Hence the components x_0, y_0 of r_0 are given by

$$\begin{aligned}
x_0 &= r_0 \cos\theta & r_0 \cos(\theta + 180°) \\
y_0 &= r_0 \sin\theta & r_0 \sin(\theta + 180°)
\end{aligned} \quad (11.32)$$

where

$$\cos 2\theta = \frac{8A}{3wz_0} \qquad \sin 2\theta = \frac{8k_2}{3wz_0}$$

By a procedure analogous to that of Sec. 11.2b, we make use of the Routh-Hurwitz criterion to obtain the stability conditions of the periodic solution. The result is

$$-k_2 A r_0{}^2 + k_1 \left[\frac{3}{8} k_1 k_2 (r_0{}^2 + 2z_0{}^2 + w^2)^2 \right.$$

$$\left. + k_2{}^2\left(4z_0{}^2 + 2w^2 + \frac{r_0{}^4}{4z_0{}^2}\right) - \frac{1}{8} A r_0{}^2(32 - 3r_0{}^2 - 3w^2)\right] > 0 \quad (11.33)$$

and

$$A(32 - 6r_0{}^2 + 6z_0{}^2 - 3w^2) + \frac{9}{2} w^2 z_0{}^2 - 4k_2{}^2 \frac{r_0{}^2}{z_0{}^2} > 0$$

The second condition may also be written as

$$\left(1 - \frac{3}{32} r_0{}^2 - \frac{3}{8} z_0{}^2\right)z_0 \frac{du_0}{dr_0{}^2} = -\frac{1}{4} A z_0 \frac{du_0}{dz_0{}^2} > 0 \quad (11.34)$$

(c) Almost Periodic Oscillations

A procedure similar to that mentioned in Sec. 11.2c is also applicable to the case where the periodic solution as determined by Eqs. (11.31) is unstable. The representative point of the system (11.29) moves, for a small value of k_1, in the neighborhood of the characteristic curve defined by $dx/d\tau = 0$ and $dy/d\tau = 0$, or

$$A^2 + k_2{}^2 = (\tfrac{3}{8} wz)^2 \quad (11.35)$$

By investigating the sign of $dz/d\tau$ along the characteristic curve of Eq. (11.35), the conclusion is reached that a limit cycle exists provided that k_1 is small.

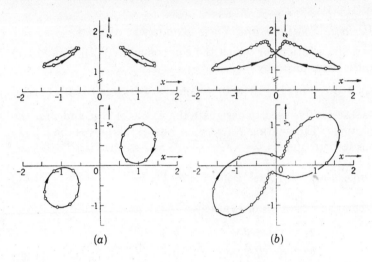

(a) (b)

FIGURE 11.13   Limit cycles correlated with almost periodic oscillations in the parametric excitation circuit of Fig. 11.12. (a) The system parameters are given by Eqs. (11.36). (b) The system parameters are given by Eqs. (11.37).

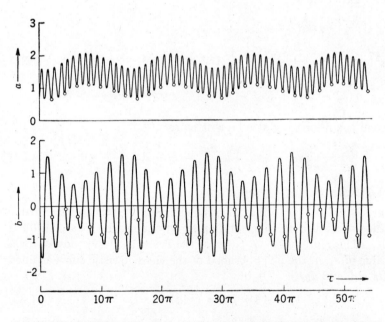

FIGURE 11.14   Waveforms of an almost periodic oscillation obtained by analog-computer analysis. The system parameters are given by Eqs. (11.36).

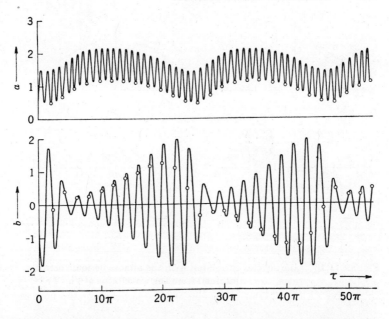

FIGURE 11.15   Waveforms of an almost periodic oscillation obtained by analog-computer analysis. The system parameters are given by Eqs. (11.37).

NUMERICAL ANALYSIS

First, let us consider an example in which the system parameters are given by

$$k_1 = 0.2 \qquad k_2 = 0.2 \qquad B = 1.0 \qquad \text{and} \qquad u_0 = 0.8 \quad (11.36)$$

A numerical computation was carried out for the system (11.29) by using these values of the parameters. The representative point keeps on moving along one of the limit cycles of Fig. 11.13a. The time interval between two successive points on the limit cycles is 2π, or equal to one cycle of the subharmonic oscillation. The period required for the representative point to complete one revolution along the limit cycle is $\pi \times 14.8 \cdots$

Second, let us consider an example in which the system parameters are given by

$$k_1 = 0.1 \qquad k_2 = 0.2 \qquad B = 1.0 \qquad \text{and} \qquad u_0 = 0.8 \quad (11.37)$$

The limit cycle calculated with these values of the parameters is shown in Fig. 11.13b. The period of one revolution along the limit cycle is $\pi \times 54.2 \cdots$.

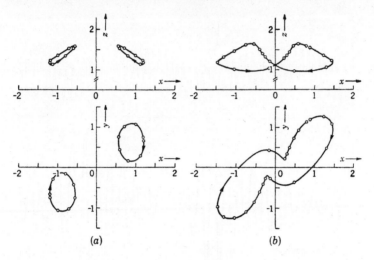

FIGURE 11.16   Limit cycles correlated with the almost periodic oscillations.   (*a*) Obtained from the waveforms of Fig. 11.14.
(*b*) Obtained from the waveforms of Fig. 11.15.

In a comparison of the two examples it is shown that there are two
distinctive types of almost periodic oscillation, as illustrated in Fig.
11.13*a* and *b*.   The former has two separate limit cycles which are symmetrically located about the *z* axis.   The projections of these limit cycles
on the *xy* plane do not contain the origin in their interiors.   In this case
the almost periodic oscillation is synchronized with the applied voltage,
even though the waveform is affected by amplitude and phase modulation.
In Fig. 11.13*b* two limit cycles are joined, resulting in a single loop; the
projection of the loop on the *xy* plane contains the origin in its interior.
In this case the almost periodic oscillation is not synchronized with the
impressed voltage, since one revolution along the limit cycle results in
the phase shift by 2π radians or two cycles of the applied voltage.

(*d*) *Analog-computer Analysis*

The computer block diagram was directly derived from Eqs. (11.24)
and (11.25).   In order to compare computer solutions with the foregoing
analysis, we use the same parameters as given by Eqs. (11.36) and (11.37).
Corresponding to these cases, the waveforms of $a\ (= v_1 + v_2)$ and
$b\ (= v_1 - v_2)$ are shown in Figs. 11.14 and 11.15, respectively.   The successive points on the curves indicate the instants when $\tau = 2n\pi$, n being
1, 2, 3, ...   By making use of Eqs. (11.28), we plot, in Fig. 11.16, the
limit cycles correlated with the almost periodic oscillations.   These limit
cycles agree well with the result obtained in the part on numerical
analysis (Fig. 11.13).

Part IV

*Self-oscillatory
Systems with
External Force*

Chapter 12

Entrainment of Frequency

12.1 Introduction

The phenomenon of frequency entrainment occurs when a periodic force is applied to a system whose free oscillation is of the self-excited type. A typical and important case is the system governed by van der Pol's equation with an additional term for periodic excitation [86]. The frequency of the self-excited oscillation falls in synchronism with the driving frequency provided that the two frequencies are not too different. If their difference is large enough, one may expect the occurrence of an almost periodic oscillation; in other words, a beat oscillation may result. However, the entrainment of frequency still occurs when the ratio between the natural frequency of the self-excited oscillation and the driving frequency is in the neighborhood of an integer (different from unity) or a fraction. Under this condition, the natural frequency of the system is entrained by a frequency which is an integral multiple or sub-multiple of the driving frequency. We shall call such entrainments *higher-harmonic* and *subharmonic entrainments*, respectively. In contrast with these types of entrainment, the phenomenon of synchronization, as mentioned earlier, will be referred to as *harmonic entrainment*.

A considerable amount of literature concerning the harmonic entrainment is available [4, 14, 18, 70, 76, 100]. However, there are not so many reports on the behavior of a self-oscillatory system under the circumstance that the driving frequency and the natural frequency of the system are far apart. Under this circumstance one may expect the occurrence of higher-harmonic or subharmonic entrainment [39, 63, 72, 77]. In this chapter we shall deal with such types of entrainment.[1] The system considered is governed by

$$\frac{d^2u}{d\tau^2} - \epsilon(1 - u^2)\frac{du}{d\tau} + u = B \cos \nu\tau + B_0 \qquad (12.1)$$

---

[1] Chapter 13 will be devoted to the study of those almost periodic oscillations which occur when the system is not entrained.

where ϵ is a small positive constant and $B \cos \nu\tau + B_0$ represents an external force containing a nonoscillatory component. In a later section of this chapter we shall also treat a system in which the restoring force is nonlinear.

12.2 Van der Pol's Equation with Forcing Term

We consider the entrainment of frequency which occurs in a system governed by Eq. (12.1). The left side of this equation takes the form of van der Pol's equation, and the forcing term appears in the right side. Introduction of a new variable defined by $v = u - B_0$ yields an alternative form of Eq. (12.1), that is,

$$\frac{d^2v}{d\tau^2} - \mu(1 - \beta v - \gamma v^2)\frac{dv}{d\tau} + v = B \cos \nu\tau$$

$$(12.2)$$

where $\mu = (1 - B_0{}^2)\epsilon$ $\beta = \dfrac{2B_0}{1 - B_0{}^2}$ and $\gamma = \dfrac{1}{1 - B_0{}^2}$

Since the system governed by Eqs. (12.2) is of the self-excited type, μ must be positive, and therefore $B_0{}^2 < 1$. One sees further that μ is also a small quantity, so that, when $B = 0$, the natural frequency of the system (12.2) is nearly equal to unity. Hence, when the driving frequency ν is in the neighborhood of unity, one may expect an entrained oscillation at the driving frequency ν, that is, the occurrence of harmonic entrainment. The entrained harmonic oscillation $v_0(\tau)$ may be expressed approximately by

$$v_0(\tau) = b_1 \sin \nu\tau + b_2 \cos \nu\tau \tag{12.3}$$

On the other hand, when the driving frequency ν is different from unity, one may expect the occurrence of higher-harmonic or subharmonic entrainment. In this case, the entrained oscillation has a frequency an integral multiple or submultiple of the driving frequency ν. An approximate solution for Eqs. (12.2) may be expected to have the form

$$v_0(\tau) = \frac{B}{1 - \nu^2} \cos \nu\tau + b_1 \sin n\nu\tau + b_2 \cos n\nu\tau$$

where $n = 2, 3, \ldots$ for higher-harmonic oscillations (12.4)
$n = \frac{1}{2}, \frac{1}{3}, \ldots$ for subharmonic oscillations

The first term represents the forced oscillation at the driving frequency ν. The second and third terms represent the entrained oscillation at the frequency $n\nu$, which is not far different from unity. It is noted that, since μ is small, terms of frequency other than ν and $n\nu$ are ignored to this order of approximation.

In the following sections, harmonic, higher-harmonic, and subharmonic oscillations as caused by frequency entrainment will be investigated.   The regions in which these different types of oscillation are sustained will be sought on the $B\nu$ plane, where the coordinates are the amplitude and frequency of the external force.

12.3 Harmonic Entrainment

(a) Periodic Solution

When the driving frequency ν is in the neighborhood of unity, the entrainment may occur at that frequency.   The resulting harmonic oscillation is then expressed by Eq. (12.3).   This solution may readily be found by the method of harmonic balance, namely, by substituting Eq. (12.3) into (12.2) and equating the coefficients of the terms containing $\sin \nu\tau$ and $\cos \nu\tau$ separately to zero.   This leads to

$$\sigma_1 x_1 + (1 - r_1{}^2)y_1 = 0 \qquad (1 - r_1{}^2)x_1 - \sigma_1 y_1 + \frac{B}{\mu\nu a_0} = 0 \quad (12.5)$$

where
$$x_1 = \frac{b_1}{a_0} \qquad y_1 = \frac{b_2}{a_0} \qquad r_1{}^2 = x_1{}^2 + y_1{}^2$$

$$a_0 = \sqrt{\frac{4}{\gamma}} \qquad \sigma_1 = \frac{1 - \nu^2}{\mu\nu} \quad \text{detuning} \tag{12.6}$$

Eliminating x_1 and y_1 from Eqs. (12.5) gives us

$$[(1 - r_1{}^2)^2 + \sigma_1{}^2]r_1{}^2 = \left(\frac{B}{\mu\nu a_0}\right)^2 \tag{12.7}$$

It is readily seen that, when the amplitude B of the external force is zero, we obtain the results $r_1 = 1$ and $\sigma_1 = 0$, so that a_0 in Eqs. (12.6) represents the ultimate amplitude of the self-excited oscillation.   The amplitudes x_1 and y_1 may also be found to be

$$x_1 = -\left(\frac{1 - r_1{}^2}{\sigma_1}\right)\frac{a_0 r_1{}^2}{A} \qquad y_1 = \frac{a_0 r_1{}^2}{A}$$

where
$$A = \frac{B}{1 - \nu^2} \tag{12.8}$$

Equation (12.7) yields what we call the response curves for the harmonic oscillation as illustrated in Fig. 12.1.   Each point on these curves yields the amplitude r_1, which is correlated with the frequency ν of a possible harmonic oscillation for a given value of the amplitude B.

(b) Stability Investigation of the Periodic Solution

The periodic solution as given by Eqs. (12.7) and (12.8) actually exists only when it is stable.   In this section the stability of the periodic

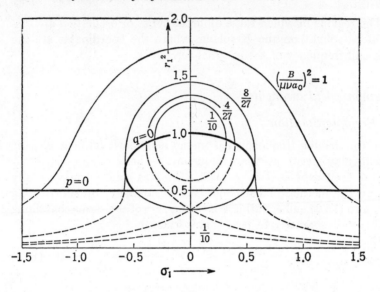

FIGURE 12.1   Response curves for the harmonic oscillation.

solution is investigated by considering the behavior of a small **variation** $\xi(\tau)$ from the periodic solution $v_0(\tau)$. If this variation $\xi(\tau)$ tends to zero with increasing τ, the periodic solution is stable; if $\xi(\tau)$ diverges, the periodic solution is unstable. The variation $\xi(\tau)$ is defined by

$$\xi(\tau) = v(\tau) - v_0(\tau)$$

where $\quad v_0(\tau) = b_1 \sin \nu\tau + b_2 \cos \nu\tau = a_0(x_1 \sin \nu\tau + y_1 \cos \nu\tau) \qquad (12.9)$

It is worth noting that $\xi(\tau)$ need not have the same frequency ν as the periodic solution.[1] Substituting Eq. (12.9) into (12.2), while bearing in mind that $\xi(\tau)$ is sufficiently small, yields the variational equation

$$\frac{d^2\xi}{d\tau^2} - \mu(1 - \beta v_0 - \gamma v_0{}^2)\frac{d\xi}{d\tau} + \left(1 + \mu\beta\frac{dv_0}{d\tau} + 2\mu\gamma v_0\frac{dv_0}{d\tau}\right)\xi = 0 \quad (12.10)$$

---

[1] The stability of the periodic solution has hitherto been discussed by considering small variations δb_1 and δb_2 from the amplitudes b_1 and b_2 defined by

$$\xi(\tau) = \delta b_1 \sin \nu\tau + \delta b_2 \cos \nu\tau$$

Since it is tacitly assumed that the variations δb_1 and δb_2 vary slowly with the time τ, $\xi(\tau)$ will have the same frequency ν as the periodic solution $v_0(\tau)$. This is a constraint imposed on the behavior of the variation $\xi(\tau)$, so that the stability condition obtained in this way is not sufficient if instability results from the variations having different frequencies. This is actually the case, as one sees shortly (Fig. 12.2).

Introducing a new variable η defined by

$$\xi(\tau) = e^{-u(\tau)}\eta(\tau)$$

where

$$u(\tau) = -\frac{\mu}{2} \int_0^\tau (1 - \beta v_0 - \gamma v_0{}^2) \, d\tau$$

$$= \delta\tau + \text{periodic terms in } \tau \tag{12.11}$$

$$\delta = \frac{\mu}{2} (2r_1{}^2 - 1)$$

and inserting the periodic solution $v_0(\tau)$ as given by the second equation of (12.9) leads to a Hill's equation of the form

$$\frac{d^2\eta}{d\tau^2} + \left[\theta_0 + 2 \sum_{m=1}^2 \theta_m \cos(m\nu\tau - \epsilon_m) \right] \eta = 0$$

where

$$\theta_0 = 1 - \frac{\mu^2}{4} \left[(1 - 2r_1{}^2)^2 + 2\left(r_1{}^2 + \frac{\beta^2}{\gamma}\right)r_1{}^2 \right] \tag{12.12}$$

$$\theta_1{}^2 = \frac{\mu^2\beta^2\nu^2}{4\gamma} r_1{}^2 \qquad \theta_2{}^2 = \mu^2\nu^2 r_1{}^4$$

In the derivation of Eqs. (12.12), the terms containing powers of μ higher than the second are ignored. From Eqs. (12.11) and (12.12) one sees that the variation $\xi(\tau)$ tends to zero with increasing τ, provided the damping δ is positive and greater than the characteristic exponent (assumed to be positive) of the solution of Eqs. (12.12). Hence the stability conditions for the periodic solution $v_0(\tau)$ are given by

$$\delta > 0 \qquad \text{or} \qquad 2r_1{}^2 - 1 > 0 \tag{12.13}$$

and, from the condition (4.9) in Sec. 4.2,

$$\left[\theta_0 - \left(\frac{m}{2}\nu\right)^2 \right]^2 + 2\left[\theta_0 + \left(\frac{m}{2}\nu\right)^2 \right]\delta^2 + \delta^4 > \theta_m{}^2 \tag{12.14}$$

From the theory of Hill's equation one may expect that, if the second condition (12.14) is not satisfied for $m = 1$ or 2, the variation $\xi(\tau)$ having the frequency $\nu/2$ or ν will build up, resulting in instability of the periodic solution $v_0(\tau)$.

By substituting the parameters θ_0, θ_1, and θ_2 as given by Eqs. (12.12) into (12.14), and bearing in mind that μ is small, we obtain, for $m = 1$,

$$(1 - 2r_1{}^2)^2 - \frac{\beta^2}{\gamma} r_1{}^2 + \sigma_{1/2}{}^2 > 0$$

where

$$\sigma_{1/2} = \frac{1 - (\nu/2)^2}{\mu(\nu/2)} \tag{12.15}$$

and, for $m = 2$,

$$(1 - r_1{}^2)(1 - 3r_1{}^2) + \sigma_1{}^2 > 0$$

where

$$\sigma_1 = \frac{1 - \nu^2}{\mu\nu} \tag{12.16}$$

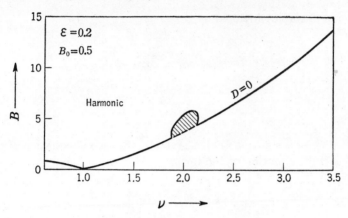

FIGURE 12.2   Region of the harmonic entrainment.

It is noted that, if the stability condition (12.15) is not satisfied, the variation $\xi(\tau)$ does not tend to zero, but gives rise to the ½-harmonic response, as we shall see in Sec. 12.5.

By making use of the stability conditions (12.13) and (12.16), namely,

$$p \equiv 2r_1^2 - 1 > 0 \qquad q \equiv (1 - r_1^2)(1 - 3r_1^2) + \sigma_1^2 > 0 \quad (12.17)$$

the unstable portions of the response curves can be identified; they are shown dashed in Fig. 12.1.   It is readily verified that the vertical tangencies of the response curves occur at the stability limit $q = 0$.   The stability condition (12.15) is to be considered when $\sigma_{\frac{1}{2}}$ is small or $\nu \cong 2$. In Fig. 12.2 is shown the region of the harmonic entrainment on the $B\nu$ plane.   The stability condition (12.15) is not satisfied in the shaded portion; thus harmonic oscillations are not sustained in this region.   As will be shown in Sec. 12.6, the entrainment at the ½-harmonic frequency occurs in this region.

(c) Complementing Remarks for Large Detuning

Since μ is small, the condition $\sigma_1^2 \gg (1 - r_1^2)^2$ is satisfied when ν is not in the neighborhood of unity.   Under this condition, we obtain, from Eq. (12.7), the approximation

$$a_0^2 r_1^2 \cong \left(\frac{B}{1 - \nu^2} \right)^2 = A^2 \qquad (12.18)$$

which is consistent with the assumption made for the amplitude of the forced oscillation in Eq. (12.4).   By using Eqs. (12.18), the first condition of (12.17) may readily be written as

$$D \equiv 1 - \frac{2A^2}{a_0^2} < 0 \tag{12.19}$$

It is clear that the second condition $q > 0$ in (12.17) is satisfied for large detuning. Therefore, (12.19) is the only condition for stability of the harmonic oscillation in the case where the driving frequency ν is not in the neighborhood of 1 or 2.

12.4 Higher-harmonic Entrainment

(a) Periodic Solution

We consider the case in which the self-excited oscillation is entrained by a frequency which is 2 or 3 times the driving frequency. The periodic solution for Eqs. (12.2) is then given by Eqs. (12.4), where n is set equal to 2 or 3 according to the ratio of the two frequencies. Substituting Eqs. (12.4) into (12.2) and equating the coefficients of the terms containing $\sin n\nu\tau$ and $\cos n\nu\tau$ separately to zero yields

$$n = 2: \qquad \sigma_2 x_2 + (D - r_2^2)y_2 - \frac{\beta}{4a_0} A^2 = 0 \tag{12.20}$$
$$(D - r_2^2)x_2 - \sigma_2 y_2 = 0$$

$$n = 3: \qquad \sigma_3 x_3 + (D - r_3^2)y_3 - \frac{\gamma}{12a_0} A^3 = 0 \tag{12.21}$$
$$(D - r_3^2)x_3 - \sigma_3 y_3 = 0$$

where
$$x_n = \frac{b_1}{a_0} \qquad y_n = \frac{b_2}{a_0} \qquad r_n^2 = x_n^2 + y_n^2$$

$$a_0 = \sqrt{\frac{4}{\gamma}} \qquad A = \frac{B}{1 - \nu^2} \qquad D = 1 - \frac{2A^2}{a_0^2} \tag{12.22}$$

$$\sigma_n = \frac{1 - (n\nu)^2}{\mu n \nu} \quad \text{detuning}$$

By eliminating x_n and y_n from Eqs. (12.20) and (12.21), one may readily find the amplitude characteristics for the higher-harmonic oscillations, that is,

$$n = 2: \qquad [(D - r_2^2)^2 + \sigma_2^2]r_2^2 = \left(\frac{\beta}{4a_0} A^2\right)^2 \tag{12.23}$$

$$n = 3: \qquad [(D - r_3^2)^2 + \sigma_3^2]r_3^2 = \left(\frac{\gamma}{12a_0} A^3\right)^2 \tag{12.24}$$

For nonzero values of D, these equations may be rewritten as

$$\left[\left(1 - \frac{r_n^2}{D}\right)^2 + \left(\frac{\sigma_n}{D}\right)^2\right]\frac{r_n^2}{D} = F_n \qquad n = 2 \text{ or } 3$$

$$\tag{12.25}$$

where $\qquad F_2 = \left(\frac{\beta}{4a_0}\right)^2\frac{A^4}{D^3} \qquad F_3 = \left(\frac{\gamma}{12a_0}\right)^2\frac{A^6}{D^3}$

Equation (12.25) has the same form as Eq. (12.7). Therefore, if the quantities r_1^2, σ_1, and $(B/\mu\nu a_0)^2$ are replaced by r_n^2/D, σ_n/D, and F_n, respectively, the response curves of Fig. 12.1 also apply to the case of higher-harmonic oscillation. As we shall verify shortly, the dashed portions of the response curves are also unstable for the higher-harmonic oscillation. It is noted that the second- and third-harmonic entrainments occur predominantly when the detuning σ_n is small, that is, when the driving frequency ν is in the neighborhoods of $\frac{1}{2}$ and $\frac{1}{3}$, respectively.

(b) Stability Investigation of the Periodic Solution

By proceeding in the same manner as in Sec. 12.3b, the stability conditions for the higher-harmonic oscillation are obtained by solving the variational equation of the Hill type. The stability conditions for the second-harmonic oscillation are as follows:

$$2r_2^2 - D > 0 \qquad (D - r_2^2)(D - 3r_2^2) + \sigma_2^2 > 0 \qquad (12.26)$$

The first inequality corresponds to the condition (12.13) for the harmonic oscillation, and the second is obtained by putting $m = 4$ in the condition (12.14). The stability conditions obtained by putting $m = 1$ to 3 were calculated, but it has turned out that these conditions are all satisfied when ν is not far different from $\frac{1}{2}$, at which the second-harmonic entrainment occurs predominantly.

The stability conditions obtained for the third-harmonic oscillation are as follows:

$$2r_3^2 - D > 0 \qquad (D - r_3^2)(D - 3r_3^2) + \sigma_3^2 > 0 \qquad (12.27)$$

The stability conditions (12.26) and (12.27) may be combined and expressed as

$$2\frac{r_n^2}{D} - 1 > 0 \qquad \left(1 - \frac{r_n^2}{D}\right)\left(1 - 3\frac{r_n^2}{D}\right) + \left(\frac{\sigma_n}{D}\right)^2 > 0 \qquad n = 2 \text{ or } 3 \quad (12.28)$$

These conditions take the same forms as (12.17), and they are useful for the investigation of stability of the response curves as given by Eqs. (12.25). Thus it follows, as mentioned earlier, that the dashed portions of the response curves in Fig. 12.1 are unstable for the higher-harmonic oscillation.

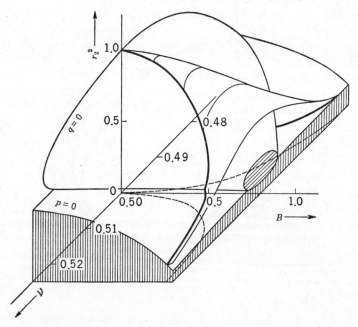

FIGURE 12.3   Response characteristic for the second-harmonic oscillation.

(c) Numerical Examples

A numerical analysis of the response characteristic for the higher-harmonic oscillation was carried out for the same values of the system parameters as those in Fig. 12.2, that is,

$$\epsilon = 0.2 \quad \text{and} \quad B_0 = 0.5$$

in Eq. (12.1).   Consequently, the parameters in Eqs. (12.2) are

$$\mu = 0.15 \quad \beta = \tfrac{1}{3} \quad \text{and} \quad \gamma = \tfrac{4}{3}$$

With these values of the parameters, and by using the relations (12.22), the response characteristics represented by Eqs. (12.23) and (12.24) are calculated and illustrated in Figs. 12.3 and 12.4, respectively.   In each figure, the surface of the response characteristic (r_n^2 versus B and ν) bordered by the thick lines shows the response of stable oscillation.   The stability limits are given by the surfaces

$$p \equiv 2r_n^2 - D = 0 \quad \text{and} \quad q \equiv (D - r_n^2)(D - 3r_n^2) + \sigma_n^2 = 0$$

which are also shown in the figures.[1]   The boundary curves of the stable

---

[1] It is not difficult to show that the vertical tangency of the response surface occurs at the stability limit $q = 0$.

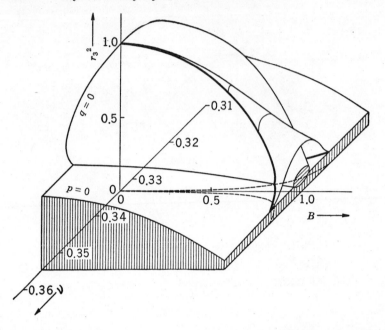

FIGURE 12.4   Response characteristic for the third-harmonic oscillation.

response are projected on the $B\nu$ coordinate plane.   Hence one sees that, when the amplitude B and the frequency ν of the external force are given at any point inside these regions, the resulting oscillation is entrained by the frequency 2 times (in Fig. 12.3) or 3 times (in Fig. 12.4) the driving frequency ν.

12.5 Subharmonic Entrainment

(a) Periodic Solution

We consider the case in which the self-excited oscillation is entrained by a frequency which is $\frac{1}{2}$ or $\frac{1}{3}$ times the driving frequency.   The periodic solution for Eqs. (12.2) is then given by Eqs. (12.4), where n is set equal to $\frac{1}{2}$ or $\frac{1}{3}$ according to the ratio of the two frequencies.   Substituting Eqs. (12.4) into (12.2) and equating the coefficients of the terms containing $\sin n\nu\tau$ and $\cos n\nu\tau$ separately to zero yields

$$n = \tfrac{1}{2}: \qquad \sigma_{\frac{1}{2}}x_{\frac{1}{2}} + \left(D - r_{\frac{1}{2}}{}^2 - \frac{\beta}{2}A \right) y_{\frac{1}{2}} = 0$$

$$\left(D - r_{\frac{1}{2}}{}^2 + \frac{\beta}{2}A \right) x_{\frac{1}{2}} - \sigma_{\frac{1}{2}}y_{\frac{1}{2}} = 0 \qquad (12.29)$$

$n = \frac{1}{3}:$
$$\sigma_{\frac{1}{3}}x_{\frac{1}{3}} + (D - r_{\frac{1}{3}}^2)y_{\frac{1}{3}} + \frac{A}{a_0}(x_{\frac{1}{3}}^2 - y_{\frac{1}{3}}^2) = 0$$

$$(D - r_{\frac{1}{3}}^2)x_{\frac{1}{3}} - \sigma_{\frac{1}{3}}y_{\frac{1}{3}} + \frac{2A}{a_0}x_{\frac{1}{3}}y_{\frac{1}{3}} = 0 \tag{12.30}$$

where
$$x_n = \frac{b_1}{a_0} \qquad y_n = \frac{b_2}{a_0} \qquad r_n^2 = x_n^2 + y_n^2$$

$$a_0 = \sqrt{\frac{4}{\gamma}} \qquad A = \frac{B}{1 - \nu^2} \qquad D = 1 - \frac{2A^2}{a_0^2} \tag{12.31}$$

$$\sigma_n = \frac{1 - (n\nu)^2}{\mu n \nu} \qquad \text{detuning}$$

Eliminating $x_{\frac{1}{2}}$ and $y_{\frac{1}{2}}$ from Eqs. (12.29) gives us

$$\left[(D - r_{\frac{1}{2}}^2)^2 + \sigma_{\frac{1}{2}}^2 - \frac{\beta^2}{4}A^2\right]r_{\frac{1}{2}}^2 = 0$$

from which the amplitude $r_{\frac{1}{2}}$ of the $\frac{1}{2}$-harmonic oscillation is found to be

$$r_{\frac{1}{2}}^2 = 0 \tag{12.32}$$

or
$$r_{\frac{1}{2}}^2 = D \pm \sqrt{\frac{\beta^2}{4}A^2 - \sigma_{\frac{1}{2}}^2} \tag{12.33}$$

Similarly, the amplitude $r_{\frac{1}{3}}$ of the $\frac{1}{3}$-harmonic oscillation is given by

$$r_{\frac{1}{3}}^2 = 0 \tag{12.34}$$

or
$$r_{\frac{1}{3}}^2 = \left(D + \frac{\gamma}{8}A^2\right) \pm \sqrt{\left(D + \frac{\gamma}{8}A^2\right)^2 - D^2 - \sigma_{\frac{1}{3}}^2} \tag{12.35}$$

When the amplitude $r_{\frac{1}{2}}$ or $r_{\frac{1}{3}}$ is zero, no subharmonic response is obtained. Consequently, the periodic solution (12.4) has only the harmonic component $B/(1 - \nu^2)$ cos $\nu\tau$. This case has already been considered in Sec. 12.3c. The amplitudes $r_{\frac{1}{2}}$ and $r_{\frac{1}{3}}$ of the subharmonic oscillations may be obtained from Eqs. (12.33) and (12.35) provided the right sides of these equations are positive. One sees from Eqs. (12.33) and (12.35) that the detunings $\sigma_{\frac{1}{2}}$ and $\sigma_{\frac{1}{3}}$ must be small enough when the entrainment at the subharmonic frequencies is expected.

(b) Stability Investigation of the Periodic Solution

The stability problem for $r_{\frac{1}{2}} = 0$ and $r_{\frac{1}{3}} = 0$ need not be considered, since it is already discussed in the case of harmonic entrainment. The stability conditions for the subharmonic oscillations as given by Eqs. (12.33) and (12.35) were sought in the same manner as in Sec. 12.3b, with the results that

$n = \frac{1}{2}:$ $\qquad 2r_{\frac{1}{2}}^2 - D > 0 \qquad r_{\frac{1}{2}}^2 - D > 0 \tag{12.36}$

$n = \frac{1}{3}:$ $\qquad 2r_{\frac{1}{3}}^2 - D > 0 \qquad r_{\frac{1}{3}}^2 - \left(D + \frac{\gamma}{8}A^2\right) > 0 \tag{12.37}$

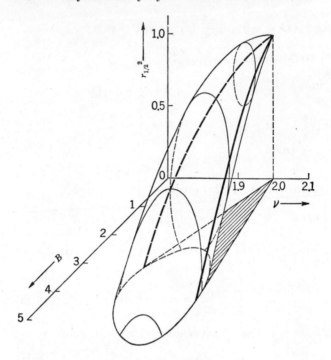

FIGURE 12.5   Response characteristic for the $\frac{1}{2}$-harmonic oscillation.

The first conditions in (12.36) and (12.37) are, as is readily seen, virtually unnecessary.   It follows from the second conditions that the amplitudes $r_{\frac{1}{2}}$ and $r_{\frac{1}{3}}$ of the stable subharmonic oscillations are given by taking the positive sign in Eqs. (12.33) and (12.35).

(c) *Numerical Examples*

A numerical analysis of the response characteristic for the subharmonic oscillation was carried out by using the same parameters as in Sec. 12.4c, that is,

$$\epsilon = 0.2 \quad \text{and} \quad B_0 = 0.5$$

so that $\quad \mu = 0.15 \quad \beta = \frac{2}{3} \quad \text{and} \quad \gamma = \frac{4}{3}$

The response characteristics calculated by using Eqs. (12.33) and (12.35) are illustrated in Figs. 12.5 and 12.6, respectively.   The thick lines in the figures are the loci of the vertical tangencies of the characteristics, and they separate the stable and unstable responses, those below the lines being unstable.   The boundaries of the stable response are projected on the $B\nu$ coordinate plane.   Hence, when the amplitude B and the fre-

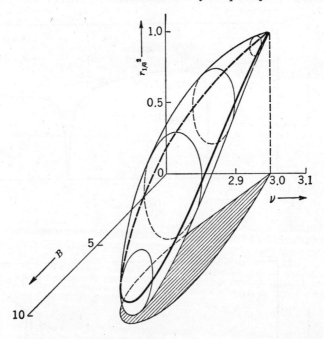

FIGURE 12.6 Response characteristic for the ⅓-harmonic oscillation.

quency ν of the external force are chosen inside these regions, the resulting oscillation is entrained by the frequency ½ times (in Fig. 12.5) or ⅓ times (in Fig. 12.6) the driving frequency ν.

12.6 Regions of Frequency Entrainment

In the preceding sections, the phenomenon of frequency entrainment at harmonic, higher-harmonic, and subharmonic frequencies has been investigated. The response characteristics of the entrained oscillation were obtained by using Eqs. (12.7), (12.23), (12.24), (12.33), and (12.35). The stability for these oscillations has also been investigated by making use of Hill's equation as the stability criterion. For convenience of comparison, the stability conditions obtained in the preceding sections are summarized in Table 12.1. As mentioned before, the higher-harmonic or subharmonic entrainment occurs within a narrow range of the driving frequency ν. On the other hand, the harmonic entrainment occurs at any driving frequency ν provided the amplitude B of the external force is sufficiently large. This fact has made it necessary to have two sets of stability conditions for the harmonic entrainment. From the results

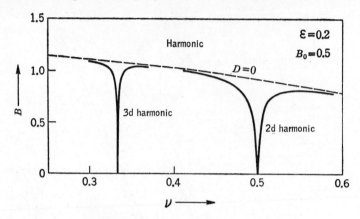

(*a*) Harmonic and higher-harmonic entrainments

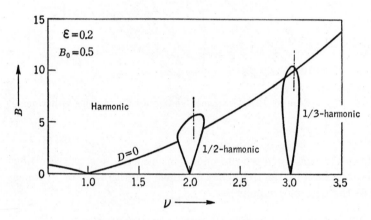

(*b*) Harmonic and subharmonic entrainments

FIGURE 12.7   Regions of frequency entrainment.

obtained in Figs. 12.2 to 12.6, the regions of frequency entrainment are reproduced on the $B\nu$ plane, as illustrated in Fig. 12.7.[1]

As mentioned in Sec. 12.3c, the stability condition for the harmonic oscillation is given by $D < 0$, provided that the driving frequency ν is not in the neighborhood of 1 or 2. The stability conditions (12.26) and (12.27) show that, if $D \leqq 0$, the higher-harmonic oscillation is stable. In Fig. 12.7a, therefore, the boundary curve $D = 0$ of the harmonic entrainment lies inside the region of the higher-harmonic entrainment. It should be noted that, since there are no abrupt changes in the ampli-

[1] The analysis to this order of approximation shows that the second- and ½-harmonic entrainments occur when the system is unsymmetrical, in other words, when the parameter β is present in the system [see Eqs. (12.23) and (12.33)].

Table 12.1 Stability Conditions for the Entrained Oscillations

Frequency	m	⅓-harmonic	m	½-harmonic	m	Harmonic	m	Second harmonic	m	Third harmonic
$\tfrac{1}{6}\nu$	1									
$\tfrac{1}{4}\nu$			1	—						
$\tfrac{1}{3}\nu$	2	$(2r_{1/3}^2 - D > 0)$ $r_{1/3}^2 - \left(D + \dfrac{\gamma}{8}A^2\right) > 0$								
$\tfrac{1}{2}\nu$	3	—	2	$(2r_{1/2}^2 - D > 0)$ $r_{1/2}^2 - D > 0$	1	$2r_1^2 - 1 > 0$ $(1-2r_1^2)^2 - \dfrac{\beta^2}{\gamma}r_1^2 + \sigma_{1/2}^2 > 0$				
$\tfrac{2}{3}\nu$	4	—	3	—						
$\tfrac{3}{4}\nu$			4	—						
ν	6				2	$2r_1^2 - 1 > 0$ $(1-r_1^2)(1-3r_1^2) + \sigma_1^2 > 0$	1	—	1	—
$\tfrac{3}{2}\nu$							2	—	2	—
2ν							3	$2r_2^2 - D > 0$ $(D - r_2^2)(D - 3r_2^2) + \sigma_2^2 > 0$	3	—
3ν							4		4	$2r_3^2 - D > 0$ $(D - r_3^2)(D - 3r_3^2) + \sigma_3^2 > 0$

Notes: 1. Only the indispensable conditions for stability are retained. Stability conditions not shown in the table (marked by —) are fulfilled when the stability conditions listed in the same column are satisfied.

2. When the second inequality in each pair of stability conditions is not satisfied, the corresponding periodic solution becomes unstable owing to the buildup of a self-excited oscillation having the frequency indicated in the first column of the table.

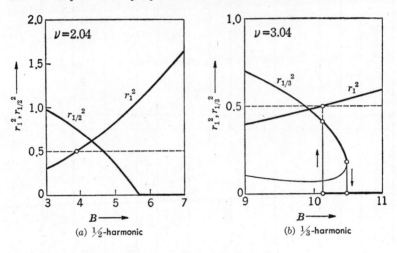

FIGURE 12.8   Response curves with varying B.

tudes of the harmonic and higher-harmonic components of an oscillation at the boundary curve $D = 0$, the boundary curve $D = 0$ has practically no significance in the figure.

In Fig. 12.7b, one sees that the continuity of the boundary curve $D = 0$ is disturbed by the intrusion of the region of the $\frac{1}{2}$-harmonic entrainment (cf. Fig. 12.2). The regions of the harmonic and $\frac{1}{3}$-harmonic entrainments, on the other hand, have an overlapping area. Thus, in this area common to the two regions, both the harmonic and $\frac{1}{3}$-harmonic oscillations are sustained. Therefore, if the driving frequency ν is kept constant and the amplitude B is varied along the chain lines of Fig. 12.7b, the response curves are subject to hysteresis for the $\frac{1}{3}$-harmonic entrainment, but no hysteresis would be observed for the $\frac{1}{2}$-harmonic entrainment. These response curves are compared in Fig. 12.8a and b.

12.7 Analog-computer Analysis

The theoretical results obtained in the preceding sections will be compared with the solutions obtained by using an analog computer. The block diagram of Fig. 12.9 shows an analog-computer setup for the solution of Eqs. (12.2), in which the parameters μ, β, and γ are set equal to the values as given in Secs. 12.3b, 12.4c, and 12.5c. The symbols in the figure follow the conventional notation.[1] The solutions of Eqs. (12.2)

---

[1] The integrating amplifiers in the block diagram integrate their inputs with respect to the machine time (in seconds), which is, in this particular case, 2 times the independent variable τ.

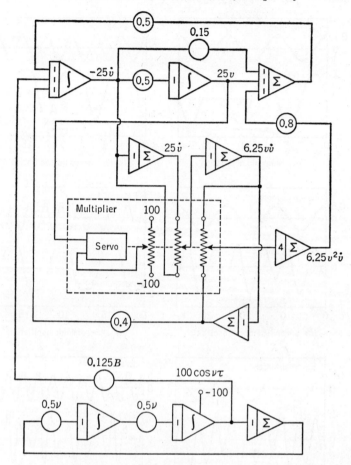

FIGURE 12.9 Block diagram of an analog-computer setup for the solution of Eqs. (12.2).

are sought for various combinations of B and ν, that is, the amplitude and frequency of the external force. Some representative waveforms of $v(\tau)$ are shown in Fig. 12.10. The points on the curves appear at the beginning of each cycle of the external force. These points are useful for distinguishing between an entrained (periodic) oscillation and a beat (nonperiodic) oscillation. The regions of entrained oscillation are illustrated in Fig. 12.11a and b. These regions agree well with the theoretical results as shown in Fig. 12.7a and b, respectively. The result concerning hysteresis derived from theory (cf. Fig. 12.8) is also confirmed by analog-computer analysis.

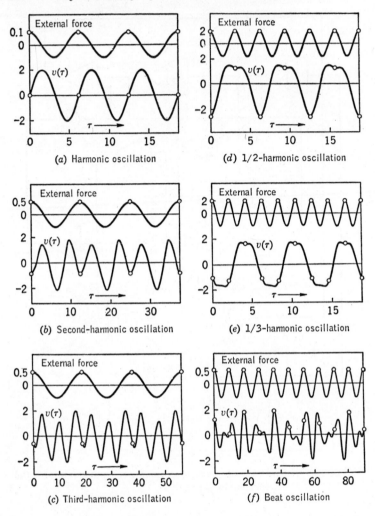

FIGURE 12.10   Waveforms of the oscillations in a system described by
Eqs. (12.2).

Thus far, the nonlinearity of the system is considered to be small; in
other words, it is assumed that $\epsilon \ll 1$ in Eq. (12.1) or $\mu \ll 1$ in Eqs. (12.2).
If the nonlinearity is large, one may expect the entrainment at frequencies other than those discussed in the preceding sections. The
theoretical analysis for such cases has not yet been accomplished, since
the algebra involved becomes very unwieldy. However, an example of
the regions of entrainment obtained by using an analog computer is

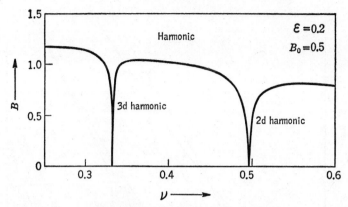

(*a*) Harmonic and higher-harmonic entrainments

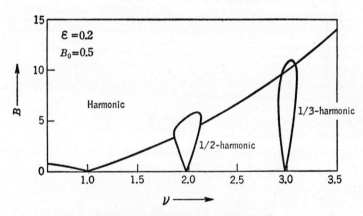

(*b*) Harmonic and subharmonic entrainments

FIGURE 12.11   Regions of frequency entrainment obtained by analog-computer analysis.

shown in Fig. 12.12.   In this case, the system is described by

$$\frac{d^2u}{d\tau^2} - \epsilon(1 - u^2)\frac{du}{d\tau} + u = B\cos\nu\tau + B_0$$

where $\epsilon = 4$ and $B_0 = 0$

One sees in the figure the regions of entrainment at the frequencies $\frac{1}{6}$, $\frac{1}{5}$, $\frac{1}{4}$, $\frac{1}{3}$, $\frac{1}{2}$, 1, 3, and 5 times the driving frequency ν.   If the amplitude B and the frequency ν of the external force are given outside these regions, that is, in the shaded portions, a beat oscillation results.

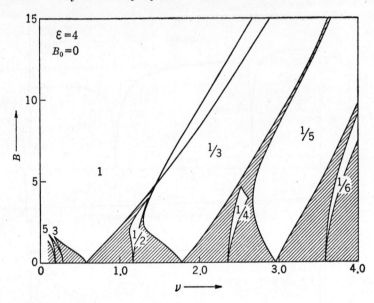

FIGURE 12.12   Regions of frequency entrainment for a system with large nonlinearity.

12.8 Self-oscillatory System with Nonlinear Restoring Force

We consider a system governed by

$$\frac{d^2v}{d\tau^2} - \mu(1 - \gamma v^2)\frac{dv}{d\tau} + v^3 = B\cos\nu\tau \tag{12.38}$$

where μ is a small positive constant and γ is positive also.   Thus the damping is negative for small values of v, so that one may expect the occurrence of frequency entrainment as before.   Furthermore, since the restoring force is nonlinear, i.e., cubic in v, the system will exhibit the phenomenon of nonlinear resonance accompanied by discontinuous jumps in amplitude.

(a) Periodic Solution

When no external force is applied to the system, a free oscillation is self-excited.   To find the ultimate amplitude and frequency of the oscillation, we put, to a first approximation,

$$v_0(\tau) = a_0\cos\omega_0\tau \tag{12.39}$$

and substitute this into Eq. (12.38) with $B = 0$. Then, equating the coefficients of the terms containing $\sin \omega_0 \tau$ and $\cos \omega_0 \tau$ separately to zero yields

$$a_0{}^2 = \frac{4}{\gamma} \qquad \omega_0{}^2 = \frac{3}{4} a_0{}^2 = \frac{3}{\gamma} \tag{12.40}$$

We see that the natural frequency ω_0 of the system is proportional to the amplitude a_0 of the free oscillation.

When the external force is present, the harmonic entrainment occurs; in it the free oscillation of the system is entrained by the driving frequency. The entrainment at other frequencies, particularly at the subharmonic frequency of order $\frac{1}{3}$, may also occur owing to the presence of the parameter γ [see Eq. (12.35)] and the restoring force in a cubic form (see Sec. 7.2). However, in this section we confine attention to the harmonic entrainment and assume the periodic solution of the form

$$v_0(\tau) = b_1 \sin \nu\tau + b_2 \cos \nu\tau \tag{12.41}$$

Substituting Eq. (12.41) into (12.38) and equating the coefficients of the terms containing $\sin \nu\tau$ and $\cos \nu\tau$ separately to zero leads to

$$\sigma_1 x_1 + (1 - r_1{}^2)y_1 = 0 \qquad (1 - r_1{}^2)x_1 - \sigma_1 y_1 + \frac{B}{\mu \nu a_0} = 0 \tag{12.42}$$

where
$$x_1 = \frac{b_1}{a_0} \qquad y_1 = \frac{b_2}{a_0} \qquad r_1{}^2 = x_1{}^2 + y_1{}^2$$

$$a_0 = \sqrt{\frac{4}{\gamma}} \qquad \sigma_1 = \frac{\omega_0{}^2 r_1{}^2 - \nu^2}{\mu \nu} \quad \text{detuning} \tag{12.43}$$

It is worth noting that the detuning is a function of both the amplitude r_1 and the frequency ν. This is a reasonable result, since the natural frequency of the system is modified to $\omega_0 r_1$. Eliminating x_1 and y_1 from Eqs. (12.42) gives us

$$[(1 - r_1{}^2)^2 + \sigma_1{}^2]r_1{}^2 = \left(\frac{B}{\mu \nu a_0}\right)^2 \tag{12.44}$$

The amplitudes x_1 and y_1 are

$$x_1 = -\left(\frac{1 - r_1{}^2}{\sigma_1}\right)\frac{a_0 r_1{}^2}{A} \qquad y_1 = \frac{a_0 r_1{}^2}{A}$$

where
$$A = \frac{B}{\omega_0{}^2 r_1{}^2 - \nu^2} \tag{12.45}$$

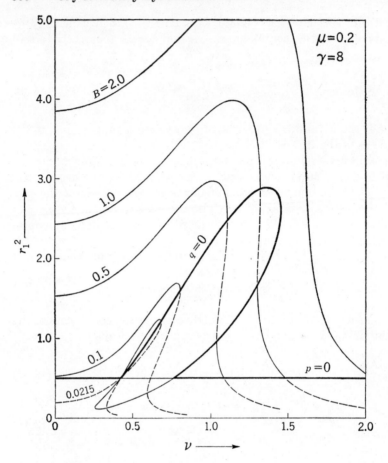

FIGURE 12.13   Response curves for the harmonic oscillation in a system
with nonlinear restoring force.

Equation (12.44) yields the response curves illustrated in Fig. 12.13.   One
sees that the response curves in the case of nonlinear restoring force could
be thought of as arising from those for the linear case (see Fig. 12.1) by
bending the latter to the right.

(b) Stability Investigation of the Periodic Solution

By a procedure analogous to that in Sec. 12.3b, we investigate the
stability of the periodic solution by considering the behavior of a small
variation defined by

$$v(\tau) = v_0(\tau) + \xi(\tau)$$

where $v_0(\tau) = b_1 \sin \nu\tau + b_2 \cos \nu\tau = a_0(x_1 \sin \nu\tau + y_1 \cos \nu\tau)$

(12.46)

If the variation $\xi(\tau)$ tends to zero with increasing τ, the periodic solution $v_0(\tau)$ is stable; otherwise, it is unstable. Substitution of Eq. (12.46) into (12.38) gives us the variational equation

$$\frac{d^2\xi}{d\tau^2} - \mu(1 - \gamma v_0{}^2)\frac{d\xi}{d\tau} + \left(3v_0{}^2 + 2\mu\gamma v_0\frac{dv_0}{d\tau}\right)\xi = 0 \qquad (12.47)$$

Introducing a new variable η defined by

$$\xi(\tau) = e^{-u(\tau)}\eta(\tau)$$

where

$$u(\tau) = -\frac{\mu}{2}\int_0^\tau (1 - \gamma v_0{}^2)\,d\tau$$
$$= \delta\tau + \text{periodic terms in } \tau \qquad (12.48)$$
$$\delta = \frac{\mu}{2}(2r_1{}^2 - 1)$$

and inserting the periodic solution $v_0(\tau)$ as given by the second equation of (12.46) leads to a Hill's equation of the form

$$\frac{d^2\eta}{d\tau^2} + \left[\theta_0 + 2\sum_{m=1}^{2}\theta_m\cos(2m\nu\tau - \epsilon_m)\right]\eta = 0$$

where

$$\theta_0 = \frac{6}{\gamma}r_1{}^2 - \mu^2\left(\frac{3}{2}r_1{}^4 - r_1{}^2 + \frac{1}{4}\right) \qquad (12.49)$$
$$\theta_1{}^2 = r_1{}^4\left[\frac{9}{\gamma^2} - \mu^2\left(\frac{6}{\gamma}r_1{}^2 - \frac{3}{\gamma} - \nu^2\right)\right] + 0(\mu^4)$$
$$\theta_2{}^2 = 0(\mu^4)$$

The stability conditions for the periodic solution $v_0(\tau)$ are given by

$$\delta > 0 \qquad \text{or} \qquad 2r_1{}^2 - 1 > 0 \qquad (12.50)$$

and, from the condition (4.6) in Sec. 4.2,

$$[\theta_0 - (m\nu)^2]^2 \pm 2[\theta_0 + (m\nu)^2]\delta^2 + \delta^4 > \theta_m{}^2 \qquad (12.51)$$

Substituting the parameters θ given by Eqs. (12.49) into (12.51) gives, for $m = 1$,

$$(1 - r_1{}^2)(1 - 3r_1{}^2) + \sigma_1{}^2 + 2\frac{\omega_0{}^2r_1{}^2}{\mu\nu}\sigma_1 > 0 \qquad (12.52)$$

In the derivation of this condition, terms containing higher order in μ are ignored. By virtue of Eq. (12.44), the condition (12.52) may be rewritten as

$$[2\nu\sigma_1{}^2 - \mu(1 - r_1{}^2)^2]\frac{d\nu}{dr_1{}^2} > 0 \qquad (12.53)$$

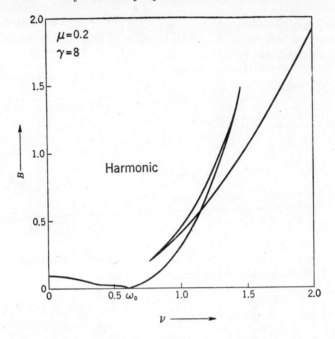

FIGURE 12.14 Region of the harmonic entrainment in a system with nonlinear restoring force.

Hence the vertical tangency of the response curve results at the stability
limit of (12.53). Since θ_2 is small, of order μ^2, the stability condition
(12.51) for $m = 2$ need not be considered. Therefore, in the end, we
have the following stability conditions for the harmonic oscillation,
that is,

$$p \equiv 2r_1^2 - 1 > 0$$

$$q \equiv (1 - r_1^2)(1 - 3r_1^2) + \sigma_1^2 + 2\frac{\omega_0^2 r_1^2}{\mu\nu}\sigma_1 > 0 \qquad (12.54)$$

The stability limits $p = 0$ and $q = 0$ are shown in Fig. 12.13. Hence the
dashed-line portions of the response curves are unstable. From this
result, the region of harmonic entrainment is reproduced on the $B\nu$ plane,
as illustrated in Fig. 12.14.[1] Two types of harmonic oscillations, i.e.,
the resonant and nonresonant oscillations, occur in the overlapped area of
entrainment. When the amplitude and frequency of the external force
are prescribed outside the region of entrainment, one may expect the
occurrence of a beat oscillation.

[1] The regions of entrainment at higher-harmonic or subharmonic frequencies may
be obtained in the same manner as in the previous case of the linear restoring force.

Chapter 13

Almost Periodic
Oscillations

13.1 Introduction

Almost periodic oscillations may occur in some sorts of nonlinear systems which are subject to periodic forces. It has been shown in Chap. 11 that oscillations of this type may arise in electrical systems which contain saturable iron cores as nonlinear elements. As mentioned in Sec. 11.1, the amplitude and phase of an almost periodic oscillation vary slowly but periodically even in the steady state. However, since the ratio between the period of the amplitude variation and that of the external force is in general incommensurable, there is no periodicity in the almost periodic oscillation.

This chapter deals with almost periodic oscillations which occur in self-oscillatory systems under periodic excitation [25, 41, 78, 93]. In the preceding chapter we have investigated the phenomenon of frequency entrainment in such systems and obtained the regions of entrainment (Fig. 12.7). If the amplitude and frequency of the external force are given inside these regions, the entrainment occurs at the harmonic (i.e., fundamental), higher-harmonic, or subharmonic frequency of the external force. If the external force is prescribed outside these regions, one may expect the occurrence of an almost periodic oscillation.

Almost periodic oscillations have been studied hitherto by assuming them as combination oscillations which consist of two simple harmonic components, one with the natural frequency of the system and the other with the driving frequency.[1] When the amplitude and frequency of the external force are given just between but not near the regions of entrainment, this approximate method of analysis furnishes a fairly good description of the almost periodic oscillation. However, if the external force is prescribed close to the regions of entrainment, the analysis[2] does

---

[1] See, for instance, Ref. 100, pp. 163–187.

[2] The accuracy of the analysis might conceivably be improved if higher-harmonic components of the almost periodic oscillation were taken into account. By using the perturbation method, computation was actually carried out along this line up to the second-order approximation. However, the results turned out to be less satisfactory than those obtained in this chapter.

not account for it very well, since the waveform of the almost periodic oscillation differs considerably from that obtained as a sum of two simple harmonic oscillations. We shall be interested in the almost periodic oscillation which occurs under such circumstances. The method of analysis in what follows is useful for the study of almost periodic oscillations as well as for the study of entrained oscillations.

13.2 Van der Pol's Equation with Forcing Term

We again consider the system governed by Eq. (12.1), that is,

$$\frac{d^2u}{dt^2} - \epsilon(1 - u^2)\frac{du}{dt} + u = B\cos vt + B_0 \tag{13.1}$$

where ϵ is a small positive constant, and $B\cos vt + B_0$ represents an external force as before. Introducing a new variable defined by $v = u - B_0$, we once more write

$$\frac{d^2v}{dt^2} - \mu(1 - \beta v - \gamma v^2)\frac{dv}{dt} + v = B\cos vt \tag{13.2}$$

where $\mu = (1 - B_0^2)\epsilon$ $\beta = \dfrac{2B_0}{1 - B_0^2}$ and $\gamma = \dfrac{1}{1 - B_0^2}$

In the preceding chapter the solution of Eqs. (13.2) was assumed, for the entrained oscillation, to take the form of Eq. (12.3) or (12.4). Then it would be natural to consider that, for the almost periodic oscillation, the coefficients b_1 and b_2 in Eqs. (12.3) and (12.4) are not constant but vary slowly with the time t. Therefore, an almost periodic oscillation which develops from the harmonic oscillation may be expressed by[1]

$$v(t) = b_1(t)\sin vt + b_2(t)\cos vt \tag{13.3}$$

Similarly, an almost periodic oscillation which develops from the higher-harmonic or subharmonic oscillation may be expressed by

$$v(t) = \frac{B}{1 - v^2}\cos vt + b_1(t)\sin nvt + b_2(t)\cos nvt \tag{13.4}$$

We now derive the relations which will determine the time-varying coefficients $b_1(t)$ and $b_2(t)$ in the above solutions. To begin with, let us consider an almost periodic oscillation which develops from the harmonic oscillation. Substituting Eq. (13.3) into (13.2) and equating the coeffi-

---

[1] This type of oscillation occurs when the amplitude B and the frequency v of the external force vary from the interior to the exterior of the region of harmonic entrainment (Fig. 12.7).

cients of the terms containing sin νt and cos νt separately to zero leads to

$$\frac{dx_1}{d\tau} = (1 - r_1{}^2)x_1 - \sigma_1 y_1 + \frac{B}{\mu\nu a_0} \equiv X_1(x_1, y_1)$$

$$\frac{dy_1}{d\tau} = \sigma_1 x_1 + (1 - r_1{}^2)y_1 \equiv Y_1(x_1, y_1)$$

(13.5)

where $\qquad x_1 = \dfrac{b_1}{a_0} \qquad y_1 = \dfrac{b_2}{a_0} \qquad r_1{}^2 = x_1{}^2 + y_1{}^2$

(13.6)

$$a_0 = \sqrt{\frac{4}{\gamma}} \qquad \tau = \frac{\mu}{2}t \qquad \sigma_1 = \frac{1 - \nu^2}{\mu\nu} \quad \text{detuning}$$

In the derivation of Eqs. (13.5) the assumptions have been made that $b_1(t)$ and $b_2(t)$ are slowly varying functions of t, so that d^2b_1/dt^2 and d^2b_2/dt^2 may be neglected, and that, since μ is a small quantity, $\mu\, db_1/dt$ and $\mu\, db_2/dt$ are also discarded. Equations (13.5) play an important role in the present investigation, since they serve as the fundamental equations in studying the almost periodic oscillation which develops from the harmonic oscillation. The entrained oscillation may also be obtained by equating $dx_1/d\tau = 0$ and $dy_1/d\tau = 0$, or by solving $X_1(x_1,y_1) = 0$ and $Y_1(x_1,y_1) = 0$ in the above equations.[1]

By the same procedure as above, almost periodic oscillations which develop from the higher-harmonic and subharmonic oscillations may also be studied. That is, substituting Eq. (13.4) into (13.2) and equating the coefficients of the terms containing sin $n\nu t$ and cos $n\nu t$ separately to zero leads to the fundamental equations for the almost periodic oscillations in what follows.

$n = 2$: $\qquad \dfrac{dx_2}{d\tau} = (D - r_2{}^2)x_2 - \sigma_2 y_2 \equiv X_2(x_2, y_2)$

$$\frac{dy_2}{d\tau} = \sigma_2 x_2 + (D - r_2{}^2)y_2 - \frac{\beta}{4a_0}A^2 \equiv Y_2(x_2, y_2)$$

(13.7)

$n = 3$: $\qquad \dfrac{dx_3}{d\tau} = (D - r_3{}^2)x_3 - \sigma_3 y_3 \equiv X_3(x_3, y_3)$

$$\frac{dy_3}{d\tau} = \sigma_3 x_3 + (D - r_3{}^2)y_3 - \frac{\gamma}{12a_0}A^3 \equiv Y_3(x_3, y_3)$$

(13.8)

$n = \tfrac{1}{2}$: $\qquad \dfrac{dx_{\frac{1}{2}}}{d\tau} = \left(D - r_{\frac{1}{2}}{}^2 + \dfrac{\beta}{2}A\right)x_{\frac{1}{2}} - \sigma_{\frac{1}{2}}y_{\frac{1}{2}} \equiv X_{\frac{1}{2}}(x_{\frac{1}{2}}, y_{\frac{1}{2}})$

$$\frac{dy_{\frac{1}{2}}}{d\tau} = \sigma_{\frac{1}{2}}x_{\frac{1}{2}} + \left(D - r_{\frac{1}{2}}{}^2 - \frac{\beta}{2}A\right)y_{\frac{1}{2}} \equiv Y_{\frac{1}{2}}(x_{\frac{1}{2}}, y_{\frac{1}{2}})$$

(13.9)

$n = \tfrac{1}{3}$: $\dfrac{dx_{\frac{1}{3}}}{d\tau} = (D - r_{\frac{1}{3}}{}^2)x_{\frac{1}{3}} - \sigma_{\frac{1}{3}}y_{\frac{1}{3}} + \dfrac{2A}{a_0}x_{\frac{1}{3}}y_{\frac{1}{3}} \equiv X_{\frac{1}{3}}(x_{\frac{1}{3}}, y_{\frac{1}{3}})$

$$\frac{dy_{\frac{1}{3}}}{d\tau} = \sigma_{\frac{1}{3}}x_{\frac{1}{3}} + (D - r_{\frac{1}{3}}{}^2)y_{\frac{1}{3}} + \frac{A}{a_0}(x_{\frac{1}{3}}{}^2 - y_{\frac{1}{3}}{}^2) \equiv Y_{\frac{1}{3}}(x_{\frac{1}{3}}, y_{\frac{1}{3}})$$

(13.10)

[1] The result, as it should, agrees with Eqs. (12.5).

where
$$x_n = \frac{b_1}{a_0} \qquad y_n = \frac{b_2}{a_0} \qquad r_n{}^2 = x_n{}^2 + y_n{}^2$$

$$a_0 = \sqrt{\frac{4}{\gamma}} \qquad A = \frac{B}{1 - \nu^2} \qquad D = 1 - \frac{2A^2}{a_0{}^2} \tag{13.11}$$

$$\tau = \frac{\mu}{2}\, t \qquad \sigma_n = \frac{1 - (n\nu)^2}{\mu n \nu} \quad \text{detuning}$$

Equations (13.7) and (13.8) determine the coefficients $b_1(t)$ and $b_2(t)$ of the almost periodic oscillations which develop from the second- and third-harmonic oscillations, respectively, while Eqs. (13.9) and (13.10) are related to the almost periodic oscillations which develop from the sub-harmonic oscillations of order ½ and of order ⅓, respectively. The entrained oscillations may readily be obtained by equating $dx_n/d\tau = 0$ and $dy_n/d\tau = 0$, or by solving $X_n(x_n,y_n) = 0$ and $Y_n(x_n,y_n) = 0$ in the above equations.

13.3 Almost Periodic Oscillations Which Develop from Harmonic Oscillations

(a) Limit Cycles Correlated with Almost Periodic Oscillations

As mentioned in Sec. 13.2, an almost periodic oscillation which develops from the harmonic oscillation may be expressed by Eq. (13.3), that is,

$$v(t) = b_1(t) \sin \nu t + b_2(t) \cos \nu t \tag{13.12}$$

where $b_1(t)$ and $b_2(t)$ are slowly varying functions of t. These time-varying coefficients, or $x_1(\tau)$ and $y_1(\tau)$ in normalized form, are to be found from Eqs. (13.5). To this end we consider the behavior of integral curves defined by

$$\frac{dy_1}{dx_1} = \frac{Y_1(x_1,y_1)}{X_1(x_1,y_1)} \tag{13.13}$$

in the $x_1 y_1$ plane. It is known that, if the external force is given just outside the region of harmonic entrainment, Eq. (13.13) has only one singularity that is unstable. It is also seen that, for large values of x_1 and y_1, integral curves are directed toward the origin for increasing τ. Therefore the existence of a stable limit cycle (that is, a limit cycle toward which neighboring integral curves tend as τ increases) may be concluded. Hence, the representative point $(x_1(\tau),y_1(\tau))$ keeps on moving along the limit cycle, resulting in the continuous variations in $b_1(t)$ and $b_2(t)$ or in the amplitude and phase of the harmonic oscillation. The period of these variations is equal to that required for the representative point to complete one revolution along the limit cycle. We shall not go into the problem in which the detuning becomes large, so that the natural fre-

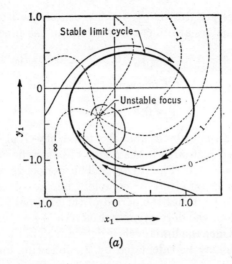

(a)

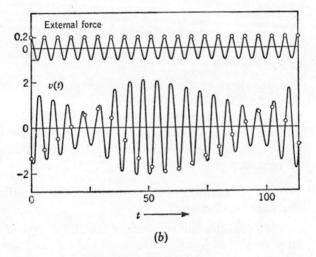

(b)

FIGURE 13.1   Almost periodic oscillation which develops
from the harmonic oscillation. (a) Limit
cycle representing the almost periodic
oscillation. (b) Waveform of the almost
periodic oscillation.

quency of the system is entrained by a higher-harmonic or subharmonic
frequency. Even though the entrainment of such a type would not
occur, an almost periodic oscillation should preferably be expressed in a
different form than Eq. (13.12), as one will see in Secs. 13.5 and 13.6.

An example of the limit cycle is shown in Fig. 13.1a. The integral curves are drawn by making use of the isocline method. The system parameters are the same as those in the preceding chapter, that is,

$$\epsilon = 0.2 \quad \text{and} \quad B_0 = 0.5$$

in Eq. (13.1). Consequently, the parameters in Eqs. (13.2) are

$$\mu = 0.15 \quad \beta = \tfrac{2}{3} \quad \text{and} \quad \gamma = \tfrac{2}{3}$$

The amplitude B and the frequency ν of the external force are prescribed just outside the region of harmonic entrainment, that is, $B = 0.2$ and $\nu = 1.1$ (Fig. 12.7b). The period required for the representative point $(x_1(\tau), y_1(\tau))$ to complete one revolution along the limit cycle is $12.8 \cdot \cdot \cdot$ times the period of the external force, so that the assumptions made in the derivation of Eqs. (13.5) are permissible.

Once the limit cycle is determined in the $x_1 y_1$ plane, the time-response curve may be calculated by the following line integral:

$$\tau = \int \frac{ds}{\sqrt{X_1{}^2(x_1, y_1) + Y_1{}^2(x_1, y_1)}}$$

where   (13.14)

$$ds = \sqrt{(dx_1)^2 + (dy_1)^2} \quad \text{line element along the limit cycle}$$

The waveform of the almost periodic oscillation obtained in this way is shown in Fig. 13.1b. The amplitude and phase of the almost periodic oscillation vary slowly but periodically. The points on the curves appear at the beginning of every cycle of the external force. One sees that the phase of the almost periodic oscillation gradually lags the external force as t increases.

(b) *Transition between Entrained Oscillations and Almost Periodic Oscillations*

We consider the behavior of integral curves of Eq. (13.13), particularly in the case where the amplitude B and the frequency ν of the external force are given near the boundary of harmonic entrainment. As mentioned in Sec. 12.3b, the boundary of harmonic entrainment is given by

$$(1 - r_1{}^2)(1 - 3r_1{}^2) + \sigma_1{}^2 = 0 \quad \text{or} \quad 2r_1{}^2 - 1 = 0 \quad (13.15)$$

The first equation applies in the case where the amplitude B and consequently the detuning σ_1 are comparatively small. Typical examples of integral curves in such a case are shown in Fig. 13.2a, b, and c. These figures show the integral curves under the conditions that B and ν of the external force are prescribed inside, on the boundary of, and outside the region of harmonic entrainment, respectively. As will be discussed in

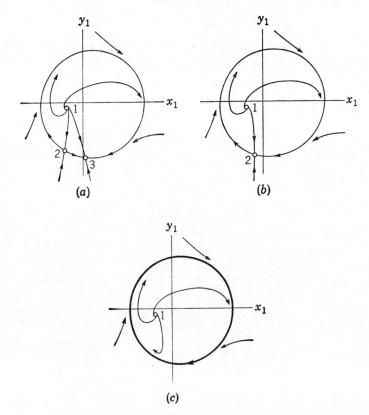

FIGURE 13.2 Transition between entrained oscillation and almost
periodic oscillation for small detuning. (*a*) Singularities representing the harmonic oscillations. (*b*)
Coalescence of singularities at the boundary of entrainment. (*c*) Limit cycle representing the almost
periodic oscillation.

Sec. 13.4*a*, the coalescence of singular points occurs at the boundary of
harmonic entrainment.

When the amplitude B and consequently the detuning σ_1 are large,
the second equation of (13.15) applies.[1] Typical examples of integral
curves in such a case are shown in Fig. 13.3. At the boundary of harmonic entrainment the only singularity is a stable focus, as will be verified
in Sec. 13.4*b*. Slight increase in the detuning beyond the boundary will

---

[1] For intermediate values of B and σ_1, some complicated phenomena may occur
[14; 100, p. 184]. In order to deal with this case precisely, we have to refer to the
original equation (13.1) instead of (13.5). However, we shall not enter into this
problem here.

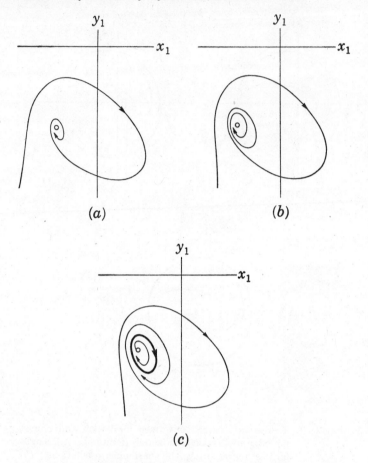

FIGURE 13.3   Transition between entrained oscillation and almost periodic oscillation for large detuning. (*a*) Singularity representing the harmonic oscillation. (*b*) Singularity of the focus type at the boundary of entrainment. (*c*) Limit cycle representing the almost periodic oscillation.

result in the occurrence of a stable limit cycle which is small in size, however.   The limit cycle grows large as the detuning increases.[1]

The difference in the behavior of integral curves between Figs. 13.2

---

[1] The shape of the limit cycle looks like a circle when the detuning is increased. Under this condition the representative point moves along the limit cycle rather quickly; therefore, the assumption made in the derivation of Eqs. (13.5) is not appropriate.   In this case the almost periodic oscillation may well be approximated by a sum of two simple harmonic oscillations [18, p. 213; 100, p. 166].

and 13.3 is explained as follows: An almost periodic oscillation may be considered as a combination of two components, i.e., the free oscillation with the natural frequency of the system and the forced oscillation with the driving frequency. When the entrainment of frequency occurs, the situation in what follows may arise. If the amplitude B of the external force is small, the forced oscillation is not predominant. Since the detuning σ_1 is also small in this case, the free oscillation is entrained by the driving frequency (Fig. 13.2). On the other hand, if B is large, the free oscillation is suppressed by the forced one (Fig. 13.3).

13.4 Geometrical Discussion of Integral Curves at the Boundary of Harmonic Entrainment

(a) Coalescence of Singular Points

Following the method of analysis due to Bendixson [6], we investigate the nature of the singular point 2 in Fig. 13.2*b*. In this case the singularity lies on the boundary $q = 0$ in Fig. 12.1; its coordinates are, from Eqs. (12.8) and the first equation of (13.15), given by

$$x_{10} = - \frac{\mu \nu a_0}{9B} (1 + 3\sigma_1{}^2 - \sqrt{1 - 3\sigma_1{}^2})$$

$$y_{10} = \frac{(1 - \nu^2)a_0}{3B} (2 + \sqrt{1 - 3\sigma_1{}^2}) \tag{13.16}$$

where the amplitude B and the frequency ν of the external force are related by

$$\left(\frac{B}{\mu \nu a_0}\right)^2 = \frac{2}{27} [1 + 9\sigma_1{}^2 - (1 - 3\sigma_1{}^2)^{\frac{3}{2}}] \tag{13.17}$$

Transferring the origin to the singular point, i.e., putting

$$x_1 = x_{10} + \xi \qquad y_1 = y_{10} + \eta$$

gives, from Eqs. (13.5),

$$\frac{d\xi}{d\tau} = a_1\xi + a_2\eta + X_1(\xi,\eta) \qquad \frac{d\eta}{d\tau} = b_1\xi + b_2\eta + Y_1(\xi,\eta)$$

where
$$\begin{aligned}
a_1 &= a_2 = 0 \\
b_1 &= -2x_{10}y_{10} + \sigma_1 \qquad b_2 = 1 - x_{10}{}^2 - 3y_{10}{}^2 \\
X_1(\xi,\eta) &= -3x_{10}\xi^2 - 2y_{10}\xi\eta - x_{10}\eta^2 - (\xi^2 + \eta^2)\xi \\
Y_1(\xi,\eta) &= -y_{10}\xi^2 - 2x_{10}\xi\eta - 3y_{10}\eta^2 - (\xi^2 + \eta^2)\eta
\end{aligned} \tag{13.18}$$

When $X_1(\xi,\eta)$ and $Y_1(\xi,\eta)$ are ignored, the characteristic equation of the system (13.18) is given by

$$\begin{vmatrix} a_1 - \lambda & a_2 \\ b_1 & b_2 - \lambda \end{vmatrix} = 0$$

from which we obtain

$$\lambda_1 = 0 \qquad \lambda_2 = b_2$$

so that the singular point with which we are dealing is not simple in character.

By proceeding in much the same way as we proceeded in Sec. 8.6b, we may determine the type of singular point and plot the integral curves near the singularity. The result is as follows: When the singularity lies on the boundary $q = 0$ of the harmonic entrainment, the coalescence of a stable node with a saddle generally occurs. An approximate solution of Eqs. (13.18) near the singularity is given by

$$\eta = \left(-\frac{b_1}{b_2} - \frac{b}{a} \xi + Ce^{-a/\xi} \right) \xi$$

with

$$a = \frac{-b_2{}^3}{b_1{}^2 x_{10} - 2b_1 b_2 y_{10} + 3b_2{}^2 x_{10}} \qquad (13.19)$$

$$b = \frac{(b_1{}^2 + b_2{}^2)(b_1 x_{10} + b_2 y_{10})}{b_2(b_1{}^2 x_{10} - 2b_1 b_2 y_{10} + 3b_2{}^2 x_{10})}$$

where C is a constant of integration. However, if the singularity is located at the points where the boundary $q = 0$ has vertical tangency, a nodal point results. This case occurs when $\sigma_1{}^2 = \frac{1}{3}$ and $(B/\mu\nu a_0)^2 = \frac{8}{27}$. An approximate solution of Eqs. (13.18) for $\sigma_1 = -1/\sqrt{3}$ is given by

$$\eta = (-\sqrt{3} + 6\sqrt{2}\,\xi - 48\sqrt{3}\,\xi^2 + C'\xi e^{-1/12\xi^2})\xi \qquad (13.20)$$

where C' is a constant of integration. Figure 13.4a and b shows the integral curves represented by Eqs. (13.19) and (13.20), respectively. The representative point moves along the integral curves in the direction of arrows. Thus the singularity is unstable in Fig. 13.4a but stable in Fig. 13.4b.

(b) Existence of a Stable Focus

Let us consider the nature of the singular point in Fig. 13.3b. In this case the singularity lies on the boundary $p = 0$ of Fig. 12.1; its coordinates are, from Eqs. (12.8) and the second equation of (13.15), given by

$$x_{10} = -\frac{\mu\nu a_0}{4B} \qquad y_{10} = \frac{(1 - \nu^2)a_0}{2B} \qquad (13.21)$$

It is noted that $|\sigma_1|$ is generally greater than 0.5. Transferring the origin

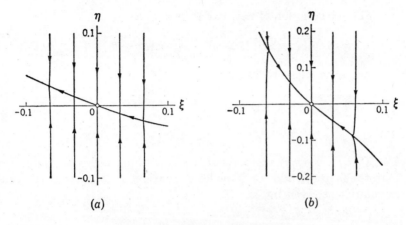

$$(a) \qquad\qquad (b)$$

FIGURE 13.4 Coalescence of singular points at the boundary of harmonic entrainment. (*a*) Node-saddle distribution of integral curves near the singularity. (*b*) Nodal distribution of integral curves near the singularity.

to the singular point, i.e., putting

$$x_1 = x_{10} + \xi \qquad y_1 = y_{10} + \eta$$

gives us, from Eqs. (13.5),

$$\frac{d\xi}{d\tau} = a_1\xi + a_2\eta + X_1(\xi,\eta) \qquad \frac{d\eta}{d\tau} = b_1\xi + b_2\eta + Y_1(\xi,\eta)$$

where
$$a_1 = -\frac{1}{2} + \frac{a_0{}^2}{2A^2} \qquad a_2 = \left(1 - \frac{a_0{}^2}{A^2}\right)\sigma_1$$

$$b_1 = \left(3 - \frac{a_0{}^2}{A^2}\right)\sigma_1 \qquad b_2 = \frac{1}{2} - \frac{a_0{}^2}{2A^2} \tag{13.22}$$

$$X_1(\xi,\eta) = -3x_{10}\xi^2 - 2y_{10}\xi\eta - x_{10}\eta^2 - (\xi^2 + \eta^2)\xi$$
$$Y_1(\xi,\eta) = -y_{10}\xi^2 - 2x_{10}\xi\eta - 3y_{10}\eta^2 - (\xi^2 + \eta^2)\eta$$

When $X_1(\xi,\eta)$ and $Y_1(\xi,\eta)$ are ignored, the characteristic equation of the system (13.22) is given by

$$\begin{vmatrix} a_1 - \lambda & a_2 \\ b_1 & b_2 - \lambda \end{vmatrix} = 0$$

Since $\qquad (a_1 - b_2)^2 + 4a_2b_1 = 1 - 4\sigma_1{}^2 < 0 \qquad a_1 + b_2 = 0$

the roots λ are imaginary, so that the singularity is either a center or a focus (Sec. 2.2*a*). Following the analysis due to Poincaré [82, 84], we investigate the type of singularity in what follows (Sec. V.2).

The introduction of new variables defined by

$$x = \sqrt{a_1 b_2 - a_2 b_1}\,(\xi + \eta) \qquad y = (a_1 + b_1)\xi + (a_2 + b_2)\eta$$

gives us, from Eqs. (13.22),

$$\frac{dx}{dz} = y + X_2(x,y) \equiv X(x,y) \qquad \frac{dy}{dz} = -x + Y_2(x,y) \equiv Y(x,y) \tag{13.23}$$

where[1]
$$z = \sqrt{a_1 b_2 - a_2 b_1}\,\tau = \sqrt{\sigma_1^2 - \tfrac{1}{4}}\,\tau$$

and $X_2(x,y)$, $Y_2(x,y)$ are polynomials containing terms of degree higher than the first in x and y. Now let us consider a closed curve around the singularity as given by

$$F(x,y) = k$$

where k is a small positive constant. If the function $F(x,y)$ could be so constructed that $\partial F/\partial x\, X + \partial F/\partial y\, Y = 0$, the singularity is a center; while if $\partial F/\partial x\, X + \partial F/\partial y\, Y$ is either positive or negative in the neighborhood of the singularity, the singularity is a focus.

In Fig. 13.3*b* the external force is so chosen that $B = 0.2$ and $\nu = 1.069$, which are located on the boundary of harmonic entrainment. The other system parameters are the same as in Sec. 13.3*a*. Then we obtain, from Eqs. (13.21),

$$x_{10} = -0.347 \qquad \text{and} \qquad y_{10} = -0.616$$

By following the procedure described in Sec. V.2, we obtain

$$F(x,y) = F_2(x,y) + F_3(x,y) + F_4(x,y) = k$$
where
$$F_2(x,y) = x^2 + y^2$$
$$F_3(x,y) = -1.26x^3 + 6.89x^2 y - 0.33xy^2 + 6.50y^3$$
$$F_4(x,y) = -11.86x^4 - 12.87x^3 y - 6.26xy^3 + 11.01y^4$$

In the neighborhood of the singularity we have

$$\frac{dF}{dz} = \frac{\partial F}{\partial x} X + \frac{\partial F}{\partial y} Y = -6.92(x^2 + y^2)^2 + \phi_5(x,y) < 0$$

where $\phi_5(x,y)$ is a polynomial containing terms of degree higher than the fourth in x and y. Hence the integral curves of Eqs. (13.23) cross the closed curve $F(x,y) = k$ from the exterior to the interior with increasing z. Therefore, we may conclude that, in the original $x_1 y_1$ plane, the singularity $x_1 = x_{10}$, $y_1 = y_{10}$ is a stable focus.

[1] It is noted that z and τ have the same sign, since $\sigma_1^2 - \tfrac{1}{4} > 0$.

13.5 Almost Periodic Oscillations Which Develop from Higher-harmonic Oscillations

(a) Limit Cycles Correlated with Almost Periodic Oscillations

When the driving frequency ν takes a value which is in the neighborhood of $\frac{1}{2}$ or $\frac{1}{3}$, the free oscillation of the system is entrained by the second- or third-harmonic frequency of the external force, respectively (Fig. 12.7a). Almost periodic oscillations which develop from these higher-harmonic oscillations may be expressed by

$$v(t) = \frac{B}{1 - \nu^2} \cos \nu t + b_1(t) \sin n\nu t + b_2(t) \cos n\nu t \qquad n = 2 \text{ or } 3 \quad (13.24)$$

where $b_1(t)$ and $b_2(t)$ are slowly varying functions of t. These coefficients, or $x_n(\tau)$ and $y_n(\tau)$ in normalized form, are to be found from Eqs. (13.7) and (13.8). We consider the behavior of integral curves in the $x_n y_n$ plane defined by

$$\frac{dy_n}{dx_n} = \frac{Y_n(x_n, y_n)}{X_n(x_n, y_n)} \qquad (13.25)$$

It is known that, if the external force is given outside the regions of higher-harmonic entrainment, Eq. (13.25) has only one unstable singularity. It is also seen that, for large values of x_n and y_n, integral curves are directed toward the origin for increasing τ. Therefore, the existence of a stable limit cycle and hence of an almost periodic oscillation may be concluded.

Figure 13.5a shows an example of the limit cycle of Eq. (13.25) for $n = 2$. In this case the corresponding almost periodic oscillation develops from the second-harmonic oscillation. The amplitude B and the frequency ν of the external force are given by $B = 0.8$ and $\nu = 0.47$, which are located just outside the region of the second-harmonic entrainment (Fig. 12.7a). The other system parameters are the same as before. The period required for the representative point to complete one revolution along the limit cycle is $7.64 \cdots$ times the period of the external force. Figure 13.5b shows the waveform of the almost periodic oscillation represented by the limit cycle. In Fig. 13.6a is plotted the limit cycle of Eq. (13.25) for $n = 3$; the corresponding almost periodic oscillation develops from the third-harmonic oscillation. The amplitude B and the frequency ν of the external force are given by $B = 1.0$ and $\nu = 0.345$, which are located just outside the region of the third-harmonic entrainment (Fig. 12.7a). The period required for the representative point to complete one revolution along the limit cycle is $10.4 \cdots$ times the period of the external force. Figure 13.6b shows the waveform of the almost periodic oscillation represented by the limit cycle.

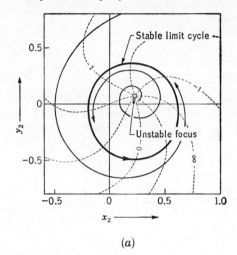

(a)

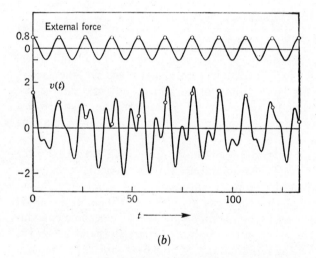

(b)

FIGURE 13.5   Almost periodic oscillation which develops
from the second-harmonic oscillation.   (a)
Limit cycle representing the almost periodic
oscillation.   (b) Waveform of the almost
periodic oscillation.

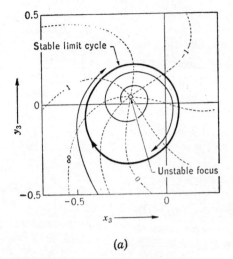

(a)

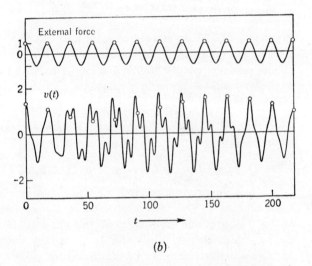

(b)

FIGURE 13.6   Almost periodic oscillation which develops
from the third-harmonic oscillation. (a)
Limit cycle representing the almost periodic
oscillation. (b) Waveform of the almost
periodic oscillation.

(b) *Transition between Entrained Oscillations and Almost Periodic Oscillations*

Equations (13.7) and (13.8) may be combined and written in the form

$$\frac{dx_n}{d\tau} = (D - r_n{}^2)x_n - \sigma_n y_n \qquad \frac{dy_n}{d\tau} = \sigma_n x_n + (D - r_n{}^2)y_n - F_n \quad (13.26)$$

where F_n is a constant dependent on the system parameters, and $n = 2$ or 3. The introduction of new variables defined by

$$x_n = \sqrt{D}\, y \qquad y_n = -\sqrt{D}\, x \qquad r_n{}^2 = Dr^2$$
$$\sigma_n = D\sigma \qquad \tau = \frac{z}{D} \tag{13.27}$$

yields, from Eqs. (13.26),

$$\frac{dx}{dz} = (1 - r^2)x - \sigma y + \frac{F_n}{D^{\frac{3}{2}}} \qquad \frac{dy}{dz} = \sigma x + (1 - r^2)y \quad (13.28)$$

Equations (13.28) have the same form as (13.5). Hence the behavior of integral curves during the transition between entrained oscillations and almost periodic oscillations is analogous to that occurring between harmonic and almost periodic oscillations.

13.6 Almost Periodic Oscillations Which Develop from Subharmonic Oscillations

(a) *Limit Cycles Correlated with Almost Periodic Oscillations*

When the driving frequency ν takes a value which is in the neighborhood of 2 or 3, the free oscillation of the system is entrained by the $\frac{1}{2}$- or $\frac{1}{3}$-harmonic frequency of the external force, respectively (Fig. 12.7b). Almost periodic oscillations which develop from these subharmonic oscillations may be expressed by

$$v(t) = \frac{B}{1 - \nu^2} \cos \nu t + b_1(t) \sin n\nu t + b_2(t) \cos n\nu t \qquad n = \tfrac{1}{2} \text{ or } \tfrac{1}{3} \tag{13.29}$$

where $b_1(t)$ and $b_2(t)$ are slowly varying functions of t. These coefficients, or $x_n(\tau)$ and $y_n(\tau)$ in normalized form, are to be found from Eqs. (13.9) and (13.10). We consider the behavior of integral curves in the $x_n y_n$ plane defined by

$$\frac{dy_n}{dx_n} = \frac{Y_n(x_n, y_n)}{X_n(x_n, y_n)} \tag{13.30}$$

Analogously as in the preceding sections, it may be concluded that, if the external force is given outside the regions of subharmonic entrainment, Eq. (13.30) possesses a limit cycle correlated with an almost periodic oscillation.

The graphical solution of Eq. (13.30) may conveniently be obtained by introducing polar coordinates $x_n = r_n \cos \theta_n$, $y_n = r_n \sin \theta_n$. Thus we write, from Eqs. (13.9) and (13.10),

$$\frac{dr_{\frac{1}{2}}}{d\tau} = (D - r_{\frac{1}{2}}{}^2)r_{\frac{1}{2}} + \frac{\beta}{2} A r_{\frac{1}{2}} \cos 2\theta_{\frac{1}{2}}$$

$$\frac{d\theta_{\frac{1}{2}}}{d\tau} = \sigma_{\frac{1}{2}} - \frac{\beta}{2} A \sin 2\theta_{\frac{1}{2}}$$

(13.31)

and

$$\frac{dr_{\frac{1}{3}}}{d\tau} = (D - r_{\frac{1}{3}}{}^2)r_{\frac{1}{3}} + \frac{A}{a_0} r_{\frac{1}{3}}{}^2 \sin 3\theta_{\frac{1}{3}}$$

$$\frac{d\theta_{\frac{1}{3}}}{d\tau} = \sigma_{\frac{1}{3}} + \frac{A}{a_0} r_{\frac{1}{3}} \cos 3\theta_{\frac{1}{3}}$$

(13.32)

Equations (13.31) are unchanged if $\theta_{\frac{1}{2}}$ is replaced by $\theta_{\frac{1}{2}} + \pi$, while Eqs. (13.32) are unchanged if $\theta_{\frac{1}{3}}$ is replaced by $\theta_{\frac{1}{3}} + \frac{2}{3}\pi$. This implies that integral curves of (13.31) are π symmetric about the origin and those of (13.32) are $\frac{2}{3}\pi$ symmetric. Let ϕ_n be the angle with which the integral curve traverses the radius vector; then

$$\tan \phi_{\frac{1}{2}} = r_{\frac{1}{2}} \frac{d\theta_{\frac{1}{2}}}{dr_{\frac{1}{2}}} = \frac{\sigma_{\frac{1}{2}} - (\beta/2)A \sin 2\theta_{\frac{1}{2}}}{D - r_{\frac{1}{2}}{}^2 + (\beta/2)A \cos 2\theta_{\frac{1}{2}}}$$

and

$$\tan \phi_{\frac{1}{3}} = r_{\frac{1}{3}} \frac{d\theta_{\frac{1}{3}}}{dr_{\frac{1}{3}}} = \frac{\sigma_{\frac{1}{3}} + (A/a_0)r_{\frac{1}{3}} \cos 3\theta_{\frac{1}{3}}}{D - r_{\frac{1}{3}}{}^2 + (A/a_0)r_{\frac{1}{3}} \sin 3\theta_{\frac{1}{3}}}$$

The graphical construction of integral curves is much facilitated by drawing iso-ϕ curves instead of isoclines.

Figure 13.7a shows an example of the limit cycle of Eqs. (13.31). Some of the iso-ϕ curves are shown dashed in the figure. The external force is prescribed just outside the region of $\frac{1}{2}$-harmonic entrainment, that is, $B = 4.0$ and $\nu = 2.15$ (Fig. 12.7b). The other system parameters are the same as before. The period T required for the representative point to complete one revolution along the limit cycle is readily obtained from the second equation of (13.31), that is,

$$T = \int_0^{2\pi} \frac{d\theta_{\frac{1}{2}}}{|\sigma_{\frac{1}{2}}| - (\beta/2)A \sin 2\theta_{\frac{1}{2}}} = \frac{2\pi}{\sqrt{\sigma_{\frac{1}{2}}{}^2 - (\beta^2/4)A^2}}$$

(13.33)

It is worth noting that in this particular case the limit cycle is also obtained analytically by constructing a Liapunov function of the form

$$V = ax_{\frac{1}{2}}{}^2 + 2bx_{\frac{1}{2}}y_{\frac{1}{2}} + cy_{\frac{1}{2}}{}^2$$

(13.34)

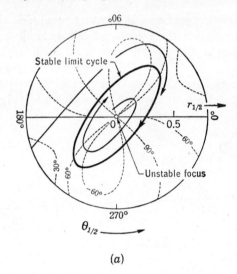

(a)

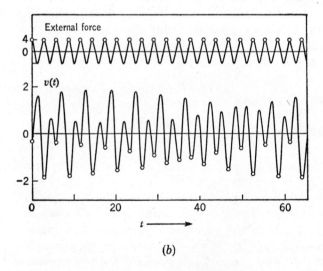

(b)

FIGURE 13.7   Almost periodic oscillation which develops
from the subharmonic oscillation of order
½.   (a) Limit cycle representing the almost
periodic oscillation.   (b) Waveform of the
almost periodic oscillation.

Following the procedure as described in Sec. 3.4, we determine the coefficients a, b, and c under the condition that $dV/d\tau = 2(x_{\frac{1}{2}}^2 + y_{\frac{1}{2}}^2)$. The result is

$$a = \frac{1}{\Delta}\left[\left(D - \frac{\beta}{2}A\right)D + \sigma_{\frac{1}{2}}^2\right]$$

$$b = -\frac{1}{\Delta}\frac{\beta}{2}A\sigma_{\frac{1}{2}}$$

$$c = \frac{1}{\Delta}\left[\left(D + \frac{\beta}{2}A\right)D + \sigma_{\frac{1}{2}}^2\right]$$

(13.35)

where

$$\Delta = \left(D^2 - \frac{\beta^2}{4}A^2 + \sigma_{\frac{1}{2}}^2\right)D$$

Then the use of Eqs. (13.9) gives us

$$\frac{dV}{d\tau} = \frac{\partial V}{\partial x_{\frac{1}{2}}}\frac{dx_{\frac{1}{2}}}{d\tau} + \frac{\partial V}{\partial y_{\frac{1}{2}}}\frac{dy_{\frac{1}{2}}}{d\tau} = 2r_{\frac{1}{2}}^2(1 - V)$$

so that the ellipse[1] $V = 1$ represents a stable limit cycle to which neighboring integral curves tend asymptotically with increasing τ. Using the values of the system parameters mentioned above, we obtain, from Eq. (13.33), $T = 10.07 \cdot\cdot\cdot$ or equal to 45.9 $\cdot\cdot\cdot$ times the period of the external force. An analytical expression for the limit cycle of Fig. 13.7a, derived from Eqs. (13.34) and (13.35), is

$$13.9x_{\frac{1}{2}}^2 - 17.9x_{\frac{1}{2}}y_{\frac{1}{2}} + 10.4y_{\frac{1}{2}}^2 = 1$$

The waveform of the almost periodic oscillation, calculated from the limit cycle, is shown in Fig. 13.7b.

In Fig. 13.8a is shown an example of the limit cycle of Eqs. (13.32). The external force is prescribed just outside the region of $\frac{1}{3}$-harmonic entrainment, that is, $B = 6.7$ and $\nu = 2.85$ (Fig. 12.7b). The period required for the representative point to complete one revolution along the limit cycle is 67.1 $\cdot\cdot\cdot$ times the period of the external force. In Fig. 13.8b is shown the waveform of the almost periodic oscillation represented by the limit cycle.

(b) *Transition between Entrained Oscillations and Almost Periodic Oscillations*

We investigate the behavior of integral curves of Eq. (13.30) when the external force is given near the boundary of subharmonic entrainment. We first deal with Eq. (13.30) for $n = \frac{1}{2}$. Three typical cases

[1] It is not difficult to show that $a > 0$, $c > 0$, and $ac > b^2$; hence V is positive definite.

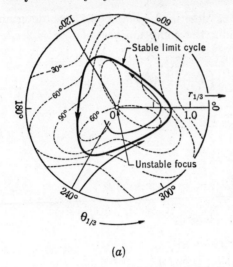

(a)

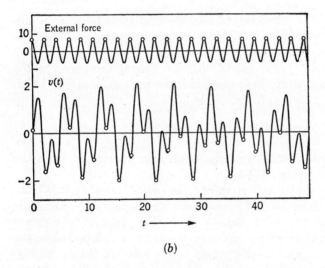

(b)

FIGURE 13.8   Almost periodic oscillation which develops
from the subharmonic oscillation of order
⅓. (a) Limit cycle representing the almost
periodic oscillation. (b) Waveform of the
almost periodic oscillation.

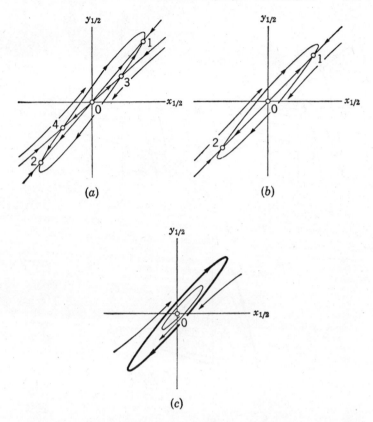

FIGURE 13.9 Transition between entrained oscillation and almost periodic oscillation. (*a*) Singularities representing the subharmonic oscillations of order ½. (*b*) Coalescence of singularities at the boundary of entrainment. (*c*) Limit cycle representing the almost periodic oscillation.

of integral curves sketched in Fig. 13.9, namely, *a*, *b*, and *c*, are such that *B* and *ν* of the external force are prescribed inside, on the boundary of, and outside the region of ½-harmonic entrainment, respectively. Singularities 1 and 2 in Fig. 13.9*a* are stable nodes correlated with entrained oscillations. Singularities 3 and 4 are saddles, which are intrinsically unstable. The origin is an unstable node. Therefore, any integral curve will tend to either 1 or 2, depending on the location of the initial point from which the integral curve has started. In Fig. 13.9*b* two singular points coalesce to form a singularity of a higher order. Such singularities, 1 and 2, are of the node-saddle type.[1] The representative

---

[1] The nature of the singularities may be investigated in the same manner as in Secs. 8.6*b* and 13.4*a*.

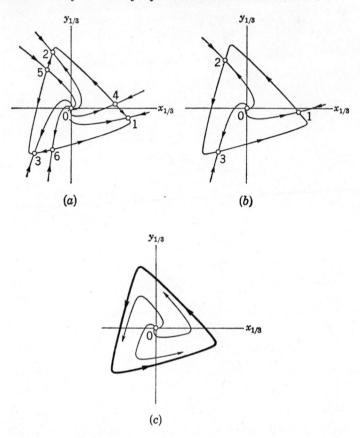

FIGURE 13.10   Transition between entrained oscillation and almost
periodic oscillation. (*a*) Singularities representing
the subharmonic oscillations of order ⅓. (*b*)
Coalescence of singularities at the boundary of
entrainment. (*c*) Limit cycle representing the almost
periodic oscillation.

point on the closed trajectory may stay for a moment at one of these
singularities, but when it is affected by a slight disturbance, the point
will move to the other singularity in the direction of the arrows. The
limit cycle in Fig. 13.9*c* is correlated with an almost periodic oscillation.

By use of the above procedure, integral curves of Eq. (13.30) for
$n = ⅓$ are plotted in Fig. 13.10. The singularities are equidistant from
and equiangular about the origin. In Fig. 13.10*a* singularities 1, 2, and 3
are correlated with entrained oscillations of the ⅓-harmonic frequency
and singularities 4, 5, 6, and the origin are unstable. In Fig. 13.10*b* the

coalescence of singular points occurs and results in the node-saddle distribution of integral curves near the singularities. The limit cycle in Fig. 13.10c is correlated with an almost periodic oscillation.

13.7 Self-oscillatory System with Nonlinear Restoring Force [25]

(a) Fundamental Equations

We consider an almost periodic oscillation which occurs in a system governed by

$$\frac{d^2v}{dt^2} - \mu(1 - \gamma v^2)\frac{dv}{dt} + v^3 = B \cos \nu t \tag{13.36}$$

In Sec. 12.8 we discussed the harmonic entrainment in the same system. For the entrained oscillation the solution was assumed to take the form of Eq. (12.41). Hence, an almost periodic oscillation which develops from the harmonic oscillation may be expressed by

$$v(t) = b_1(t) \sin \nu t + b_2(t) \cos \nu t \tag{13.37}$$

where the coefficients $b_1(t)$ and $b_2(t)$ are slowly varying functions of the time t.

By a procedure analogous to that of Sec. 13.2, we derive the relations which determine the time-varying coefficients $b_1(t)$ and $b_2(t)$ in the above solution. We thus substitute Eq. (13.37) into (13.36) and equate the coefficients of the terms containing $\sin \nu t$ and $\cos \nu t$ separately to zero. The result is

$$\frac{dx_1}{d\tau} = (1 - r_1{}^2)x_1 - \sigma_1 y_1 + \frac{B}{\mu\nu a_0} \equiv X_1(x_1, y_1)$$

$$\frac{dy_1}{d\tau} = \sigma_1 x_1 + (1 - r_1{}^2)y_1 \equiv Y_1(x_1, y_1) \tag{13.38}$$

where
$$x_1 = \frac{b_1}{a_0} \qquad y_1 = \frac{b_2}{a_0} \qquad r_1{}^2 = x_1{}^2 + y_1{}^2$$

$$a_0 = \sqrt{\frac{4}{\gamma}} \qquad \omega_0 = \sqrt{\frac{3}{\gamma}} \qquad \tau = \frac{\mu}{2}t \tag{13.39}$$

$$\sigma_1 = \frac{\omega_0{}^2 r_1{}^2 - \nu^2}{\mu\nu} \quad \text{detuning}$$

It should, however, be remembered that the assumptions of Sec. 13.2 must be made for the derivation of Eqs. (13.38). These equations serve as fundamental equations in studying the almost periodic oscillation which develops from the harmonic oscillation. The entrained oscillation may also be obtained by equating $dx_1/d\tau = 0$ and $dy_1/d\tau = 0$ in Eqs. (13.38).

(b) *Limit Cycle Correlated with Almost Periodic Oscillations*

We consider the behavior of integral curves of

$$\frac{dy_1}{dx_1} = \frac{Y_1(x_1,y_1)}{X_1(x_1,y_1)} \tag{13.40}$$

in the x_1y_1 plane, where $X_1(x_1,y_1)$ and $Y_1(x_1,y_1)$ are given by Eqs. (13.38). We are concerned, for the time being, with the singular points of Eq. (13.40) which are correlated with entrained oscillations. The nature of the singularity (x_{10},y_{10}) may be found by solving the characteristic equation (see Sec. 2.2a)

$$\begin{vmatrix} a_1 - \lambda & a_2 \\ b_1 & b_2 - \lambda \end{vmatrix} = 0$$

with
$$a_1 = \left(\frac{\partial X_1}{\partial x_1}\right)_0 = 1 - 3x_{10}^2 - y_{10}^2 - 2\frac{\omega_0^2}{\mu\nu}x_{10}y_{10}$$

$$a_2 = \left(\frac{\partial X_1}{\partial y_1}\right)_0 = -2x_{10}y_{10} - \sigma_1 - 2\frac{\omega_0^2}{\mu\nu}y_{10}^2 \tag{13.41}$$

$$b_1 = \left(\frac{\partial Y_1}{\partial x_1}\right)_0 = -2x_{10}y_{10} + \sigma_1 + 2\frac{\omega_0^2}{\mu\nu}x_{10}^2$$

$$b_2 = \left(\frac{\partial Y_1}{\partial y_1}\right)_0 = 1 - x_{10}^2 - 3y_{10}^2 + 2\frac{\omega_0^2}{\mu\nu}x_{10}y_{10}$$

where $(\partial X_1/\partial x_1)_0, \ldots, (\partial Y_1/\partial y_1)_0$ denote the values of $\partial X_1/\partial x_1, \ldots,$ $\partial Y_1/\partial y_1$ at $x_1 = x_{10}$ and $y_1 = y_{10}$. By making use of the Routh-Hurwitz criterion, the conditions for stability of the singular point are given by

$$a_1 + b_2 < 0 \qquad a_1b_2 - a_2b_1 > 0 \tag{13.42}$$

Substituting Eqs. (13.41) into (13.42) gives the same result as already obtained in Sec. 12.8b.

We now turn to our present investigation. When an almost periodic oscillation occurs in the system described by Eq. (13.36), the coefficients $b_1(t)$ and $b_2(t)$ in Eq. (13.37), or $x_1(\tau)$ and $y_1(\tau)$ in Eqs. (13.39), vary periodically with a large period. Thus, as before, the limit cycle of Eq. (13.40), if it exists, is correlated with an almost periodic oscillation. It depends upon the system parameters whether Eq. (13.40) has a limit cycle or not.[1] We show in what follows an example of the integral curves which tend either to a singular point or to a limit cycle.

---

[1] The occurrence and extinction of a limit cycle, as the system parameters vary, will be discussed in Sec. 13.7c.

Table 13.1   Singular Points in Fig. 13.11

Singular point	x_{10}	y_{10}	λ_1, λ_2	μ_1, μ_2*	Classification
1	-0.086	-0.546	$0.39 \pm 3.80i$		Unstable focus
2	0.536	-1.212	$0.87, -5.89$	$2.80,\ \ 0.09$	Saddle (unstable)
3	1.571	-0.255	$-1.11, -7.02$	$4.77, -2.54$	Stable node

* μ_1, μ_2 are the tangential directions of the integral curves at the singular points (Sec. 2.2a).

NUMERICAL EXAMPLE

Let us consider a case in which the parameters of Eq. (13.36) are given by

$$\mu = 0.2 \qquad \gamma = 8 \qquad B = 0.35 \qquad \text{and} \qquad \nu = 1$$

Substituting these values into Eqs. (12.44) and (12.45) locates the singular points in the $x_1 y_1$ plane. Their stability may be decided by the use of (12.54). The result is listed in Table 13.1.

Contrary to the case illustrated in Fig. 8.2 (Sec. 8.4), the singularity 1 becomes unstable, because, as one sees from the condition (12.50), the system has a negative damping for values of $r_1{}^2$ less than 0.5. By making use of Eqs. (2.28) and (2.29), we obtain, for the integral curves in the neighborhood of the singularity (in polar coordinates r_1 and θ_1),

$$r_1(\theta_1 + 2\pi) = 0.525 r_1(\theta_1)$$

From the conditions (2.30) we see that $d\theta_1/d\tau < 0$. Hence the representative point $(r_1(\tau), \theta_1(\tau))$ moves along the integral curve in the clockwise direction and leaves the singularity as τ increases.

By making use of the isocline method, we seek the integral curves of Eq. (13.40) for the above values of the system parameters. The result is shown in Fig. 13.11 in terms of the original variables b_1 and b_2.* We see in the figure a stable limit cycle which encircles the unstable focus 1. One of the integral curves (drawn by thick line in the figure) that contains the saddle point 2, that is, the separatrix, divides the whole plane into two regions; in one of them all integral curves tend to the limit cycle, in the other, to the nodal point 3. Thus both the almost periodic and entrained oscillations may occur in this system; which ones do occur will depend on the initial condition.

Figure 13.12 shows the waveform of the almost periodic oscillation represented by the limit cycle of Fig. 13.11. The period required for

* It is noted that $x_1 = \sqrt{2}\, b_1$ and $y_1 = \sqrt{2}\, b_2$.

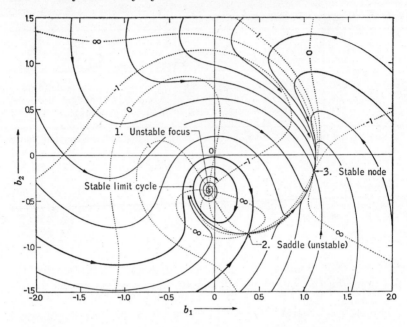

FIGURE 13.11   Integral curves for the harmonic oscillation with limit cycle.
Full line: integral curves.   Dotted line: isoclines.

the representative point to complete one revolution along the limit cycle, that is, the period of amplitude variation of the almost periodic oscillation, is $2\pi \times 3.426 \cdots$. Since the limit cycle does not contain the origin in its interior, the oscillation is synchronized with the external force.

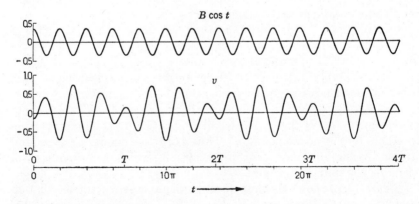

FIGURE 13.12   Almost periodic oscillation represented by the limit cycle of Fig. 13.11.

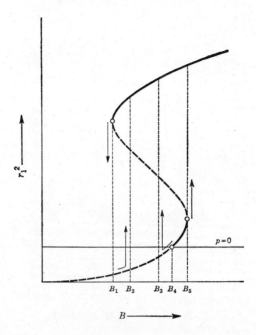

FIGURE 13.13   Amplitude characteristic of the harmonic oscillation, showing the transition between entrained and almost periodic oscillations.

(c) *Transition between Entrained Oscillations and Almost Periodic Oscillations*

As mentioned before, the existence of a limit cycle of Eq. (13.40) depends on the system parameters. However, the situation is much more complicated than that in a system with linear restoring force. In the preceding investigation, where the system is described by Eqs. (13.2), an almost periodic oscillation generally occurs when the system parameters are prescribed outside the region of entrainment.[1] On the other hand, in a system with nonlinear restoring force, both entrained and almost periodic oscillations may occur for certain values of the system

---

[1] For certain values of the system parameters, particularly in the neighborhood of $(B/\mu\nu a_0)^2 = \frac{8}{27}$ and $|\sigma_1| = 0.5$ (Fig. 12.1), there exists a harmonic oscillation as well as an almost periodic oscillation (for the same values of the system parameters). However, such a range of the system parameters is extremely limited; therefore, it is overlooked in the preceding analysis.

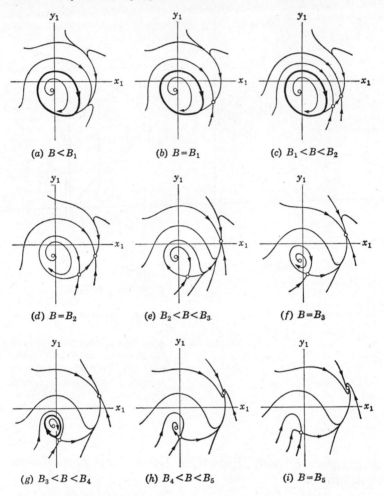

FIGURE 13.14   Integral curves of Eq. (13.40) for various values of B.

parameters, as one might expect from the skewed form of the resonance curves (see Fig. 12.13).[1]

Equation (12.44), which may also be derived from Eqs. (13.38) by equating both $dx_1/d\tau$ and $dy_1/d\tau$ to zero, furnishes the relation between B, ν, and r_1. If ν is kept constant, we obtain the amplitude characteristic (B versus r_1^2) as illustrated in Fig. 13.13. From the stability conditions (12.54) for the harmonic oscillation, one readily sees that the dashed portions of the characteristic curve are unstable.

[1] An example of such a system is given in Fig. 13.11.

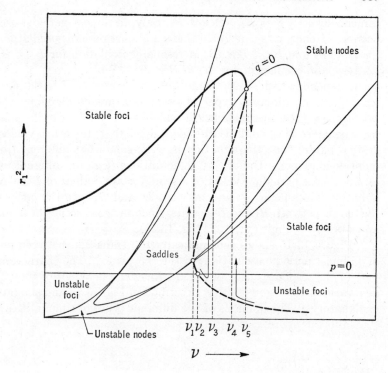

FIGURE 13.15 Response curve of the harmonic oscillation, showing the transition between entrained and almost periodic oscillations.

The occurrence and extinction of a limit cycle, as B varies, is most easily understood by plotting the integral curves of Eq. (13.40) for various values of B. The result is shown in Fig. 13.14. When B is given below B_1 of Fig. 13.13, there is a limit cycle which encircles an unstable focal point. This state is shown in Fig. 13.14a. A coalescence of singularities occurs at $B = B_1$. Further increase in B results in the coexistence of a stable limit cycle and a stable node. When $B = B_2$, there is a closed integral curve starting from the saddle point and coming back to the same point. The limit cycle disappears when B is increased beyond B_2. However, when B reaches B_3, the integral curves show the same behavior as in Fig. 13.14d (see Fig. 13.14f). A limit cycle appears once more for $B_3 < B < B_4$ and is reduced to a stable focus at $B = B_4$ (cf. Fig. 13.3b). For values of B between B_4 and B_5, there are two types of harmonic oscillations, i.e., the resonant and nonresonant oscillations, but there is no limit cycle. The coalescence of singularities occurs at $B = B_5$. When B is decreased, the resonant oscillation is sustained down to $B = B_1$.   Below B_1, only an almost periodic oscillation occurs.   Hence, it is concluded that an almost periodic oscillation occurs for $B < B_2$ and $B_3 < B < B_4$, a resonant oscillation for $B_1 < B$, and a nonresonant oscillation for $B_4 < B < B_5$.   Which types of oscillation occur depends on the initial condition.   The region of initial conditions leading to the different types of oscillation are bordered by the separatrices, i.e., the integral curves which tend to the saddle point with increasing τ.   It should also be mentioned that, for other values of the system parameters, the situation may be somewhat different from that mentioned above.   Under certain conditions (e.g., for different values of ν), an almost periodic oscillation occurs for any values of B below B_4 or only for values below B_1 and between B_3 and B_4.   In the latter case no hysteresis between B_1 and B_2 results, since the coalescence of singularities appears on the limit cycle (cf. Fig. 13.2b).

Figure 13.15 shows schematically the transition between entrained and almost periodic oscillations in the $\nu r_1{}^2$ plane.   The characteristic is a part of the response curve of Fig. 12.13.   The behavior of the transition as ν varies is analogous to that described in Fig. 13.13 and should be self-explanatory.   The domains of the different types of singularities are also shown in the figure.

Appendix I: **Expansions of the Mathieu Functions**

Mathieu's equation [see Eq. (3.60)] has the periodic solutions of period π or 2π corresponding to the characteristic number of a for a given q. Following Mathieu [73, 74] and Whittaker [112, 113] these periodic solutions, i.e., the Mathieu functions, are developed into power series of q when $|q|$ is small. Thus we write some expansions for the Mathieu functions in what follows.

First, for $ce_0(\tau,q)$, we have

$$ce_0(\tau,q) = 1 + 4q\cos 2\tau + 2q^2\cos 4\tau + q^3(-28\cos 2\tau + \tfrac{4}{9}\cos 6\tau)$$
$$+ q^4(-\tfrac{160}{9}\cos 4\tau + \tfrac{1}{18}\cos 8\tau) + \cdots$$

for which the associated characteristic number is given by

$$a_{c0} = -32q^2 + 224q^4 - \frac{2^{10}\cdot 29}{9}q^6 + \cdots$$

Similarly, for other Mathieu functions, we have

$$ce_1(\tau,q) = \cos\tau + q\cos 3\tau + q^2(-\cos 3\tau + \tfrac{1}{3}\cos 5\tau)$$
$$+ q^3(\tfrac{1}{3}\cos 3\tau - \tfrac{4}{9}\cos 5\tau + \tfrac{1}{18}\cos 7\tau)$$
$$+ q^4(1\tfrac{1}{9}\cos 3\tau + \tfrac{1}{6}\cos 5\tau - \tfrac{1}{12}\cos 7\tau + \tfrac{1}{180}\cos 9\tau) + \cdots$$

$$se_1(\tau,q) = \sin\tau + q\sin 3\tau + q^2(\sin 3\tau + \tfrac{1}{3}\sin 5\tau)$$
$$+ q^3(\tfrac{1}{3}\sin 3\tau + \tfrac{4}{9}\sin 5\tau + \tfrac{1}{18}\sin 7\tau)$$
$$+ q^4(-1\tfrac{1}{9}\sin 3\tau + \tfrac{1}{6}\sin 5\tau + \tfrac{1}{12}\sin 7\tau + \tfrac{1}{180}\sin 9\tau) + \cdots$$

$$ce_2(\tau,q) = \cos 2\tau + q(-2 + \tfrac{2}{3}\cos 4\tau) + \tfrac{1}{6}q^2\cos 6\tau$$
$$+ q^3(4\tfrac{0}{3} + \tfrac{43}{27}\cos 4\tau + \tfrac{1}{45}\cos 8\tau)$$
$$+ q^4(293\tfrac{3}{540}\cos 6\tau + \tfrac{1}{540}\cos 10\tau) + \cdots$$

$$se_2(\tau,q) = \sin 2\tau + \tfrac{2}{3}q\sin 4\tau + \tfrac{1}{6}q^2\sin 6\tau - q^3(\tfrac{5}{27}\sin 4\tau - \tfrac{1}{45}\sin 8\tau)$$
$$+ q^4(-\tfrac{37}{540}\sin 6\tau + \tfrac{1}{540}\sin 10\tau) + \cdots$$

$$ce_3(\tau,q) = \cos 3\tau - q(\cos \tau - \tfrac{1}{2} \cos 5\tau) + q^2(\cos \tau + \tfrac{1}{10} \cos 7\tau)$$
$$- q^3(\tfrac{1}{2} \cos \tau - \tfrac{7}{40} \cos 5\tau - \tfrac{1}{90} \cos 9\tau)$$
$$+ q^4(- \cos \tau - \tfrac{1}{4} \cos 5\tau + \tfrac{17}{360} \cos 7\tau + \tfrac{1}{1260} \cos 11\tau) + \cdots$$
$$se_3(\tau,q) = \sin 3\tau - q(\sin \tau - \tfrac{1}{2} \sin 5\tau) + q^2(- \sin \tau + \tfrac{1}{10} \sin 7\tau)$$
$$- q^3(\tfrac{1}{2} \sin \tau - \tfrac{7}{40} \sin 5\tau - \tfrac{1}{90} \sin 9\tau)$$
$$+ q^4(\sin \tau + \tfrac{1}{4} \sin 5\tau + \tfrac{17}{360} \sin 7\tau + \tfrac{1}{1260} \sin 11\tau) + \cdots$$

The characteristic numbers a_{cn} and a_{sn} associated, respectively, with the Mathieu functions $ce_n(\tau,q)$ and $se_n(\tau,q)$ were previously given by Eqs. (3.64).

Appendix II: **Unstable Solutions of Hill's Equation**

As mentioned in Sec. 4.2, the stability conditions (4.6) and (4.9) were calculated to the first approximation, since the characteristic exponent μ was given by Eq. (3.86). However, a closer evaluation of the characteristic exponent will be necessary when the stability condition of a higher-order approximation is desired. Therefore, by proceeding in the same manner as described in the case of Mathieu's equation (Sec. 3.6b), we shall seek the unstable solutions for Hill's equation of the form

$$\frac{d^2x}{d\tau^2} + \left(\theta_0 + 2 \sum_{\nu=1}^{4} \theta_\nu \cos 2\nu\tau \right) x = 0 \qquad (\text{II.1})$$

Following Whittaker [112; 113, p. 424], the general solution of Eq. (II.1) is given by

$$x = c_1 e^{\mu\tau} \phi(\tau,\sigma) + c_2 e^{-\mu\tau} \phi(\tau,-\sigma) \qquad (\text{II.2})$$

where c_1 and c_2 are arbitrary constants. Then, we obtain the expansions for μ and θ_0 in what follows.[1]

II.1 *Unbounded Solution Associated with First Unstable Region*

The characteristic exponent μ is given by

$$
\begin{aligned}
\mu = {}& \tfrac{1}{2}\theta_1 \sin 2\sigma + \tfrac{1}{8}\theta_1\theta_2 \sin 2\sigma + \tfrac{1}{24}\theta_2\theta_3 \sin 2\sigma - \tfrac{3}{128}\theta_1{}^3 \sin 2\sigma \\
& + \tfrac{1}{128}\theta_1{}^2\theta_2 \sin 4\sigma + \tfrac{5}{384}\theta_1{}^2\theta_3 \sin 2\sigma + \tfrac{7}{288}\theta_1\theta_2{}^2 \sin 2\sigma \\
& + \tfrac{17}{4608}\theta_1\theta_3{}^2 \sin 2\sigma + \tfrac{1}{96}\theta_2{}^2\theta_3 \sin 2\sigma \\
& \qquad\qquad\qquad\qquad + \tfrac{1}{1152}\theta_1\theta_2\theta_3 \sin 4\sigma + \cdots \qquad (\text{II.3})
\end{aligned}
$$

---

[1] The computation is carried out up to terms of third degree in θ's.

in which the parameter σ is to be determined by

$$\theta_0 = 1 + \theta_1 \cos 2\sigma + (-\tfrac{1}{4} + \tfrac{1}{8} \cos 4\sigma)\theta_1{}^2 - \tfrac{1}{6}\theta_2{}^2 - \tfrac{1}{16}\theta_3{}^2$$
$$+ \tfrac{1}{4}\theta_1\theta_2 \cos 2\sigma + \tfrac{1}{12}\theta_2\theta_3 \cos 2\sigma - \tfrac{1}{64}\theta_1{}^3 \cos 2\sigma$$
$$+ (-\tfrac{11}{192} + \tfrac{5}{64} \cos 4\sigma)\theta_1{}^2\theta_2 + \tfrac{5}{192}\theta_1{}^2\theta_3 \cos 2\sigma$$
$$- \tfrac{1}{144}\theta_1\theta_2{}^2 \cos 2\sigma - \tfrac{1}{2304}\theta_1\theta_3{}^2 \cos 2\sigma$$
$$+ \tfrac{1}{48}\theta_2{}^2\theta_3 \cos 2\sigma + (-\tfrac{13}{192} + \tfrac{13}{576} \cos 4\sigma)\theta_1\theta_2\theta_3 + \cdots \quad \text{(II.4)}$$

II.2 Unbounded Solution Associated with Second Unstable Region

Similarly, the expansions for μ and θ_0 are given by

$$\mu = \tfrac{1}{4}\theta_2 \sin 2\sigma - \tfrac{1}{16}\theta_1{}^2 \sin 2\sigma + \tfrac{1}{24}\theta_1\theta_3 \sin 2\sigma + \tfrac{1}{64}\theta_2\theta_4 \sin 2\sigma$$
$$- \tfrac{3}{4096}\theta_2{}^3 \sin 2\sigma + (-\tfrac{1}{72} \sin 2\sigma + \tfrac{1}{128} \sin 4\sigma)\theta_1{}^2\theta_2$$
$$+ \tfrac{7}{2304}\theta_1{}^2\theta_4 \sin 2\sigma + \tfrac{1}{4096}\theta_2{}^2\theta_4 \sin 4\sigma + \tfrac{7}{1800}\theta_2\theta_3{}^2 \sin 2\sigma$$
$$+ \tfrac{7}{9216}\theta_2\theta_4{}^2 \sin 2\sigma + \tfrac{5}{2304}\theta_3{}^2\theta_4 \sin 2\sigma$$
$$+ (-\tfrac{5}{384} \sin 2\sigma + \tfrac{1}{576} \sin 4\sigma)\theta_1\theta_2\theta_3$$
$$- \tfrac{11}{1920}\theta_1\theta_3\theta_4 \sin 2\sigma + \cdots \quad \text{(II.5)}$$

$$\theta_0 = 4 + \theta_2 \cos 2\sigma + (\tfrac{1}{6} - \tfrac{1}{4} \cos 2\sigma)\theta_1{}^2$$
$$+ (-\tfrac{1}{16} + \tfrac{1}{32} \cos 4\sigma)\theta_2{}^2 - \tfrac{1}{10}\theta_3{}^2 - \tfrac{1}{24}\theta_4{}^2 + \tfrac{1}{6}\theta_1\theta_3 \cos 2\sigma$$
$$+ \tfrac{1}{16}\theta_2\theta_4 \cos 2\sigma - \tfrac{1}{1024}\theta_2{}^3 \cos 2\sigma$$
$$+ (\tfrac{1}{12} - \tfrac{1}{9} \cos 2\sigma + \tfrac{1}{64} \cos 4\sigma)\theta_1{}^2\theta_2 + \tfrac{7}{576}\theta_1{}^2\theta_4 \cos 2\sigma$$
$$+ (-\tfrac{11}{3072} + \tfrac{5}{1024} \cos 4\sigma)\theta_2{}^2\theta_4 - \tfrac{1}{225}\theta_2\theta_3{}^2 \cos 2\sigma$$
$$- \tfrac{1}{2304}\theta_2\theta_4{}^2 \cos 2\sigma + \tfrac{5}{576}\theta_3{}^2\theta_4 \cos 2\sigma$$
$$+ (\tfrac{1}{20} - \tfrac{5}{96} \cos 2\sigma + \tfrac{5}{288} \cos 4\sigma)\theta_1\theta_2\theta_3$$
$$+ (\tfrac{1}{360} - \tfrac{11}{480} \cos 2\sigma)\theta_1\theta_3\theta_4 + \cdots \quad \text{(II.6)}$$

II.3 Unbounded Solution Associated with Third Unstable Region

$$\mu = \tfrac{1}{6}\theta_3 \sin 2\sigma - \tfrac{1}{24}\theta_1\theta_2 \sin 2\sigma + \tfrac{1}{48}\theta_1\theta_4 \sin 2\sigma + \tfrac{1}{384}\theta_1{}^3 \sin 2\sigma$$
$$- \tfrac{1}{10368}\theta_3{}^3 \sin 2\sigma - \tfrac{7}{1536}\theta_1{}^2\theta_3 \sin 2\sigma - \tfrac{1}{2400}\theta_2{}^2\theta_3 \sin 2\sigma$$
$$+ \tfrac{23}{18816}\theta_3\theta_4{}^2 \sin 2\sigma + \tfrac{1}{384}\theta_1\theta_2\theta_3 \sin 4\sigma - \tfrac{1}{320}\theta_1\theta_2\theta_4 \sin 2\sigma$$
$$+ \tfrac{1}{1536}\theta_1\theta_3\theta_4 \sin 4\sigma - \tfrac{5}{1728}\theta_2\theta_3\theta_4 \sin 2\sigma + \cdots \quad \text{(II.7)}$$

$$\theta_0 = 9 + \theta_3 \cos 2\sigma + \tfrac{1}{16}\theta_1{}^2 + \tfrac{1}{10}\theta_2{}^2 + (-\tfrac{1}{36} + \tfrac{1}{72} \cos 4\sigma)\theta_3{}^2$$
$$- \tfrac{1}{14}\theta_4{}^2 - \tfrac{1}{4}\theta_1\theta_2 \cos 2\sigma + \tfrac{1}{8}\theta_1\theta_4 \cos 2\sigma + \tfrac{1}{64}\theta_1{}^3 \cos 2\sigma$$
$$- \tfrac{1}{5184}\theta_3{}^3 \cos 2\sigma - \tfrac{3}{160}\theta_1{}^2\theta_2 - \tfrac{9}{256}\theta_1{}^2\theta_3 \cos 2\sigma$$
$$- \tfrac{9}{400}\theta_2{}^2\theta_3 \cos 2\sigma + \tfrac{3}{140}\theta_2{}^2\theta_4 - \tfrac{9}{3136}\theta_3\theta_4{}^2 \cos 2\sigma$$
$$+ (\tfrac{121}{2880} + \tfrac{5}{576} \cos 4\sigma)\theta_1\theta_2\theta_3 - \tfrac{3}{160}\theta_1\theta_2\theta_4 \cos 2\sigma$$
$$+ (\tfrac{293}{16128} + \tfrac{17}{2304} \cos 4\sigma)\theta_1\theta_3\theta_4$$
$$- \tfrac{5}{288}\theta_2\theta_3\theta_4 \cos 2\sigma + \cdots \quad \text{(II.8)}$$

The periodic functions $\phi(\tau,\sigma)$ and $\phi(\tau,-\sigma)$ in Eq. (II.2) are also obtained for the respective unstable regions. They have, however, no direct concern with the stability criterion, so that the full details of the expansions of these periodic functions are omitted here.[1]

[1] It is noted that the expansions of the periodic functions up to terms of second degree in θ's may be obtained by putting $\epsilon_\nu = 0$ in Appendix III.

Appendix III: **Unstable Solutions for the Extended Form of Hill's Equation**

We shall calculate the unstable solutions of Eq. (3.83). Taking the alternative form

$$\frac{d^2x}{d\tau^2} + \left[\theta_0 + 2 \sum_{\nu=1}^{4} \theta_\nu \cos (2\nu\tau - \epsilon_\nu) \right] x = 0 \qquad \text{(III.1)}$$

we write the general solution as

$$x = c_1 e^{\mu\tau} \phi(\tau,\sigma_1) + c_2 e^{-\mu\tau} \phi(\tau,\sigma_2) \qquad \text{(III.2)}$$

where c_1 and c_2 are arbitrary constants. The parameters σ_1, σ_2 will be determined from the expansions of θ_0 in what follows.

III.1 Unbounded Solution Associated with First Unstable Region

$$\begin{aligned}
\mu = \tfrac{1}{2}\theta_1 \sin 2\sigma &+ \tfrac{1}{8}\theta_1\theta_2 \sin (2\sigma + 2\epsilon_1 - \epsilon_2) \\
&+ \tfrac{1}{24}\theta_2\theta_3 \sin (2\sigma + \epsilon_1 + \epsilon_2 - \epsilon_3) \\
&+ \tfrac{1}{48}\theta_3\theta_4 \sin (2\sigma + \epsilon_1 + \epsilon_3 - \epsilon_4) + \cdots \quad \text{(III.3)}
\end{aligned}$$

$$\begin{aligned}
\theta_0 = 1 + \theta_1 \cos 2\sigma &+ (-\tfrac{1}{4} + \tfrac{1}{8} \cos 4\sigma)\theta_1{}^2 - \tfrac{1}{6}\theta_2{}^2 - \tfrac{1}{16}\theta_3{}^2 \\
&- \tfrac{1}{30}\theta_4{}^2 + \tfrac{1}{4}\theta_1\theta_2 \cos (2\sigma + 2\epsilon_1 - \epsilon_2) \\
&+ \tfrac{1}{12}\theta_2\theta_3 \cos (2\sigma + \epsilon_1 + \epsilon_2 - \epsilon_3) \\
&+ \tfrac{1}{24}\theta_3\theta_4 \cos (2\sigma + \epsilon_1 + \epsilon_3 - \epsilon_4) + \cdots \quad \text{(III.4)}[1]
\end{aligned}$$

$$\begin{aligned}
\phi(\tau,\sigma) = \sin (\tau - \tfrac{1}{2}\epsilon_1 - \sigma) &+ A_1\theta_1 + A_2\theta_2 + A_3\theta_3 + A_4\theta_4 \\
&+ B_1\theta_1{}^2 + B_2\theta_2{}^2 + B_3\theta_3{}^2 + B_4\theta_4{}^2 + B_{12}\theta_1\theta_2 + B_{13}\theta_1\theta_3 \\
&+ B_{14}\theta_1\theta_4 + B_{23}\theta_2\theta_3 + B_{24}\theta_2\theta_4 + B_{34}\theta_3\theta_4 + \cdots \quad \text{(III.5)}
\end{aligned}$$

[1] Solving for σ, we obtain two principal values σ_1 and σ_2.

where

$A_1 = \frac{1}{8} \sin (3\tau - \frac{3}{2}\epsilon_1 - \sigma)$

$A_2 = -\frac{1}{8} \cos 2\sigma \sin (3\tau + \frac{1}{2}\epsilon_1 - \epsilon_2 - \sigma)$
$\quad - \frac{1}{8} \sin 2\sigma \cos (3\tau + \frac{1}{2}\epsilon_1 - \epsilon_2 - \sigma) + \frac{1}{24} \sin (5\tau - \frac{1}{2}\epsilon_1 - \epsilon_2 - \sigma)$

$A_3 = -\frac{1}{24} \cos 2\sigma \sin (5\tau + \frac{1}{2}\epsilon_1 - \epsilon_3 - \sigma)$
$\quad - \frac{1}{24} \sin 2\sigma \cos (5\tau + \frac{1}{2}\epsilon_1 - \epsilon_3 - \sigma) + \frac{1}{48} \sin (7\tau - \frac{1}{2}\epsilon_1 - \epsilon_3 - \sigma)$

$A_4 = -\frac{1}{48} \cos 2\sigma \sin (7\tau + \frac{1}{2}\epsilon_1 - \epsilon_4 - \sigma)$
$\quad - \frac{1}{48} \sin 2\sigma \cos (7\tau + \frac{1}{2}\epsilon_1 - \epsilon_4 - \sigma) + \frac{1}{80} \sin (9\tau - \frac{1}{2}\epsilon_1 - \epsilon_4 - \sigma)$

$B_1 = \frac{1}{64} \cos 2\sigma \sin (3\tau - \frac{3}{2}\epsilon_1 - \sigma) + \frac{3}{64} \sin 2\sigma \cos (3\tau - \frac{3}{2}\epsilon_1 - \sigma)$
$\quad\quad\quad\quad\quad\quad\quad + \frac{1}{192} \sin (5\tau - \frac{5}{2}\epsilon_1 - \sigma)$

$B_2 = -\frac{1}{384} \cos 2\sigma \sin (7\tau + \frac{1}{2}\epsilon_1 - 2\epsilon_2 - \sigma)$
$\quad\quad\quad - \frac{1}{384} \sin 2\sigma \cos (7\tau + \frac{1}{2}\epsilon_1 - 2\epsilon_2 - \sigma)$
$\quad\quad\quad\quad\quad + \frac{1}{1920} \sin (9\tau - \frac{1}{2}\epsilon_1 - 2\epsilon_2 - \sigma)$

$B_3 = -\frac{1}{2880} \cos 2\sigma \sin (11\tau + \frac{1}{2}\epsilon_1 - 2\epsilon_3 - \sigma)$
$\quad\quad\quad - \frac{1}{2880} \sin 2\sigma \cos (11\tau + \frac{1}{2}\epsilon_1 - 2\epsilon_3 - \sigma)$
$\quad\quad\quad\quad\quad + \frac{1}{8064} \sin (13\tau - \frac{1}{2}\epsilon_1 - 2\epsilon_3 - \sigma)$

$B_4 = -\frac{1}{10752} \cos 2\sigma \sin (15\tau + \frac{1}{2}\epsilon_1 - 2\epsilon_4 - \sigma)$
$\quad\quad\quad - \frac{1}{10752} \sin 2\sigma \cos (15\tau + \frac{1}{2}\epsilon_1 - 2\epsilon_4 - \sigma)$
$\quad\quad\quad\quad\quad + \frac{1}{23040} \sin (17\tau - \frac{1}{2}\epsilon_1 - 2\epsilon_4 - \sigma)$

$B_{12} = (\frac{1}{48} - \frac{1}{32} \cos 4\sigma) \sin (3\tau + \frac{1}{2}\epsilon_1 - \epsilon_2 - \sigma)$
$\quad\quad - \frac{1}{32} \sin 4\sigma \cos (3\tau + \frac{1}{2}\epsilon_1 - \epsilon_2 - \sigma)$
$\quad\quad - \frac{1}{288} \cos 2\sigma \sin (5\tau - \frac{1}{2}\epsilon_1 - \epsilon_2 - \sigma)$
$\quad\quad + \frac{1}{288} \sin 2\sigma \cos (5\tau - \frac{1}{2}\epsilon_1 - \epsilon_2 - \sigma)$
$\quad\quad\quad\quad\quad + \frac{1}{288} \sin (7\tau - \frac{3}{2}\epsilon_1 - \epsilon_2 - \sigma)$

$B_{13} = -\frac{1}{48} \cos 2\sigma \sin (3\tau + \frac{3}{2}\epsilon_1 - \epsilon_3 - \sigma)$
$\quad\quad - \frac{1}{48} \sin 2\sigma \cos (3\tau + \frac{3}{2}\epsilon_1 - \epsilon_3 - \sigma)$
$\quad\quad + (\frac{5}{1152} - \frac{1}{192} \cos 4\sigma) \sin (5\tau + \frac{1}{2}\epsilon_1 - \epsilon_3 - \sigma)$
$\quad\quad - \frac{1}{192} \sin 4\sigma \cos (5\tau + \frac{1}{2}\epsilon_1 - \epsilon_3 - \sigma)$
$\quad\quad - \frac{1}{2304} \cos 2\sigma \sin (7\tau - \frac{1}{2}\epsilon_1 - \epsilon_3 - \sigma)$
$\quad\quad + \frac{5}{2304} \sin 2\sigma \cos (7\tau - \frac{1}{2}\epsilon_1 - \epsilon_3 - \sigma)$
$\quad\quad\quad\quad\quad + \frac{7}{3840} \sin (9\tau - \frac{3}{2}\epsilon_1 - \epsilon_3 - \sigma)$

$B_{14} = -\frac{7}{1152} \cos 2\sigma \sin (5\tau + \frac{3}{2}\epsilon_1 - \epsilon_4 - \sigma)$
$\quad\quad - \frac{7}{1152} \sin 2\sigma \cos (5\tau + \frac{3}{2}\epsilon_1 - \epsilon_4 - \sigma)$
$\quad\quad + (\frac{1}{640} - \frac{1}{576} \cos 4\sigma) \sin (7\tau + \frac{1}{2}\epsilon_1 - \epsilon_4 - \sigma)$
$\quad\quad - \frac{1}{576} \sin 4\sigma \cos (7\tau + \frac{1}{2}\epsilon_1 - \epsilon_4 - \sigma)$
$\quad\quad - \frac{1}{9600} \cos 2\sigma \sin (9\tau - \frac{1}{2}\epsilon_1 - \epsilon_4 - \sigma)$
$\quad\quad + \frac{11}{9600} \sin 2\sigma \cos (9\tau - \frac{1}{2}\epsilon_1 - \epsilon_4 - \sigma)$
$\quad\quad\quad\quad\quad + \frac{11}{9600} \sin (11\tau - \frac{3}{2}\epsilon_1 - \epsilon_4 - \sigma)$

$B_{23} = \frac{7}{384} \sin (3\tau - \frac{1}{2}\epsilon_1 + \epsilon_2 - \epsilon_3 - \sigma)$
$\quad\quad - \frac{1}{480} \cos 2\sigma \sin (9\tau + \frac{1}{2}\epsilon_1 - \epsilon_2 - \epsilon_3 - \sigma)$
$\quad\quad - \frac{1}{480} \sin 2\sigma \cos (9\tau + \frac{1}{2}\epsilon_1 - \epsilon_2 - \epsilon_3 - \sigma)$
$\quad\quad\quad\quad\quad + \frac{1}{1920} \sin (11\tau - \frac{1}{2}\epsilon_1 - \epsilon_2 - \epsilon_3 - \sigma)$

$$B_{24} = -\tfrac{1}{128} \cos 2\sigma \sin (3\tau + \tfrac{1}{2}\epsilon_1 + \epsilon_2 - \epsilon_4 - \sigma)$$
$$- \tfrac{1}{128} \sin 2\sigma \cos (3\tau + \tfrac{1}{2}\epsilon_1 + \epsilon_2 - \epsilon_4 - \sigma)$$
$$+ \tfrac{11}{1920} \sin (5\tau - \tfrac{1}{2}\epsilon_1 + \epsilon_2 - \epsilon_4 - \sigma)$$
$$- \tfrac{7}{5760} \cos 2\sigma \sin (11\tau + \tfrac{1}{2}\epsilon_1 - \epsilon_2 - \epsilon_4 - \sigma)$$
$$- \tfrac{7}{5760} \sin 2\sigma \cos (11\tau + \tfrac{1}{2}\epsilon_1 - \epsilon_2 - \epsilon_4 - \sigma)$$
$$+ \tfrac{13}{40320} \sin (13\tau - \tfrac{1}{2}\epsilon_1 - \epsilon_2 - \epsilon_4 - \sigma)$$

$$B_{34} = \tfrac{13}{1920} \sin (3\tau - \tfrac{1}{2}\epsilon_1 + \epsilon_3 - \epsilon_4 - \sigma)$$
$$- \tfrac{1}{2688} \cos 2\sigma \sin (13\tau + \tfrac{1}{2}\epsilon_1 - \epsilon_3 - \epsilon_4 - \sigma)$$
$$- \tfrac{1}{2688} \sin 2\sigma \cos (13\tau + \tfrac{1}{2}\epsilon_1 - \epsilon_3 - \epsilon_4 - \sigma)$$
$$+ \tfrac{1}{6720} \sin (15\tau - \tfrac{1}{2}\epsilon_1 - \epsilon_3 - \epsilon_4 - \sigma)$$

III.2 Unbounded Solution Associated with Second Unstable Region

$$\mu = \tfrac{1}{4}\theta_2 \sin 2\sigma - \tfrac{1}{16}\theta_1{}^2 \sin (2\sigma - 2\epsilon_1 + \epsilon_2)$$
$$+ \tfrac{1}{24}\theta_1\theta_3 \sin (2\sigma + \epsilon_1 + \epsilon_2 - \epsilon_3)$$
$$+ \tfrac{1}{64}\theta_2\theta_4 \sin (2\sigma + 2\epsilon_2 - \epsilon_4) + \cdots \quad \text{(III.6)}$$

$$\theta_0 = 4 + \theta_2 \cos 2\sigma + [\tfrac{1}{6} - \tfrac{1}{4} \cos (2\sigma - 2\epsilon_1 + \epsilon_2)]\theta_1{}^2$$
$$+ (-\tfrac{1}{16} + \tfrac{1}{32} \cos 4\sigma)\theta_2{}^2 - \tfrac{1}{10}\theta_3{}^2 - \tfrac{1}{24}\theta_4{}^2$$
$$+ \tfrac{1}{6}\theta_1\theta_3 \cos (2\sigma + \epsilon_1 + \epsilon_2 - \epsilon_3)$$
$$+ \tfrac{1}{16}\theta_2\theta_4 \cos (2\sigma + 2\epsilon_2 - \epsilon_4) + \cdots \quad \text{(III.7)}$$

$$\phi(\tau,\sigma) = \sin (2\tau - \tfrac{1}{2}\epsilon_2 - \sigma) + A_1\theta_1 + A_2\theta_2 + A_3\theta_3 + A_4\theta_4$$
$$+ B_1\theta_1{}^2 + B_2\theta_2{}^2 + B_3\theta_3{}^2 + B_4\theta_4{}^2 + B_{12}\theta_1\theta_2 + B_{13}\theta_1\theta_3$$
$$+ B_{14}\theta_1\theta_4 + B_{23}\theta_2\theta_3 + B_{24}\theta_2\theta_4 + B_{34}\theta_3\theta_4 + \cdots \quad \text{(III.8)}$$

where

$$A_1 = -\tfrac{1}{4} \sin (\epsilon_1 - \tfrac{1}{2}\epsilon_2 - \sigma) + \tfrac{1}{12} \sin (4\tau - \epsilon_1 - \tfrac{1}{2}\epsilon_2 - \sigma)$$
$$A_2 = \tfrac{1}{32} \sin (6\tau - \tfrac{3}{2}\epsilon_2 - \sigma)$$
$$A_3 = -\tfrac{1}{12} \cos 2\sigma \sin (4\tau + \tfrac{1}{2}\epsilon_2 - \epsilon_3 - \sigma)$$
$$- \tfrac{1}{12} \sin 2\sigma \cos (4\tau + \tfrac{1}{2}\epsilon_2 - \epsilon_3 - \sigma)$$
$$+ \tfrac{1}{60} \sin (8\tau - \tfrac{1}{2}\epsilon_2 - \epsilon_3 - \sigma)$$
$$A_4 = -\tfrac{1}{32} \cos 2\sigma \sin (6\tau + \tfrac{1}{2}\epsilon_2 - \epsilon_4 - \sigma)$$
$$- \tfrac{1}{32} \sin 2\sigma \cos (6\tau + \tfrac{1}{2}\epsilon_2 - \epsilon_4 - \sigma)$$
$$+ \tfrac{1}{96} \sin (10\tau - \tfrac{1}{2}\epsilon_2 - \epsilon_4 - \sigma)$$

$$B_1 = \tfrac{1}{384} \sin (6\tau - 2\epsilon_1 - \tfrac{1}{2}\epsilon_2 - \sigma)$$
$$B_2 = \tfrac{1}{1024} \cos 2\sigma \sin (6\tau - \tfrac{3}{2}\epsilon_2 - \sigma)$$
$$+ \tfrac{3}{1024} \sin 2\sigma \cos (6\tau - \tfrac{3}{2}\epsilon_2 - \sigma) + \tfrac{1}{3072} \sin (10\tau - \tfrac{5}{2}\epsilon_2 - \sigma)$$
$$B_3 = -\tfrac{1}{1152} \cos 2\sigma \sin (10\tau + \tfrac{1}{2}\epsilon_2 - 2\epsilon_3 - \sigma)$$
$$- \tfrac{1}{1152} \sin 2\sigma \cos (10\tau + \tfrac{1}{2}\epsilon_2 - 2\epsilon_3 - \sigma)$$
$$+ \tfrac{1}{11520} \sin (14\tau - \tfrac{1}{2}\epsilon_2 - 2\epsilon_3 - \sigma)$$
$$B_4 = -\tfrac{1}{6144} \cos 2\sigma \sin (14\tau + \tfrac{1}{2}\epsilon_2 - 2\epsilon_4 - \sigma)$$
$$- \tfrac{1}{6144} \sin 2\sigma \cos (14\tau + \tfrac{1}{2}\epsilon_2 - 2\epsilon_4 - \sigma)$$
$$+ \tfrac{1}{30720} \sin (18\tau - \tfrac{1}{2}\epsilon_2 - 2\epsilon_4 - \sigma)$$

$$B_{12} = \tfrac{5}{96} \sin (\sigma + \epsilon_1 - \tfrac{1}{2}\epsilon_2) - \tfrac{1}{32} \sin (3\sigma - \epsilon_1 + \tfrac{1}{2}\epsilon_2)$$
$$- \tfrac{7}{384} \sin (4\tau + \epsilon_1 - \tfrac{3}{2}\epsilon_2 - \sigma)$$
$$+ \tfrac{1}{36} \cos 2\sigma \sin (4\tau - \epsilon_1 - \tfrac{1}{2}\epsilon_2 - \sigma)$$
$$+ \tfrac{5}{144} \sin 2\sigma \cos (4\tau - \epsilon_1 - \tfrac{1}{2}\epsilon_2 - \sigma)$$
$$+ \tfrac{11}{5760} \sin (8\tau - \epsilon_1 - \tfrac{3}{2}\epsilon_2 - \sigma)$$

$$B_{13} = -\tfrac{7}{960} \sin (6\tau + \epsilon_1 - \tfrac{1}{2}\epsilon_2 - \epsilon_3 - \sigma)$$
$$+ \tfrac{1}{192} \cos 2\sigma \sin (6\tau - \epsilon_1 + \tfrac{1}{2}\epsilon_2 - \epsilon_3 - \sigma)$$
$$+ \tfrac{1}{192} \sin 2\sigma \cos (6\tau - \epsilon_1 + \tfrac{1}{2}\epsilon_2 - \epsilon_3 - \sigma)$$
$$+ \tfrac{1}{960} \sin (10\tau - \epsilon_1 - \tfrac{1}{2}\epsilon_2 - \epsilon_3 - \sigma)$$

$$B_{14} = -\tfrac{11}{1152} \cos 2\sigma \sin (4\tau + \epsilon_1 + \tfrac{1}{2}\epsilon_2 - \epsilon_4 - \sigma)$$
$$- \tfrac{11}{1152} \sin 2\sigma \cos (4\tau + \epsilon_1 + \tfrac{1}{2}\epsilon_2 - \epsilon_4 - \sigma)$$
$$- \tfrac{23}{5760} \sin (8\tau + \epsilon_1 - \tfrac{1}{2}\epsilon_2 - \epsilon_4 - \sigma)$$
$$+ \tfrac{7}{1920} \cos 2\sigma \sin (8\tau - \epsilon_1 + \tfrac{1}{2}\epsilon_2 - \epsilon_4 - \sigma)$$
$$+ \tfrac{7}{1920} \sin 2\sigma \cos (8\tau - \epsilon_1 + \tfrac{1}{2}\epsilon_2 - \epsilon_4 - \sigma)$$
$$+ \tfrac{3}{4480} \sin (12\tau - \epsilon_1 - \tfrac{1}{2}\epsilon_2 - \epsilon_4 - \sigma)$$

$$B_{23} = \tfrac{11}{384} \sin (\sigma + \tfrac{3}{2}\epsilon_2 - \epsilon_3)$$
$$+ (\tfrac{7}{1440} - \tfrac{1}{96} \cos 4\sigma) \sin (4\tau + \tfrac{1}{2}\epsilon_2 - \epsilon_3 - \sigma)$$
$$- \tfrac{1}{96} \sin 4\sigma \cos (4\tau + \tfrac{1}{2}\epsilon_2 - \epsilon_3 - \sigma)$$
$$- \tfrac{1}{900} \cos 2\sigma \sin (8\tau - \tfrac{1}{2}\epsilon_2 - \epsilon_3 - \sigma)$$
$$- \tfrac{1}{3600} \sin 2\sigma \cos (8\tau - \tfrac{1}{2}\epsilon_2 - \epsilon_3 - \sigma)$$
$$+ \tfrac{23}{67200} \sin (12\tau - \tfrac{3}{2}\epsilon_2 - \epsilon_3 - \sigma)$$

$$B_{24} = (\tfrac{1}{768} - \tfrac{1}{512} \cos 4\sigma) \sin (6\tau + \tfrac{1}{2}\epsilon_2 - \epsilon_4 - \sigma)$$
$$- \tfrac{1}{512} \sin 4\sigma \cos (6\tau + \tfrac{1}{2}\epsilon_2 - \epsilon_4 - \sigma)$$
$$- \tfrac{1}{4608} \cos 2\sigma \sin (10\tau - \tfrac{1}{2}\epsilon_2 - \epsilon_4 - \sigma)$$
$$+ \tfrac{1}{4608} \sin 2\sigma \cos (10\tau - \tfrac{1}{2}\epsilon_2 - \epsilon_4 - \sigma)$$
$$+ \tfrac{1}{4608} \sin (14\tau - \tfrac{3}{2}\epsilon_2 - \epsilon_4 - \sigma)$$

$$B_{34} = \tfrac{23}{1920} \sin (\sigma + \tfrac{1}{2}\epsilon_2 + \epsilon_3 - \epsilon_4)$$
$$+ \tfrac{1}{128} \sin (4\tau - \tfrac{1}{2}\epsilon_2 + \epsilon_3 - \epsilon_4 - \sigma)$$
$$- \tfrac{11}{13440} \cos 2\sigma \sin (12\tau + \tfrac{1}{2}\epsilon_2 - \epsilon_3 - \epsilon_4 - \sigma)$$
$$- \tfrac{11}{13440} \sin 2\sigma \cos (12\tau + \tfrac{1}{2}\epsilon_2 - \epsilon_3 - \epsilon_4 - \sigma)$$
$$+ \tfrac{13}{120960} \sin (16\tau - \tfrac{1}{2}\epsilon_2 - \epsilon_3 - \epsilon_4 - \sigma)$$

III.3 Unbounded Solution Associated with Third Unstable Region

$$\mu = \tfrac{1}{6}\theta_3 \sin 2\sigma - \tfrac{1}{24}\theta_1\theta_2 \sin (2\sigma - \epsilon_1 - \epsilon_2 + \epsilon_3)$$
$$+ \tfrac{1}{48}\theta_1\theta_4 \sin (2\sigma + \epsilon_1 + \epsilon_3 - \epsilon_4) + \cdots \quad \text{(III.9)}$$

$$\theta_0 = 9 + \theta_3 \cos 2\sigma + \tfrac{1}{16}\theta_1{}^2 + \tfrac{1}{10}\theta_2{}^2$$
$$+ (-\tfrac{1}{36} + \tfrac{1}{72} \cos 4\sigma)\theta_3{}^2 - \tfrac{1}{14}\theta_4{}^2$$
$$- \tfrac{1}{4}\theta_1\theta_2 \cos (2\sigma - \epsilon_1 - \epsilon_2 + \epsilon_3)$$
$$+ \tfrac{1}{8}\theta_1\theta_4 \cos (2\sigma + \epsilon_1 + \epsilon_3 - \epsilon_4) + \cdots \quad \text{(III.10)}$$

$$\phi(\tau,\sigma) = \sin (3\tau - \tfrac{1}{2}\epsilon_3 - \sigma) + A_1\theta_1 + A_2\theta_2 + A_3\theta_3 + A_4\theta_4$$
$$+ B_1\theta_1{}^2 + B_2\theta_2{}^2 + B_3\theta_3{}^2 + B_4\theta_4{}^2 + B_{12}\theta_1\theta_2$$
$$+ B_{13}\theta_1\theta_3 + B_{14}\theta_1\theta_4 + B_{23}\theta_2\theta_3 + B_{24}\theta_2\theta_4 + B_{34}\theta_3\theta_4 + \cdots \quad \text{(III.11)}$$

where

$$A_1 = -\tfrac{1}{8}\sin (\tau + \epsilon_1 - \tfrac{1}{2}\epsilon_3 - \sigma) + \tfrac{1}{16}\sin (5\tau - \epsilon_1 - \tfrac{1}{2}\epsilon_3 - \sigma)$$

$$A_2 = \tfrac{1}{8}\cos 2\sigma \sin (\tau - \epsilon_2 + \tfrac{1}{2}\epsilon_3 - \sigma)$$
$$+ \tfrac{1}{8}\sin 2\sigma \cos (\tau - \epsilon_2 + \tfrac{1}{2}\epsilon_3 - \sigma) + \tfrac{1}{40}\sin (7\tau - \epsilon_2 - \tfrac{1}{2}\epsilon_3 - \sigma)$$

$$A_3 = \tfrac{1}{72}\sin (9\tau - \tfrac{3}{2}\epsilon_3 - \sigma)$$

$$A_4 = -\tfrac{1}{16}\cos 2\sigma \sin (5\tau + \tfrac{1}{2}\epsilon_3 - \epsilon_4 - \sigma)$$
$$- \tfrac{1}{16}\sin 2\sigma \cos (5\tau + \tfrac{1}{2}\epsilon_3 - \epsilon_4 - \sigma)$$
$$+ \tfrac{1}{112}\sin (11\tau - \tfrac{1}{2}\epsilon_3 - \epsilon_4 - \sigma)$$

$$B_1 = -\tfrac{1}{64}\cos 2\sigma \sin (\tau - 2\epsilon_1 + \tfrac{1}{2}\epsilon_3 - \sigma)$$
$$- \tfrac{1}{64}\sin 2\sigma \cos (\tau - 2\epsilon_1 + \tfrac{1}{2}\epsilon_3 - \sigma)$$
$$+ \tfrac{1}{640}\sin (7\tau - 2\epsilon_1 - \tfrac{1}{2}\epsilon_3 - \sigma)$$

$$B_2 = \tfrac{1}{128}\cos 2\sigma \sin (5\tau - 2\epsilon_2 + \tfrac{1}{2}\epsilon_3 - \sigma)$$
$$+ \tfrac{1}{128}\sin 2\sigma \cos (5\tau - 2\epsilon_2 + \tfrac{1}{2}\epsilon_3 - \sigma)$$
$$+ \tfrac{1}{4480}\sin (11\tau - 2\epsilon_2 - \tfrac{1}{2}\epsilon_3 - \sigma)$$

$$B_3 = \tfrac{1}{5184}\cos 2\sigma \sin (9\tau - \tfrac{3}{2}\epsilon_3 - \sigma)$$
$$+ \tfrac{1}{1728}\sin 2\sigma \cos (9\tau - \tfrac{3}{2}\epsilon_3 - \sigma)$$
$$+ \tfrac{1}{15552}\sin (15\tau - \tfrac{5}{2}\epsilon_3 - \sigma)$$

$$B_4 = -\tfrac{1}{2560}\cos 2\sigma \sin (13\tau + \tfrac{1}{2}\epsilon_3 - 2\epsilon_4 - \sigma)$$
$$- \tfrac{1}{2560}\sin 2\sigma \cos (13\tau + \tfrac{1}{2}\epsilon_3 - 2\epsilon_4 - \sigma)$$
$$+ \tfrac{1}{39424}\sin (19\tau - \tfrac{1}{2}\epsilon_3 - 2\epsilon_4 - \sigma)$$

$$B_{12} = \tfrac{1}{128}\sin (\tau - \epsilon_1 + \epsilon_2 - \tfrac{1}{2}\epsilon_3 - \sigma)$$
$$- \tfrac{1}{160}\sin (5\tau + \epsilon_1 - \epsilon_2 - \tfrac{1}{2}\epsilon_3 - \sigma)$$
$$+ \tfrac{7}{5760}\sin (9\tau - \epsilon_1 - \epsilon_2 - \tfrac{1}{2}\epsilon_3 - \sigma)$$

$$B_{13} = \tfrac{3}{128}\cos 2\sigma \sin (\tau + \epsilon_1 - \tfrac{1}{2}\epsilon_3 - \sigma)$$
$$+ \tfrac{5}{384}\sin 2\sigma \cos (\tau + \epsilon_1 - \tfrac{1}{2}\epsilon_3 - \sigma)$$
$$+ \tfrac{3}{256}\cos 2\sigma \sin (5\tau - \epsilon_1 - \tfrac{1}{2}\epsilon_3 - \sigma)$$
$$+ \tfrac{11}{768}\sin 2\sigma \cos (5\tau - \epsilon_1 - \tfrac{1}{2}\epsilon_3 - \sigma)$$
$$- \tfrac{1}{360}\sin (7\tau + \epsilon_1 - \tfrac{3}{2}\epsilon_3 - \sigma)$$
$$+ \tfrac{11}{16128}\sin (11\tau - \epsilon_1 - \tfrac{3}{2}\epsilon_3 - \sigma)$$

$$B_{14} = \tfrac{1}{640}\cos 2\sigma \sin (7\tau - \epsilon_1 + \tfrac{1}{2}\epsilon_3 - \epsilon_4 - \sigma)$$
$$+ \tfrac{1}{640}\sin 2\sigma \cos (7\tau - \epsilon_1 + \tfrac{1}{2}\epsilon_3 - \epsilon_4 - \sigma)$$
$$- \tfrac{13}{8064}\sin (9\tau + \epsilon_1 - \tfrac{1}{2}\epsilon_3 - \epsilon_4 - \sigma)$$
$$+ \tfrac{1}{2240}\sin (13\tau - \epsilon_1 - \tfrac{1}{2}\epsilon_3 - \epsilon_4 - \sigma)$$

$$B_{23} = -(\tfrac{1}{120} + \tfrac{1}{96}\cos 4\sigma) \sin (\tau - \epsilon_2 + \tfrac{1}{2}\epsilon_3 - \sigma)$$
$$- \tfrac{1}{96}\sin 4\sigma \cos (\tau - \epsilon_2 + \tfrac{1}{2}\epsilon_3 - \sigma)$$
$$- \tfrac{1}{144}\sin (5\tau + \epsilon_2 - \tfrac{3}{2}\epsilon_3 - \sigma)$$
$$+ \tfrac{3}{800}\cos 2\sigma \sin (7\tau - \epsilon_2 - \tfrac{1}{2}\epsilon_3 - \sigma)$$
$$+ \tfrac{11}{2400}\sin 2\sigma \cos (7\tau - \epsilon_2 - \tfrac{1}{2}\epsilon_3 - \sigma)$$
$$+ \tfrac{7}{28800}\sin (13\tau - \epsilon_2 - \tfrac{3}{2}\epsilon_3 - \sigma)$$

$$B_{24} = \tfrac{7}{640} \cos 2\sigma \sin (\tau + \epsilon_2 + \tfrac{1}{2}\epsilon_3 - \epsilon_4 - \sigma)$$
$$+ \tfrac{7}{640} \sin 2\sigma \cos (\tau + \epsilon_2 + \tfrac{1}{2}\epsilon_3 - \epsilon_4 - \sigma)$$
$$- \tfrac{13}{4480} \sin (7\tau + \epsilon_2 - \tfrac{1}{2}\epsilon_3 - \epsilon_4 - \sigma)$$
$$+ \tfrac{1}{1152} \cos 2\sigma \sin (9\tau - \epsilon_2 + \tfrac{1}{2}\epsilon_3 - \epsilon_4 - \sigma)$$
$$+ \tfrac{1}{1152} \sin 2\sigma \cos (9\tau - \epsilon_2 + \tfrac{1}{2}\epsilon_3 - \epsilon_4 - \sigma)$$
$$+ \tfrac{19}{120960} \sin (15\tau - \epsilon_2 - \tfrac{1}{2}\epsilon_3 - \epsilon_4 - \sigma)$$

$$B_{34} = - \tfrac{11}{1152} \sin (\tau - \tfrac{3}{2}\epsilon_3 + \epsilon_4 - \sigma)$$
$$+ (\tfrac{5}{2688} - \tfrac{1}{192} \cos 4\sigma) \sin (5\tau + \tfrac{1}{2}\epsilon_3 - \epsilon_4 - \sigma)$$
$$- \tfrac{1}{192} \sin 4\sigma \cos (5\tau + \tfrac{1}{2}\epsilon_3 - \epsilon_4 - \sigma)$$
$$- \tfrac{3}{6272} \cos 2\sigma \sin (11\tau - \tfrac{1}{2}\epsilon_3 - \epsilon_4 - \sigma)$$
$$- \tfrac{5}{18816} \sin 2\sigma \cos (11\tau - \tfrac{1}{2}\epsilon_3 - \epsilon_4 - \sigma)$$
$$+ \tfrac{23}{282240} \sin (17\tau - \tfrac{3}{2}\epsilon_3 - \epsilon_4 - \sigma)$$

Stability Criterion Obtained by Using the Perturbation Method

The stability of periodic solutions in nonlinear systems has been discussed in many scientific and technical journals. We refer to a paper by Mandelstam and Papalexi [72] and compare their stability criterion with the result obtained by our method in Chap. 4. They have discussed the subharmonic oscillations in vacuum-tube circuits described by the differential equation

$$\frac{d^2v}{d\tau^2} + v = \lambda F\left(v, \frac{dv}{d\tau}\right) + B \cos \nu\tau \qquad \text{with} \qquad \nu = 2, 3, 4, \ldots \quad \text{(IV.1)}$$

in which the parametric coefficient λ of the nonlinear function $F(v, dv/d\tau)$ is assumed to be small. They have treated the problem by the perturbation method and have obtained the following periodic solution for the subharmonic oscillation of order $1/\nu$, that is,

$$v(\tau) = x \sin \tau + y \cos \tau + w \cos \nu\tau \qquad \text{with} \qquad w = \frac{B}{1 - \nu^2} \quad \text{(IV.2)}$$

in which the amplitudes x and y are determined by

$$\int_0^{2\pi} \psi(x,y,\tau) \sin \tau \, d\tau = 0 \qquad \int_0^{2\pi} \psi(x,y,\tau) \cos \tau \, d\tau = 0$$

where $\quad \psi(x,y,\tau) = F(x \sin \tau + y \cos \tau + w \cos \nu\tau,$
$$x \cos \tau - y \sin \tau - \nu w \sin \nu\tau) \qquad \text{(IV.3)}$$

This procedure leads to the same result as that obtained by substituting Eq. (IV.2) into (IV.1) and equating the coefficients of the terms containing $\sin \tau$ and $\cos \tau$ separately to zero (see Sec. 1.5). Then, as for the stability, they have obtained the condition

$$\left| \begin{array}{cc} \int_0^{2\pi} \dfrac{\partial \psi}{\partial x} \sin \tau \, d\tau & \int_0^{2\pi} \dfrac{\partial \psi}{\partial x} \cos \tau \, d\tau \\[2ex] \int_0^{2\pi} \dfrac{\partial \psi}{\partial y} \sin \tau \, d\tau & \int_0^{2\pi} \dfrac{\partial \psi}{\partial y} \cos \tau \, d\tau \end{array} \right| > 0 \qquad (IV.4)$$

which has been derived from the consideration that the variations of the amplitudes x and y of the subharmonic oscillation tend to zero with the increase of time τ.

On the other hand, in our investigation, the subharmonic oscillations described by Eq. (7.1) are studied. The variational equation (3.55) becomes in this case

$$\frac{d^2\eta}{d\tau^2} + \left[\left(\frac{df}{dv} \right)_0 - \delta^2 \right] \eta = 0 \qquad (IV.5)$$

where the subscript 0 denotes the insertion of the periodic solution after the differentiation.

We now proceed to show that the stability condition (IV.4) corresponds either to our condition (4.6) for the first unstable region ($n = 1$) or to (4.9) for the second unstable region ($n = 2$) in Chap. 4.[1]

CASE 1

We first consider the case in which the variational equation (IV.5) leads to Hill's equation of the form (4.4). Then $(df/dv)_0$ in Eq. (IV.5) may be expanded in a Fourier series as

$$\left(\frac{df}{dv} \right)_0 = a_0 + a_1 \cos 2\tau + a_2 \cos 4\tau + \cdots + b_1 \sin 2\tau + b_2 \sin 4\tau + \cdots$$

where
$$a_0 = \frac{1}{2\pi} \int_0^{2\pi} \left(\frac{df}{dv} \right)_0 d\tau$$

$$a_\nu = \frac{1}{\pi} \int_0^{2\pi} \left(\frac{df}{dv} \right)_0 \cos 2\nu\tau \, d\tau \qquad\qquad (IV.6)$$

$$b_\nu = \frac{1}{\pi} \int_0^{2\pi} \left(\frac{df}{dv} \right)_0 \sin 2\nu\tau \, d\tau \qquad \nu = 1, 2, 3, \ldots$$

Substituting Eqs. (IV.6) into (IV.5) and comparing the equation so formed with (4.4) gives us

$$\theta_0 = a_0 - \delta^2 \qquad 2\theta_\nu = \sqrt{a_\nu{}^2 + b_\nu{}^2} \qquad \epsilon_\nu = \tan^{-1} \frac{b_\nu}{a_\nu} \qquad (IV.7)$$

---

[1] Which condition it will be depends on the nonlinear characteristic and the order of the subharmonic oscillation (Chap. 7).

Hence, the stability condition (4.6) may be written as

$$(a_0 - n^2)^2 + 4n^2\delta^2 > \tfrac{1}{4}(a_n^2 + b_n^2) \qquad n = 1, 2, 3, \ldots$$

and further, substituting for a_0, a_n, and b_n their values as given by Eqs. (IV.6), leads to

$$\left\{ \int_0^{2\pi} \left[\left(\frac{df}{dv}\right)_0 - n^2 \right] d\tau \right\}^2 + 16\pi^2 n^2 \delta^2 > \left[\int_0^{2\pi} \left(\frac{df}{dv}\right)_0 \cos 2n\tau \, d\tau \right]^2$$

$$+ \left[\int_0^{2\pi} \left(\frac{df}{dv}\right)_0 \sin 2n\tau \, d\tau \right]^2$$

or $\displaystyle \int_0^{2\pi} \left[\left(\frac{df}{dv}\right)_0 - n^2 \right] \sin^2 n\tau \, d\tau \cdot \int_0^{2\pi} \left[\left(\frac{df}{dv}\right)_0 - n^2 \right] \cos^2 n\tau \, d\tau$

$$- \left\{ \int_0^{2\pi} \left[\left(\frac{df}{dv}\right)_0 - n^2 \right] \sin n\tau \cos n\tau \, d\tau \right\}^2 + 4\pi^2 n^2 \delta^2 > 0$$

This stability condition may be rewritten in a form similar to (IV.4), that is,

$$\begin{vmatrix} \displaystyle\int_0^{2\pi} \Psi_x \sin n\tau \, d\tau & \displaystyle\int_0^{2\pi} \Psi_x \cos n\tau \, d\tau \\[2mm] \displaystyle\int_0^{2\pi} \Psi_y \sin n\tau \, d\tau & \displaystyle\int_0^{2\pi} \Psi_y \cos n\tau \, d\tau \end{vmatrix} > 0$$

in which
$$\Psi_x = \left[\left(\frac{df}{dv}\right)_0 - n^2 \right] \sin n\tau + 2n\delta \cos n\tau \qquad \text{(IV.8)}$$

$$\Psi_y = \left[\left(\frac{df}{dv}\right)_0 - n^2 \right] \cos n\tau - 2n\delta \sin n\tau$$

Now the stability condition (IV.4) may be derived by putting $n = 1$ in (IV.8); namely, the comparison of Eq. (IV.1) with (7.1) gives

$$F\left(v, \frac{dv}{d\tau}\right) = \frac{1}{\lambda}\left[v - f(v) - 2\delta \frac{dv}{d\tau} \right]$$

and hence, by virtue of Eq. (IV.2),

$$\frac{\partial \psi}{\partial x} = \frac{1}{\lambda}\left[1 - \left(\frac{df}{dv}\right)_0 \right] \sin \tau - \frac{2\delta}{\lambda} \cos \tau = -\frac{1}{\lambda}[\Psi_x]_{n=1}$$

$$\frac{\partial \psi}{\partial y} = \frac{1}{\lambda}\left[1 - \left(\frac{df}{dv}\right)_0 \right] \cos \tau + \frac{2\delta}{\lambda} \sin \tau = -\frac{1}{\lambda}[\Psi_y]_{n=1}$$

The substitution of these relations into (IV.8) yields the stability condition (IV.4).

Thus we see that, from the condition (IV.4), no information is available concerning the stability related to the unstable region of order $n \geqq 2$. We explain this in the following. In the derivation of the condition (IV.4), only a variation in the amplitude of the subharmonic oscillation is considered; hence a constraint results on the variation that it must have the same frequency as the subharmonic frequency. Consequently, a variation which may have a different frequency from the subharmonic frequency is not taken into account. In our investigation, there is no such constraint. The generalized stability condition (4.6) for the nth unstable region furnishes the criterion for testing the stability against the buildup of the nth harmonic of the subharmonic oscillation. Thus, for example, when we discuss the subharmonic oscillation of order $\frac{1}{3}$, the stability condition takes the form of inequality (4.6). It is mentioned in Sec. 7.4 that the condition (4.6) for $n = 2$ as well as for $n = 1$ must be considered in the case when the nonlinearity is characterized by a quintic function.

CASE 2

We have also to discuss the case in which the variational equation (IV.5) leads to Hill's equation of the form (4.7). Proceeding analogously to the description in Case 1, we may conclude that the stability condition (IV.4) is obtained from our condition (4.9) by putting $n = 2$. For the subharmonic oscillation of order $\frac{1}{2}$, the stability condition leads to inequality (4.9); the condition for the nth unstable region ascertains the stability against the buildup of the $n/2$th harmonic of the subharmonic oscillation. As explained in Sec. 7.6b, the unstable oscillations not only of order $\frac{1}{2}$ but also of orders $\frac{1}{4}$ and $\frac{3}{4}$ occur in certain situations, so that the conditions (IV.8) must be considered for $n = 1, 2,$ and 3.

To summarize the foregoing two cases, it may be concluded that the condition (IV.4) ascertains only the stability against the buildup of an unstable oscillation whose period is the same as that of the subharmonic oscillation $x \sin \tau + y \cos \tau$ in Eq. (IV.2) and that the condition (IV.4) is included in the generalized stability condition (4.6) or (4.9).

Appendix V: **Remarks Concerning Integral Curves and Singular Points**

In Chap. 2 of the text we have discussed the fundamental properties of integral curves and singular points defined by a set of first-order differential equations. This appendix supplements the text by topological studies of integral curves, as done by Poincaré and Bendixson.

V.1 Theorems of Bendixson

The following theorems[1] are due to Bendixson and are cited in our investigation of integral curves in Chaps. 8 and 13.

(a) First Theorem of Bendixson

We consider a system of differential equations

$$\frac{dx}{d\tau} = X_m + X_{m+1} \qquad \frac{dy}{d\tau} = Y_m + Y_{m+1} \tag{V.1}$$

where X_m and Y_m are polynomials which are homogeneous and of degree m in x and y; X_{m+1} and Y_{m+1} are analytic functions containing terms of degree higher than m.

The theorem of Bendixson states that the integral curves of the system (V.1) tend to the origin either without any definite direction, thus in the manner of spirals, or with the definite direction as determined by the equation

$$xY_m - yX_m = 0 \tag{V.2}$$

---

[1] For the proofs of these theorems see Bendixson [6, pp. 34, 45].

(b) Second Theorem of Bendixson

We deal with the differential equation

$$x^m \frac{dy}{dx} = ay + bx + B(x,y) \tag{V.3}$$

where $B(x,y)$ is a polynomial containing terms of degree higher than the first in x and y and we assume that $a \neq 0$.

Bendixson has obtained the following conclusions.

CASE 1. $a > 0$, m IS AN EVEN NUMBER

There is one and only one branch of integral curves tending to the origin on the left side of the y axis, while integral curves on the other side constitute a nodal distribution, i.e., they form a nodal region for $x > 0$. These integral curves are schematically illustrated in Fig. V.1.

CASE 2. $a < 0$, m IS AN EVEN NUMBER

By the substitution $x = -x'$, this case may be transformed to the foregoing case; thus we have a nodal distribution of integral curves for $x < 0$ and a saddle distribution for $x > 0$.

CASE 3. $a > 0$, m IS AN ODD NUMBER

The origin is a nodal point.

CASE 4. $a < 0$, m IS AN ODD NUMBER

The origin is a saddle point.

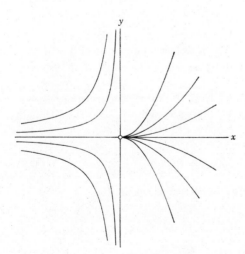

FIGURE V.1   Coalescence of node and saddle.

V.2 Theory of Centers

The types of singular points have been classified according to the roots of the characteristic equation (see Sec. 2.2a). If, in particular, the roots are imaginary, so that

$$a_1 + b_2 = 0 \qquad (a_1 - b_2)^2 + 4a_2 b_1 < 0 \tag{V.4}$$

the singularity will be either a center or a focus. The following procedure due to Poincaré [82; 84, pp. 95–114] will distinguish between these two types of singularity when the conditions (V.4) are satisfied.

We consider a set of differential equations

$$\begin{aligned}
\frac{dx}{d\tau} &= X_1 + X_2 + \cdots + X_n \equiv X \\
\frac{dy}{d\tau} &= Y_1 + Y_2 + \cdots + Y_n \equiv Y
\end{aligned} \tag{V.5}$$

where X_m and Y_m $(m = 1, 2, \ldots, n)$ are homogeneous polynomials of degree m in x and y. Since both X and Y vanish for $x = y = 0$, the origin is a singular point of the system (V.5).

In order to investigate the behavior of integral curves near the singularity, we consider a family of closed curves around the origin defined by

$$F(x,y) = C \tag{V.6}$$

where $F(x,y)$ is a positive definite function of x and y and C is a sufficiently small positive constant. We write

$$F(x,y) = F_2 + F_3 + F_4 + \cdots \tag{V.7}$$

where F_i $(i = 2, 3, 4, \ldots)$ is a homogeneous polynomial of degree i in x and y. Then, using Eqs. (V.5) and (V.7), we obtain

$$\begin{aligned}
\frac{dF}{d\tau} &= \frac{\partial F}{\partial x}\frac{dx}{d\tau} + \frac{\partial F}{\partial y}\frac{dy}{d\tau} \\
&= \left(\frac{\partial F_2}{\partial x} + \frac{\partial F_3}{\partial x} + \cdots \right)(X_1 + X_2 + \cdots) \\
&\qquad + \left(\frac{\partial F_2}{\partial y} + \frac{\partial F_3}{\partial y} + \cdots \right)(Y_1 + Y_2 + \cdots)
\end{aligned}$$

If we write

$$\phi_{ik} = \frac{\partial F_i}{\partial x} X_k + \frac{\partial F_i}{\partial y} Y_k \tag{V.8}$$

then

$$\frac{dF}{d\tau} = \phi_{21} + (\phi_{31} + \phi_{22}) + (\phi_{41} + \phi_{32} + \phi_{23}) + \cdots \tag{V.9}$$

Following the analysis due to Poincaré, the unknown functions F_i will be sought by equating to zero the polynomials of the same degree in the right side of Eq. (V.9); namely,

$$\phi_{21} = 0$$
$$\phi_{31} = -\phi_{22}$$
$$\phi_{41} = -\phi_{32} - \phi_{23} \tag{V.10}$$
$$\cdots \cdots \cdots$$

When the conditions (V.4) are satisfied, an appropriate linear transformation of the variables will enable choosing X_1 and Y_1 in Eqs. (V.5) such that

$$X_1 = y \quad \text{and} \quad Y_1 = -x$$

Then $F_2 = x^2 + y^2$, and Eqs. (V.10) may be written in the form

$$\frac{\partial F_q}{\partial x} y - \frac{\partial F_q}{\partial y} x = H_q \qquad q = 2, 3, 4, \ldots \tag{V.11}$$

Introducing polar coordinates $x = r \cos \varphi$ and $y = r \sin \varphi$ gives us

$$\frac{\partial F_q}{\partial x} y - \frac{\partial F_q}{\partial y} x = -\frac{\partial F_q}{\partial \varphi}$$

Equation (V.11) may be rewritten as

$$\frac{df_q}{d\varphi} = -h_q(\varphi) \qquad q = 2, 3, 4, \ldots$$

where
$$F_q(r \cos \varphi, r \sin \varphi) = r^q F_q(\cos \varphi, \sin \varphi) = r^q f_q(\varphi)$$
$$H_q(r \cos \varphi, r \sin \varphi) = r^q H_q(\cos \varphi, \sin \varphi) = r^q h_q(\varphi) \tag{V.12}$$

A necessary condition for the solution of this equation is

$$\int_0^{2\pi} h_q(\varphi) \, d\varphi = 0 \tag{V.13}$$

It is clear that this condition is also sufficient because, if it holds, we have

$$\frac{df_q(\varphi)}{d\varphi} = -h_q(\varphi) = \sum_{\nu=1}^{q} (a_\nu \cos \nu\varphi + b_\nu \sin \nu\varphi)$$

and, therefore,

$$f_q(\varphi) = \sum_{\nu=1}^{q} \left(\frac{a_\nu}{\nu} \sin \nu\varphi - \frac{b_\nu}{\nu} \cos \nu\varphi \right) + \text{const}$$

If q is odd, the condition (V.13) is obviously satisfied. If, however, q is even, it may happen that the integral does not vanish and thus we are stopped in our procedure. We consider the following two cases.

CASE 1.   WHEN THE CONDITION (V.13) CEASES TO HOLD AS q INCREASES

Suppose that $f_2, f_3, \ldots, f_{q-1}$ are determined such that they satisfy the first $q - 2$ equations of (V.12), but f_q does not satisfy the $(q - 1)$st equation. In this case we replace the $(q-1)$st equation by

$$\frac{df_q(\varphi)}{d\varphi} = -h_q(\varphi) + c_0$$

$$(V.14)$$

where

$$c_0 = \frac{1}{2\pi} \int_0^{2\pi} h_q(\varphi) \, d\varphi$$

and thus achieve the integrability for f_q. Returning to the cartesian coordinates, we obtain, after multiplication by r^q,

$$\frac{\partial F_q}{\partial x} y - \frac{\partial F_q}{\partial y} x = H_q(x,y) - c_0(x^2 + y^2)^{q/2} \qquad (V.15)$$

It is noted that q is even in this case. We stop our procedure at this point and consider the polynomial

$$F(x,y) = F_2 + F_3 + \cdots + F_{q-1} + F_q \qquad (V.16)$$

Then, for a sufficiently small value of C, $F = C$ represents a closed curve encircling the singularity. If c_0 is negative, we see, from Eqs. (V.9), (V.15), and (V.16), that

$$\frac{dF}{d\tau} = \frac{\partial F}{\partial x} X + \frac{\partial F}{\partial y} Y > 0$$

so that the representative point on the integral curve will cross the closed curve from the inside to the outside as τ increases. Hence the singularity is an unstable focus. If c_0 is positive, on the other hand, the representative point will cross the closed curve in the reversed direction, so that the singularity is a stable focus.

CASE 2. WHEN THE CONDITION (V.13) HOLDS FOR ALL VALUES OF q

In this case Eqs. (V.11) are solved for all values of q; thus we obtain

$$F(x,y) = F_2 + F_3 + \cdots + F_q + \cdots \qquad (V.17)$$

which satisfies

$$\frac{\partial F}{\partial x} X + \frac{\partial F}{\partial y} Y = 0 \qquad (V.18)$$

The solution of Eqs. (V.5) is given by

$$F(x,y) = C \qquad (V.19)$$

which represents a family of closed curves encircling the singularity. Hence the singularity is a center.

In order to illustrate the use of the foregoing procedure, we investigate the type of the singularity at the origin of the system of Eqs. (8.25) in Chap. 8; namely,

$$\frac{d\xi}{d\tau} = X_1 + X_2 + X_3 \equiv X \qquad \frac{d\eta}{d\tau} = Y_1 + Y_2 + Y_3 \equiv Y$$

where
$$
\begin{aligned}
X_1 &= a_2\eta & Y_1 &= b_1\xi \\
X_2 &= -\tfrac{3}{8}y_0\xi^2 - \tfrac{9}{8}y_0\eta^2 & Y_2 &= \tfrac{3}{4}y_0\xi\eta \\
X_3 &= -\tfrac{3}{8}\xi^2\eta - \tfrac{3}{8}\eta^3 & Y_3 &= \tfrac{3}{8}\xi^3 + \tfrac{3}{8}\xi\eta^2
\end{aligned}
\qquad \text{(V.20)}
$$

It is clear that, if $a_2 b_1 > 0$, the singularity is a saddle point. We shall show that, if $a_2 b_1 < 0$, the singularity is a center.

We put

$$F_2(\xi,\eta) = a\xi^2 + 2b\xi\eta + c\eta^2$$

and by substituting this into

$$\phi_{21} = \frac{\partial F_2}{\partial \xi} X_1 + \frac{\partial F_2}{\partial \eta} Y_1 = 0$$

we obtain

$$b_1 b\xi^2 + (a_2 a + b_1 c)\xi\eta + a_2 b\eta^2 = 0 \qquad \text{(V.21)}$$

If we choose a, b, and c such that[1]

$$a = -b_1 \qquad b = 0 \qquad \text{and} \qquad c = a_2$$

Eq. (V.21) is identically satisfied and $F_2(\xi,\eta)$ becomes

$$F_2(\xi,\eta) = -b_1\xi^2 + a_2\eta^2 \qquad \text{(V.22)}$$

Since a_2 and b_1 have opposite signs, the equation $F_2(\xi,\eta) = \text{constant}$ represents a family of concentric ellipses around the origin. By substituting Eqs. (V.20) and (V.22) into

$$\frac{\partial F_3}{\partial \xi} X_1 + \frac{\partial F_3}{\partial \eta} Y_1 = -\frac{\partial F_2}{\partial \xi} X_2 - \frac{\partial F_2}{\partial \eta} Y_2$$

we obtain

$$F_3(\xi,\eta) = -\tfrac{3}{4}y_0\eta(\xi^2 + \eta^2)$$

By proceeding analogously, we obtain

$$F_4(\xi,\eta) = -\tfrac{3}{16}(\xi^2 + \eta^2)^2 \qquad F_5 = F_6 = \cdots = 0$$

Therefore Eq. (V.18) is satisfied by

$$F(\xi,\eta) = -b_1\xi^2 + a_2\eta^2 - \tfrac{3}{4}y_0\eta(\xi^2 + \eta^2) - \tfrac{3}{16}(\xi^2 + \eta^2)^2$$

and the solution of Eqs. (V.20), that is, $F(\xi,\eta) = C$, represents a family of closed curves encircling the origin. Hence it is concluded that the singularity is a center.[2]

---

[1] We assume that $a_2 > 0$. If $a_2 < 0$, we put

$$a = b_1 \qquad b = 0 \qquad \text{and} \qquad c = -a_2$$

Then $F_2(\xi,\eta) = b_1\xi^2 - a_2\eta^2$ is positive definite.

[2] An example is also given in Sec. 13.4b, where the singularity with the conditions (V.4) turns out to be a focus.

As mentioned in Secs. 8.7 and 9.3, we made use of an electronic synchronous switch in order to prescribe the initial condition θ_0, that is, the phase angle of the applied voltage at which the circuit was closed. An ordinary mechanical switch (timed mechanically or even electronically) is not adequate for this purpose, because it has a time lag which may not be constant for every operation.

Figure VI.1 shows the circuit diagram of an electronic switch which is used in our experiments. The main tubes V_1 and V_2 are thyratrons connected in inverse-parallel, thus allowing the alternating current to flow in both directions. These tubes are electronically controlled by auxiliary thyratrons V_3 and V_4 incorporated with a peaking transformer PT and a phase shifter PS.

The sequence of operation is as follows: First, the manual switch SW is open. Then the magnetic relay MC is not energized; its contacts a, b are also open. Therefore, the tubes V_3 and V_4 are nonconductive. Since no currents are flowing through the resistors R_1 and R_2, there are no voltage drops across R_1 and R_2 and the grids of V_1 and V_2 are kept negative by d-c biases E_{21} and E_{22}, respectively. Hence, the tubes V_1 and V_2 are also blocked.[1]

Next, suppose that the switch SW is closed at the instant marked in Fig. VI.2a. Then the relay MC is energized and the contacts a, b close; but this would not give rise to the simultaneous firing of V_3 and V_4. These tubes will fire at the instant when the positive peaks from the

---

[1] A thyratron fires (under positive anode voltage) when the grid voltage is raised over the critical grid potential. After breakdown, the anode current cannot be stopped or influenced by grid control. Conduction can be stopped only by reducing the current to zero long enough for the tube to deionize. During conduction, the arc drop (about 10 volts in our case) is slightly affected by a change in the anode current.

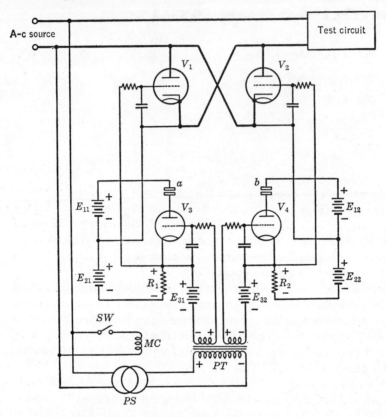

Figure VI.1　Connection diagram of an electronic switch. V_1, V_2: main thyratron tubes. V_3, V_4: thyratron tubes for control. PT: peaking transformer. PS: phase shifter. MC: magnetic relay.

peaking transformer PT are impressed on their grids against the negative biases E_{31} and E_{32} (Fig. VI.2b).　After breakdown of V_3 and V_4, the voltage drops which develop across R_1 and R_2 exceed the biases E_{21} and E_{22}, respectively, and the grid potentials of V_1 and V_2 become positive as shown in Fig. VI.2c.　As a consequence, V_1 and V_2 fire alternately with the alternation of the main current.　An example of the waveform of the main current is illustrated in Fig. VI.2d.

　　We thus see that, regardless of the time of closure of the manual switch SW, the main current always starts at a predetermined phase angle, that is, θ_0, of the voltage wave as determined by the instant at which the peaked positive voltages are impressed on the grids of V_3 and V_4.　This phase angle may arbitrarily be chosen by the setting of the phase shifter PS which energizes the peaking transformer PT.

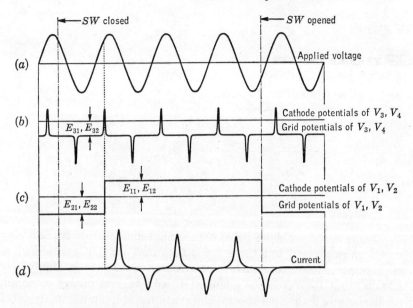

FIGURE VI.2   Waveforms of the voltage and current, illustrating operation of the electronic switch.

It is also mentioned that when the manual switch *SW* is opened, the relay *MC* is deenergized and the contacts *a, b* become open, stopping the firing of V_3 and V_4. Simultaneously, the negative biases are impressed on the grids of V_1 and V_2, but the discharge continues until the anode voltage is reduced to zero (Fig. VI.2*d*).

The accuracy of timing of the electronic synchronous switch is determined by the ionization time of thyratrons as well as the error in indication of the phase shifter. The former is of order of 10 microseconds, which corresponds to 0.2 degree in the phase angle[1] (1 cycle of a-c voltage corresponds to 360 degrees), and the latter may be kept within 0.3 degree if the phase shifter is carefully constructed in this respect. Consequently, the total error in the phase angle θ_0 may be less than 0.5 degree.

[1] The supply frequency is 60 cps.

Problems

Some of the problems listed here are not difficult to solve and may be useful as exercises,[1] but there are others that may be considered more appropriate as subject matter for research work. It is highly recommended that the reader be acquainted with various phases of nonlinear oscillations by both experimental observation and computer analysis before he gets into an elaborate mathematical treatment. For this purpose a number of problems pertaining to experimental study and computer analysis are also included in the list.

Chapter 1

1. Using the perturbation method, obtain the periodic solution of Rayleigh's equation

$$\frac{d^2x}{dt^2} - \mu\left[1 - \frac{1}{3}\left(\frac{dx}{dt}\right)^2\right]\frac{dx}{dt} + x = 0$$

to the second-order approximation, that is, to the order of μ^2.

2. Consider a spring which has the characteristic expressed by

$$\text{Spring force} = \kappa(1 \pm a^2x^2)x$$

where x is the deflection of the spring and κ and a^2 are positive constants. The plus sign applies to a "hard" spring, and the minus sign to a "soft" spring. If a mass M is attached to such a spring, the equation applying to the system is

$$M\frac{d^2x}{dt^2} + \kappa(1 \pm a^2x^2)x = 0$$

---

[1] Since the mathematical treatment in the text is terse, it is suggested that the reader follow it by his own derivation.

Assume that a^2 is small. Then, by using the perturbation method, solve this equation with the initial conditions

$$x(0) = A \qquad \text{and} \qquad \left(\frac{dx}{dt}\right)_{t=0} = 0$$

An exact solution can be obtained in terms of the elliptic function (see Refs. 18, p. 75, and 70, p. 24). Compare the result obtained by the perturbation method with the exact solution.

3. The averaging method is sometimes used to replace a certain type of nonlinear differential equation with a so-called "equivalent linear equation" which has the same solution to the first-order approximation. Using the relations given by Eqs. (1.101), show that an equivalent linear equation for Eq. (1.92) is given by

$$\ddot{x} - \frac{2\dot{r}}{r}\,\dot{x} + (1 - 2\theta)x = 0$$

Apply this procedure to van der Pol's equation (1.31) and obtain the equivalent linear equation

$$\ddot{x} - \mu\left(1 - \frac{r^2}{4}\right)\dot{x} + x = 0$$

4. By using the method of harmonic balance, obtain the harmonic solution of Duffing's equation:

$$\frac{d^2x}{d\tau^2} + 0.2\frac{dx}{d\tau} + x^3 = B\cos\tau$$

for values of B between 0 and 1.0. Calculate the amplitudes of the fundamental and third-harmonic components, and plot the result against B.

Chapter 2

5. Classify the singular points of the following differential equations:

(a) $\dfrac{dy}{dx} = \dfrac{2x}{y}$ (b) $\dfrac{dy}{dx} = -\dfrac{y}{x}$

(c) $\dfrac{dy}{dx} = \dfrac{x + y}{x - y}$ (d) $\dfrac{dy}{dx} = \dfrac{y - xy}{x - xy}$

(e) $\dfrac{dy}{dx} = \dfrac{y}{x^2}$ (f) $\dfrac{dy}{dx} = \dfrac{-x + 2x^2 - x^3}{y}$

For the equations (e) and (f) follow a procedure similar to that described in Secs. 8.5b and 8.6b.

6. By using the isocline method, construct a phase-plane diagram for the following equations:

Van der Pol's equation: $\dfrac{d^2x}{dt^2} - (1 - x^2)\dfrac{dx}{dt} + x = 0$

Rayleigh's equation: $\dfrac{d^2x}{dt^2} - \left[1 - \dfrac{1}{3}\left(\dfrac{dx}{dt}\right)^2\right]\dfrac{dx}{dt} + x = 0$

Mass on hard spring: $\dfrac{d^2x}{dt^2} + (1 + 0.2x^2)x = 0$

Mass on soft spring: $\dfrac{d^2x}{dt^2} + (1 - 0.2x^2)x = 0$

Is there any other method which is more convenient for obtaining solution curves of each equation on the $x\ dx/dt$ plane?

7. A pendulum with viscous damping is governed by the differential equation

$$\frac{d^2x}{dt^2} + k\frac{dx}{dt} + \sin x = 0$$

where k is the damping coefficient. By using the isocline method, investigate the behavior of the motion of the pendulum on the $x\ dx/dt$ plane. The damping coefficient k is chosen as 0.5, 2, and 4.

8. The operation of a synchronous motor is governed, to a first approximation, by a differential equation similar to that for the pendulum with viscous damping, that is,

$$\frac{d^2x}{dt^2} + k\frac{dx}{dt} + c\sin x = L$$

in which the term d^2x/dt^2 represents the inertia torque due to the mass of the rotor and its connected load, the term $k\ dx/dt$ arises from the torque due to the slip of the damper winding relative to the rotating field, the term $c\sin x$ is the torque due to the angle between the fields of the rotor and armature, and L represents the torque of the external load on the motor. The angle x is measured from an axis which rotates with the synchronous speed of the rotating electric field. Assuming that $c > L$, plot solution curves of this equation on the $x\ dx/dt$ plane and discuss the pullout problem of the synchronous motor (see Refs. 70, p. 160, and 100, p. 70).

9. Two identical d-c series generators are connected in parallel. The field winding of each generator has a resistance r and an inductance L. The armature resistance is negligible. The two generators in parallel supply current to a load of resistance R. The relation between the generated voltage e and the current i of each machine is $e = a \tan^{-1} bi$, where a and b are positive constants.

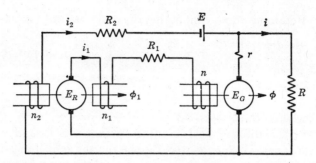

FIGURE P.1

Find the equilibrium values of the currents, i_1 and i_2, of these machines. Discuss the stability of the equilibrium by plotting trajectories of the representative point on the i_1i_2 plane.

If the field winding of either generator is connected in series with the armature of the other generator, one has $e_1 = a \tan^{-1} bi_2$, and vice versa. Determine equilibrium conditions for this mode of operation and discuss their stability.

Note: See Refs. 3, p. 233, and 76, p. 82.

10. The schematic diagram of a rototrol-controlled d-c generator is shown in Fig. P.1, for which the circuit equations may be written as

$$n \frac{d\phi}{dt} + R_1 i_1 + n_1 \frac{d\phi_1}{dt} = E_R$$

$$n_2 \frac{d\phi_1}{dt} + R_2 i_2 = E - Ri$$

$$k_1 i_1 + k_2 i_2 = E_R = S_1 \phi_1$$
$$(R + r)i = E_G = S\phi$$

where E_R and E_G are the induced voltages of the rototrol and the generator and are proportional to the field fluxes ϕ_1 and ϕ, respectively. Since the rototrol is usually designed to operate within the linear portion of the magnetization curve, the induced voltage E_R is assumed to be proportional to the field currents i_1 and i_2. However, the magnetic saturation of the generator field is taken into account; its characteristic is assumed to be

$$i_1 = a_1 \phi + a_3 \phi^3$$

Since the rototrol is operated under a "tuned" condition, we have the relationship

$$R_1 = k_1$$

Retaining ϕ and eliminating the other variables in these equations gives us

$$\frac{nL_2}{k_2}\frac{d^2\phi}{dt^2} + \left[\frac{nR_2}{k_2} - \frac{R}{R+r}\frac{SM}{k_2} + \left(\frac{L_1R_2}{k_2} + M\right)(a_1 + 3a_3\phi^2)\right]\frac{d\phi}{dt}$$
$$+ \frac{R}{R+r}S\phi = E$$

where $L_1 \,(= n_1k_1/S_1)$ and $L_2 \,(= n_2k_2/S_1)$ are the self-inductances of the rototrol field windings with turns n_1 and n_2, respectively, and $M \,(= n_1k_2/S_1 = n_2k_1/S_1)$ is the mutual inductance between these windings. Because of the presence of the mutual inductance M, the coefficient of $d\phi/dt$ can become negative for $\phi = (R + r)E/RS$; and an undamped oscillation, i.e., a hunting oscillation, results in this case.

Show that the condition for the occurrence of hunting is given by

$$\left(\frac{nR_2}{S} - \frac{MR}{R+r}\right)\left(\frac{dE_G}{di_1}\right)_0 + L_1R_2 + k_2M < 0$$

where $(dE_G/di_1)_0$ denotes the value of dE_G/di_1 at $E_G = (R + r)E/R$. By applying the phase-plane analysis, plot a limit cycle representing such a hunting oscillation in the $\phi\, d\phi/dt$ plane.

Note: For further details, see Chihiro Hayashi, Hunting Oscillation in Nonlinear Control Systems and Its Stabilization, *Fachtagung Regelungstechnik*, Beitr. Nr. 35, Heidelberg, 1956.

11. A mechanical system consisting of a mass mounted on a linear spring and sliding over a smooth surface with Coulomb friction is governed by the equation

$$\frac{d^2x}{dt^2} + h\,\text{sgn}\,\frac{dx}{dt} + x = 0$$

where h is a positive constant and the function sgn dx/dt has the value of $+1$ for positive dx/dt and -1 for negative dx/dt.

Construct a phase-plane diagram for this system by the δ method, with use of the initial conditions that $dx/dt = 0$ and $x = 20h$ at $t = 0$. Sketch a curve for x as a function of t until motion ceases.

12. Consider the vacuum-tube oscillator shown in Fig. P.2. The grid and plate circuits are inductively coupled. The differential equation of the oscillating circuit is

$$L\frac{di}{dt} + Ri + \frac{1}{C}\int (i - i_p)\, dt = 0$$

The grid potential e_g is given by

$$e_g = M\frac{di}{dt}$$

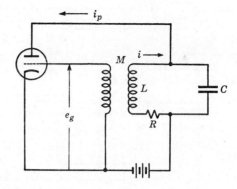

FIGURE P.2

It is assumed that the plate current i_p depends only upon the grid potential e_g and that the relation between these two quantities is given by

$$i_p = f(e_g) = a_1 e_g + a_3 e_g{}^3 + a_5 e_g{}^5$$

where a_1, a_3, and a_5 are constants determined by the tube characteristic. From the above equations one readily obtains

$$\frac{d^2 e_g}{d\tau^2} - \omega_0(Ma_1 - RC + 3Ma_3 e_g{}^2 + 5Ma_5 e_g{}^4)\frac{de_g}{d\tau} + e_g = 0$$

where $\omega_0 = 1/\sqrt{LC}$ and $\tau = \omega_0 t$.

Apply the phase-plane analysis to this equation, and discuss the behavior of the oscillator for varying M. Plot phase-plane diagrams for the following two cases:

(i) $a_1 > 0$ $a_3 < 0$ $a_5 = 0$
(ii) $a_1 > 0$ $a_3 > 0$ $a_5 < 0$

Note: It will be observed in case i that, by increasing the mutual inductance M between the grid and plate circuits, the self-excitation starts smoothly as soon as a critical value $M = M_0$ of this parameter is reached. For $M > M_0$, the amplitude of oscillation steadily increases with increasing M. For decreasing M the phenomenon is reversible; the self-excitation disappears at $M = M_0$. This kind of self-excitation is known as a "soft" excitation.

On the other hand, case ii leads to a "hard" excitation. In this case, by increasing M, the self-excitation starts abruptly with a finite amplitude at $M = M_0$ and increases smoothly for $M > M_0$. For decreasing M the self-excitation does not disappear at $M = M_0$; it disappears abruptly at a certain value of M which is lower than M_0. Thus the phenomenon exhibits a kind of hysteresis.

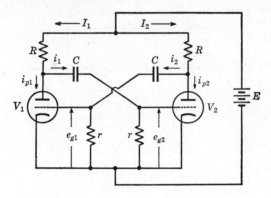

Figure P.3

13. The circuit diagram of Fig. P.3 shows an example of a symmetrical multivibrator. In this connection we neglect the grid current and the plate reaction. By use of the notations in the figure, the following equations are obtained:

$$RI_1 + \frac{1}{C} \int i_1 \, dt + ri_1 = E \qquad\qquad RI_2 + \frac{1}{C} \int i_2 \, dt + ri_2 = E$$

$$i_{p1} + i_1 = I_1 \qquad\qquad i_{p2} + i_2 = I_2$$

$$i_{p1} = \phi(e_{g1}) = \phi(ri_2) \qquad\qquad i_{p2} = \phi(e_{g2}) = \phi(ri_1)$$

where $i_p = \phi(e_g)$ is the nonlinear characteristic of the vacuum tubes V_1 and V_2. Setting up the differential equations for i_1 and i_2 and solving for di_1/dt and di_2/dt leads to

$$\frac{di_1}{dt} = \frac{P(i_1, i_2)}{U(i_1, i_2)} \qquad\qquad \frac{di_2}{dt} = \frac{Q(i_1, i_2)}{U(i_1, i_2)}$$

where

$$P(i_1, i_2) = (R + r)\frac{i_1}{C} - Rr\phi'(ri_2)\frac{i_2}{C}$$

$$Q(i_1, i_2) = (R + r)\frac{i_2}{C} - Rr\phi'(ri_1)\frac{i_1}{C}$$

$$U(i_1, i_2) = R^2 r^2 \phi'(ri_1)\phi'(ri_2) - (R + r)^2$$

The prime denotes the differentiation with respect to the argument ri.

Carry out the phase-plane analysis for $di_1/di_2 = P(i_1, i_2)/Q(i_1, i_2)$. The function $\phi'(ri)$ is assumed by the form $a \exp [-(bri)^2]$, where a and b are positive constants. When $U(i_1, i_2) = 0$, di_1/dt and di_2/dt become infinite, and the currents i_1 and i_2 vary abruptly. Hence, a discontinuous periodic solution is expected.

Note: For further details, see Refs. 3, p. 274, and 76, p. 627. The effect of the stray capacitance in a multivibrator circuit is discussed in Ref. 3 (Russian ed.), pp. 846–855.

14. An oscillatory system is governed by the equation

$$\frac{d^2x}{d\tau^2} + 0.2\frac{dx}{d\tau} + |x|x = 1.5 \cos 2\tau + 0.5$$

By using the δ method, construct a phase-plane diagram for this system with the initial conditions that $x = 0$ and $dx/d\tau = 0$ at $\tau = 0$. Compare the result of this analysis with that of Fig. 2.20.

15. By using the slope-line method, construct a phase-plane diagram for the differential equation

$$\frac{d^2x}{d\tau^2} + 0.1\frac{dx}{d\tau} + x^3 = 0.15 \cos \tau$$

The initial conditions are given as follows:

(i) $x = -0.15$ $(dx/d\tau)_{\tau=0} = 0$
(ii) $x = 0.65$ $(dx/d\tau)_{\tau=0} = 1.0$
(iii) $x = 0.18$ $(dx/d\tau)_{\tau=0} = 0.05$

For each of these cases, plot the solution curve until it practically retraces a closed trajectory which represents a periodic solution.

Chapter 3

16. Consider the system

$$\frac{dx}{dt} = f(x) + by \qquad f(0) = 0$$
$$\frac{dy}{dt} = cx + dy$$

where b, c, and d are constants and $bc + d^2 \neq 0$.* Show that the equilibrium, $x = y = 0$, of the system is asymptotically stable provided that

$$x[f(x) + dx] < 0 \qquad \text{and} \qquad x[df(x) - bcx] > 0 \qquad x \neq 0$$

Apply this criterion to the following equation:

$$\frac{d^2x}{dt^2} + \mu\left[1 - \frac{1}{3}\left(\frac{dx}{dt}\right)^2\right]\frac{dx}{dt} + x = 0 \qquad \mu > 0$$

and discuss the region of asymptotic stability.

Note: The proof is accomplished by use of the positive definite Liapunov function

$$V(x,y) = \int_0^x [df(\xi) - bc\xi]\, d\xi + \frac{1}{2}(dx - by)^2$$

* See Ref. 29 (English ed.), p. 43.

17. By using the perturbation method, obtain the expansions for the periodic solutions, $ce_0(\tau,q)$, $ce_1(\tau,q)$, . . . , $se_3(\tau,q)$, of Mathieu's equation.

Note: The results are given in Appendix I.

18. By following Whittaker's method of solution explained in Sec. 3.6b, derive the unstable solutions, Eqs. (3.69) to (3.72), of Mathieu's equation associated with the second and third unstable regions.

Chapter 4

19. Consider the differential equation

$$\frac{d^2v}{d\tau^2} + 2\delta \frac{dv}{d\tau} + f(v) = B \cos \tau$$

where $f(v)$ is a polynomial in v, and assume a subharmonic solution of the form

$$v(\tau) = x_{\frac{1}{3}} \sin \tfrac{1}{3}\tau + y_{\frac{1}{3}} \cos \tfrac{1}{3}\tau + x_1 \sin \tau + y_1 \cos \tau$$

By applying the stability condition (4.18), investigate the stability of this solution. Show that the vertical tangency of the characteristic curve occurs at the stability limit $\Delta(\delta) = 0$.

Chapter 5

20. A system governed by

$$\frac{d^2v}{d\tau^2} + k \frac{dv}{d\tau} + v^3 = B \cos \tau$$

is considered. Show that a harmonic oscillation having the amplitude r defined by $v = x \sin \tau + y \cos \tau$ and $r^2 = x^2 + y^2$ is unstable provided that

$$8 - 4\sqrt{1 - 3k^2} < 9r^2 < 8 + 4\sqrt{1 - 3k^2}$$

21. By a procedure analogous to that of Sec. 5.1b, draw frequency-response curves of the harmonic oscillation of a system governed by

$$\frac{d^2v}{d\tau^2} + k \frac{dv}{d\tau} + v^3 = B \cos \nu\tau + B_0$$

where $k = 0.1$ and $B_0 = 0.36$ (cf. Fig. 5.3).

Chapter 6

22. By making use of an analog-computer setup for Eq. (6.1), investigate the self-excitation of higher-harmonic oscillations when the

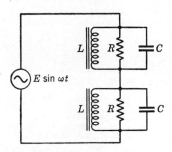

amplitude B of the external force is exceedingly raised. Observe the waveform of $v(\tau)$ for the values of B between 0 and 25.

Note: The second-harmonic oscillation is accompanied with subharmonic oscillations in a certain interval of B.

23. An electric circuit is shown in Fig. P.4, in which two parallel-resonance circuits are connected in series, each having equal values of L, R, and C. With reference to the study in Sec. 6.4, find the final amplitude of the self-excited oscillations in the steady state.

Note: Assume the fundamental component for the oscillations in the first unstable region, the nonoscillatory and second-harmonic components in the second unstable region, and the fundamental and third-harmonic components in the third unstable region.   Then, by using the method of harmonic balance, determine each component.

Chapter 7

24. Find the subharmonic solution of order $\frac{1}{3}$ for the differential equation

$$\frac{d^2v}{d\tau^2} + k\frac{dv}{d\tau} + v^3 = B\cos 3\tau$$

and test its stability by using Hill's equation as a stability criterion. The periodic solution is assumed by the form

$$v(\tau) = x_1\sin\tau + y_1\cos\tau + x_3\sin 3\tau + y_3\cos 3\tau$$

Show that the amplitude of the harmonic component, that is, $(x_3{}^2 + y_3{}^2)^{\frac{1}{2}}$, may be approximated by $B/8$, as it was in Sec. 7.3c (see Fig. 7.6).

25. A nondissipative system governed by the differential equation

$$\frac{d^2v}{d\tau^2} + v^3 = B\cos 3\tau$$

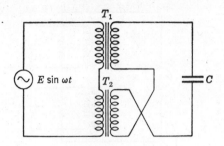

F<small>IGURE</small> P.5

is considered.   Show that the periodic solution $v_0(\tau) = y \cos \tau + w \cos 3\tau$ is stable provided that

$$(y + 4w)\, B\, \frac{dB}{dy} = -(2y + w)B\, \frac{dB}{dw} > 0$$

Thus one sees that the vertical tangency of the characteristic curve (y,B relation or w,B relation) results at the stability limit

$$\frac{dB}{dy} = \frac{dB}{dw} = 0$$

Note: Apply the stability condition (4.18).

26. In Prob. 25 replace the cubic function v^3 by a quintic function v^5.   Then show that the stability condition for the periodic solution $v_0(\tau)$ is given by

$$(5y^3 + 24y^2w + 18yw^2 + 24w^3)B\, \frac{dB}{dy}$$
$$= -(8y^3 + 15y^2w + 24yw^2 + 6w^3)B\, \frac{dB}{dw} > 0$$

27. Perform an experiment on the circuit of Fig. P.5 and seek periodic oscillations under various combinations of the voltage E and the capacitance C. T_1 and T_2 are transformers of identical rating.   One will readily obtain the subharmonic response of orders $\frac{1}{2}$ and $\frac{1}{3}$ by starting the oscillation with appropriate initial conditions.   It is to be noted that, under certain conditions, subharmonic components of orders $\frac{2}{5}$ and $\frac{3}{7}$ may predominantly appear in the secondary circuit.

Note: The same circuit appears in Fig. 6.11, where the self-excitation of higher harmonics is shown.   Subharmonic oscillations are likely to occur for comparatively low values of the applied voltage and large values of the capacitance.

Chapter 8

28. In the phase-plane diagram of Fig. 8.2, there are three singularities of points 1 to 3: points 1 and 3 are stable and point 2 is unstable. As the amplitude B of the external force increases, points 1 and 2 approach each other and finally coalesce. An abrupt increase in the amplitude of the harmonic oscillation results in this case. An integral curve, issuing from the point of coalescence and arriving at the singularity of point 3, represents the transient state during the jump.

Determine the value of B for which the jump occurs and draw a phase-plane diagram of Eq. (8.12) for this value of B. Also, by using Eqs. (8.3) and (8.13), plot the time-response curve of $v(\tau)$ during this transient.

Note: A jump in the amplitude from a higher value to a lower value occurs when B is decreased to a certain value which is lower than before.

29. Design a synchronous switch which employs silicon-controlled rectifiers instead of the thyratrons V_1 and V_2 in Fig. VI.1. This modification serves to reduce the voltage drop across the switch under conducting conditions.[1] When this new switch is available, perform an experiment as described in Sec. 8.7 and plot the regions of initial conditions which lead to the resonant and nonresonant oscillations.

Chapter 9

30. The phase-plane analysis of a subharmonic oscillation may be facilitated by using polar coordinates. Express Eqs. (9.5), (9.12), and (9.21) in terms of r and θ, where $x = r \cos \theta$ and $y = r \sin \theta$; then reproduce Figs. 9.1, 9.7, and 9.10 in polar coordinates. Make use of the method of iso-ϕ curves described in Sec. 2.4.

31. It was mentioned in Sec. 7.7 that the subharmonic oscillation of order $\frac{1}{2}$ occurs in a system governed by the differential equation

$$\frac{d^2v}{d\tau^2} + k\frac{dv}{d\tau} + |v|v = B \cos 2\tau + B_0$$

The phase-plane analysis of the subharmonic oscillation is described in Secs. 9.5 and 9.6, where the nonlinear term $|v|v$ is approximated by $c_1v + c_3v^3$ [Eq. (9.30)] for the convenience of analytical treatment.

By proceeding analogously, plot the integral curves of Eq. (9.21) for the following two cases:

(i) $k = 0.2$ $B = 1.5$ $B_0 = 0.5$
(ii) $k = 0.2$ $B = 0.6$ $B_0 = 0.4$

---

[1] A silicon-controlled rectifier has characteristics similar to those of a gas thyratron except that the forward drop is about one-tenth that of a thyratron.

Plot also, on the $v\dot{v}$ plane, the trajectories of periodic solutions and the regions of initial conditions which lead to different kinds of response.

Chapter 10

32. A specific example of Duffing's equation, that is,

$$\frac{d^2v}{d\tau^2} + 0.2\frac{dv}{d\tau} + v^3 = 0.3\cos\tau$$

is considered in Sec. 8.4 and also in Sec. 10.3a. By use of the relation given by Eq. (8.20), the singular points and integral curves of Fig. 8.2 can be reproduced on the $v(0)\dot{v}(0)$ plane. Compare the result obtained in this manner with that of Fig. 10.4, which gives more exact information than Fig. 8.2.

33. By using the mapping method, determine the regions of initial conditions which lead to different kinds of periodic solutions for the differential equation

$$\frac{d^2v}{d\tau^2} + k\frac{dv}{d\tau} + |v|v = B\cos 2\tau + B_0$$

The values of the system parameters are the same as in Prob. 31.

Note: Make use of the result of Prob. 31 to locate the fixed points (which are correlated with the periodic solutions) on the $v\dot{v}$ plane.

Chapter 11

34. Using Eqs. (11.12), calculate the amplitude r_0 and the non-oscillatory component z_0 of the periodic solution, Eqs. (11.9), and plot against B_0 for values between 0 and 0.06 of this parameter. The other system parameters are chosen as follows:

$$k_1 = 0.20 \qquad k_2 = 0.03 \qquad B = 0.50$$

Also, by using the stability conditions (11.18) and (11.19), test the stability of the periodic solution.

35. By making use of the block diagram illustrated in Fig. 11.4, investigate the occurrence of almost periodic oscillations of a system described by Eqs. (11.22). Use the same system parameters as given in Prob. 34.

36. By using the circuit of Fig. 11.12 and a procedure analogous to that of Sec. 11.3, perform an experiment to obtain almost periodic oscillations which develop from the subharmonic oscillation of order $\frac{1}{2}$. Compare the results obtained experimentally with those of the analytical study in Sec. 11.4c. Also, investigate the region of the system parameters in which almost periodic oscillations occur.

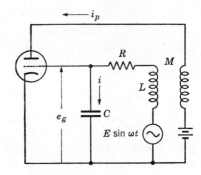

FIGURE P.6

Chapter 12

37. It has been shown in Sec. 12.6 that the regions of harmonic and ⅓-harmonic entrainments have an overlapping area (Fig. 12.7*b*). If the amplitude *B* and the frequency *ν* are chosen inside this region, either a harmonic or a ⅓-harmonic oscillation occurs depending on different values of initial conditions.

Carry out the phase-plane analysis for Eqs. (13.10), and determine the region of initial conditions which lead to the ⅓-harmonic response. Use the values of *B* and *ν* as chosen above. Also, by using the relation given by Eq. (13.4), reproduce this region on the *v dv/dτ* plane.

Note: The phase-plane analysis for Eqs. (13.10) is facilitated by changing the variables $x_{\frac{1}{3}}$ and $y_{\frac{1}{3}}$ into polar coordinates. Apply the method of iso-ϕ curves described in Sec. 2.4.

It is interesting to note that all initial conditions lead to the ½-harmonic response when Eqs. (13.9) are treated.

38. By making use of the block diagram of Fig. 12.9, carry out an analog-computer analysis and determine the regions of initial conditions, $v(0)$ and $(dv/d\tau)_{\tau=0}$, which lead to the harmonic and ⅓-harmonic responses. Use the same values of *B* and *ν* as in Prob. 37. Compare the result with that of Prob. 37.

39. Figure P.6 shows a typical physical system in which the phenomenon of entrainment occurs. This vacuum-tube oscillator contains a source *E* sin *ωt* of alternating voltage in the grid circuit, as indicated in the figure. Thus, the frequency of the oscillator is entrained by the frequency *ω* of the external source.

The differential equations for the system are

$$L \frac{di}{dt} + Ri + e_g - M \frac{di_p}{dt} = E \sin \omega t$$

$$C \frac{de_g}{dt} = i$$

The current in the grid itself is ignored in deriving these equations.   It is assumed that the plate current i_p depends only upon the grid potential e_g and that the relation between these two quantities is given by

$$i_p = f(e_g) = Se_g \left(1 - \frac{e_g^2}{3V_s^2} \right)$$

where S and V_s are positive constants.   The quantity S is called the steepness of the characteristic, and V_s is called the saturation potential. These quantities should be appropriately chosen in order to fit the function $i_p = f(e_g)$ to the experimental curve.   It is convenient to introduce the following quantities:

$$v = \frac{e_g}{V_s} \qquad \mu = \frac{MS - RC}{\sqrt{LC}} \qquad \gamma = \frac{MS}{MS - RC}$$

$$B = \frac{E}{V_s} \qquad \nu = \sqrt{LC}\,\omega \qquad \tau = \frac{1}{\sqrt{LC}}\, t$$

Then the differential equation for the system becomes

$$\frac{d^2v}{d\tau^2} - \mu(1 - \gamma v^2)\frac{dv}{d\tau} + v = B \sin \nu\tau$$

In order that self-excited oscillations can occur, μ must be positive.   It is to be noted that $1/\sqrt{LC}$ is the natural frequency of the system.

Prepare such an oscillator circuit and investigate the entrainment of self-excited oscillations by changing the amplitude E and the frequency ω of the external source.   Plot the regions of harmonic, higher-harmonic, and subharmonic entrainments on the $B\nu$ plane.

40. Consider a system governed by the differential equation

$$\frac{d^2v}{d\tau^2} - \mu(1 - \gamma v^2)\frac{dv}{d\tau} + v^3 = B \cos \nu\tau$$

where μ is a small positive quantity and γ is positive also.   To investigate the subharmonic entrainment at one-third the driving frequency ν, assume a periodic solution of the form

$$v_0(\tau) = \frac{B}{\omega_0^2 r_1^2 - \nu^2} \cos \nu\tau + b_1 \sin \frac{1}{3} \nu\tau + b_2 \cos \frac{1}{3} \nu\tau$$

where the amplitude of the harmonic component, $B/(\omega_0^2 r_1^2 - \nu^2)$, is given by Eqs. (12.44) and (12.45).[1]   Determine the amplitudes b_1 and b_2 and plot the region of subharmonic entrainment on the $B\nu$ plane.

[1] Since $(1 - r_1^2)^2 \ll \sigma_1^2$ in the region where the subharmonic entrainment occurs, we neglect a term that contains $\sin \nu\tau$.

41. By proceeding analogously to Sec. 12.8, investigate the phenomenon of harmonic entrainment which occurs in a system governed by the differential equation

$$\frac{d^2v}{d\tau^2} - \mu\left[1 - \gamma\left(\frac{dv}{d\tau}\right)^2\right]\frac{dv}{d\tau} + v^3 = B\cos\nu\tau$$

where μ and γ are positive constants and μ is assumed to be small.

Chapter 13

42. By proceeding analogously to Secs. 8.6*b* and 13.4*a*, investigate the integral curves near the singularity, 1 or 2, of Fig. 13.9*b*.

43. It was mentioned in Prob. 39 that entrained oscillations of a forced oscillator circuit occur at the harmonic, higher-harmonic, and subharmonic frequencies provided the parameters B and ν (or E and ω) are given inside the regions of entrainment.

Obtain experimentally almost periodic oscillations which occur in the same circuit when the amplitude E and the frequency ω of the external source are chosen outside these regions.

44. By making use of an analog computer, set up a block diagram for the solution of Eq. (13.36) and investigate the occurrence of almost periodic oscillations as the amplitude B and the frequency ν of the external force vary. Compare the results of the computer analysis with those obtained by solving the autonomous system of Eqs. (13.38).

Note: The region of entrainment at one-third the driving frequency ν is analytically sought in Prob. 40.

45. By a procedure analogous to that of Sec. 13.7, investigate the occurrence of almost periodic oscillations in a system described by the same differential equation as in Prob. 41.

Bibliography

The literature on nonlinear differential equations and their applications to physical systems is extensive. The list of books and papers given here is not intended to be exhaustive. Rather, it is a collection of material which has been referred to in the text and which may well be helpful to the inquiring reader.[1]

1. Actes du colloque international des vibrations non linéaires (Ile de porquerolles, 1951), *Publ. Sci. Tech. Min. Air (France)*, **281** (1953).
2. Alimov, Yu. I.: On the Construction of Liapunov Functions for Systems of Linear Differential Equations with Constant Coefficients, *Siberian Math. J.*, **2**: 3–6 (1961) (in Russian).
3. Andronow, A. A., and S. E. Chaikin: "Theory of Oscillations," Princeton University Press, Princeton, N.J., 1949; enlarged 2d ed. by A. A. Andronow, A. A. Witt, and S. E. Chaikin, Fizmatgiz, Moscow, 1959 (in Russian).
4. Andronow, A. A., and A. A. Witt: Zur Theorie des Mitnehmens von van der Pol, *Arch. Elektrotech.*, **24**: 99–110 (1930).
5. Appleton, E. V., and B. van der Pol: On a Type of Oscillation-hysteresis in a Simple Triode Generator, *Phil. Mag.*, **6-43**: 177–193 (1922).
6. Bendixson, I.: Sur les courbes définies par des équations différentielles, *Acta Math.*, **24**: 1–88 (1901).
7. Besicovitch, A. S.: "Almost Periodic Functions," Cambridge University Press, London, 1938; reprint, Dover Publications, Inc., New York, 1954.
8. Bessonov, L. A.: "Auto-oscillations in Electric Circuits Containing Iron Cores," Gosenergoizdat, Moscow, 1958 (in Russian).
9. Blair, K. W., and W. S. Loud: Periodic Solutions of $x'' + cx' + g(x) = Ef(t)$ under Variation of Certain Parameters, *J. Soc. Ind. Appl. Math.*, **8**: 74–101 (1960).
10. Bogoliuboff, N. N., and Yu. A. Mitropolsky: "Asymptotic Methods in the Theory of Nonlinear Oscillations," Fizmatgiz, Moscow, 1963 (in Russian);

[1] A good summary of the literature on nonlinear mechanics and nonlinear control theory, existing in 1956, is given by Higgins [42]. See also Ref. 43, by the same author. A number of collected papers are devoted to the study of nonlinear oscillations and nonlinear circuit analysis; see Refs. 1, 59, 102, 103, 104, and 106.

English translation, Gordon and Breach, Science Publishers, Inc., New York, 1961.

11. Bohr, H.: Zur Theorie der fastperiodischen Funktionen, *Acta Math.*, **45**: 29–127 (1924); **46**: 101–214 (1925); **47**: 237–281 (1926).

12. Brillouin, L.: A Practical Method for Solving Hill's Equation, *Quart. Appl. Math.*, **6**: 167–178 (1948).

13. Buland, R. N.: Analysis of Nonlinear Servos by Phase-plane-delta Method, *J. Franklin Inst.*, **257**: 37–48 (1954).

14. Cartwright, M. L.: Forced Oscillations in Nearly Sinusoidal Systems, *J. Inst. Elec. Engrs. (London)*, **95**(3): 88–96 (1948).

15. Cartwright, M. L.: Nonlinear Vibrations, *Brit. Assoc. Advan. Sci.*, **6**(21): April, 1949.

16. Coddington, E. A., and N. Levinson: "Theory of Ordinary Differential Equations," McGraw-Hill Book Company, New York, 1955.

17. Cosgriff, R. L.: "Nonlinear Control Systems," McGraw-Hill Book Company, New York, 1958.

18. Cunningham, W. J.: "Introduction to Nonlinear Analysis," McGraw-Hill Book Company, New York, 1958.

19. Den Hartog, J. P.: "Mechanical Vibrations," 4th ed., McGraw-Hill Book Company, New York, 1956.

20. Den Hartog, J. P., and S. J. Mikina: Forced Vibrations with Nonlinear Spring Constants, *Trans. ASME*, **54**: 157–164 (1932).

21. Duffing, G.: "Erzwungene Schwingungen bei veränderlicher Eigenfrequenz und ihre technische Bedeutung," Sammlung Vieweg, Braunschweig, 1918.

22. Floquet, G.: Sur les équations différentielles linéaires à coefficients périodiques, *Ann. Ecole Norm. Super.*, **2-12**: 47–88 (1883).

23. Gille, J. C., P. Decaulne, and M. Pélegrin: "Méthodes modernes d'étude des systèmes asservis," Dunod, Paris, 1960.

24. Gille, J. C., M. J. Pélegrin, and P. Decaulne: "Feedback Control Systems," McGraw-Hill Book Company, New York, 1959.

25. Gillies, A. W.: The Singularities and Limit Cycles of an Autonomous System of Differential Equations of the Second Order Associated with Nonlinear Oscillations, *Symp. Nonlinear Oscillations (Intern. Union Theoret. Appl. Mech.)*, Kiev, 1961.

26. Goldstein, S.: Mathieu Functions, *Trans. Cambridge Phil. Soc.*, **23**: 303–336 (1927).

27. Goto, E.: The Parametron, A Digital Computing Element Which Utilizes Parametric Oscillation, *Proc. IRE*, **47**: 1304–1316 (1959).

28. Hahn, W.: Behandlung von Stabilitätsproblemen mit der zweiten Methode von Ljapunov, *Nichtlineare Regelungsvorgänge*, Verlag R. Oldenbourg, München, 1956.

29. Hahn, W.: Theorie und Anwendung der direkten Methode von Ljapunov, *Ergeb. Math. Grenzg.*, Neue Folge, **22**, Springer-Verlag, Berlin, 1959; English translation, Prentice-Hall, Inc., Englewood Cliffs, N.J., 1963.

30. Hámos, L. V.: Beitrag zur Frequenzanalyse von nichtlinearen Systemen, *Fachtagung Regelungstechnik*, Beitr. 65, Heidelberg, 1956.

31. Hayashi, C.: Studies on Low Frequency Oscillations in Transformer Cir-

cuits, *Elec. Rev.*, **29**: 599–608, 670–677, 732–740 (1941); **30**: 418–425, 479–495, 551–560, 597–609 (1942) (in Japanese).

32. Hayashi, C.: "Forced Oscillations in Nonlinear Systems," Nippon Printing and Publishing Co., Ltd., Osaka, Japan, 1953.

33. Hayashi, C.: Forced Oscillations with Nonlinear Restoring Force, *J. Appl. Phys.*, **24**: 198–207 (1953).

34. Hayashi, C.: Stability Investigation of the Nonlinear Periodic Oscillations, *J. Appl. Phys.*, **24**: 344–348 (1953).

35. Hayashi, C.: Subharmonic Oscillations in Nonlinear Systems, *J. Appl. Phys.*, **24**: 521–529 (1953).

36. Hayashi, C.: Initial Conditions for Certain Types of Nonlinear Oscillations, *Symp. Nonlinear Circuit Analysis*, **6**: 63–92 (1956), Polytechnic Institute of Brooklyn, New York.

37. Hayashi, C.: Quasi-periodic Oscillations in Nonlinear Control Systems, *Autom. Remote Control (First Intern. Congr. IFAC)*, vol. 2, 889–894, Butterworth Scientific Publications, London, 1961.

38. Hayashi, C., Y. Nishikawa, and M. Abe: Subharmonic Oscillations of Order One-half, *Trans. IRE Circuit Theory*, **CT-7**: 102–111 (1960).

39. Hayashi, C., H. Shibayama, and Y. Nishikawa: Frequency Entrainment in a Self-oscillatory System with External Force, *Trans. IRE Circuit Theory*, **CT-7**: 413–422 (1960).

40. Hayashi, C., and Y. Nishikawa: Initial Conditions Leading to Different Types of Periodic Solutions for Duffing's Equation, *Symp. Nonlinear Oscillations (Intern. Union Theoret. Appl. Mech.)*, Kiev, 1961.

41. Hayashi, C., H. Shibayama, and Y. Ueda: Quasi-periodic Oscillations in a Self-oscillatory System with External Force, *Symp. Nonlinear Oscillations (Intern. Union Theoret. Appl. Mech.)*, Kiev, 1961.

42. Higgins, T. J.: A Resume of the Development and Literature of Nonlinear Control-system Theory, *Trans. ASME*, **79**: 445–449 (1957).

43. Higgins, T. J.: A Resume of the Basic Literature on Nonlinear System Theory (with Particular Reference to Lyapunov's Methods), *Workshop on Lyapunov's Direct Method (Joint Autom. Control Conf.)*, M.I.T., Cambridge, 1960.

44. Hill, G. W.: On the Part of the Motion of the Lunar Perigee, *Acta Math.*, **8**: 1–36 (1886).

45. Hurwitz, A.: Über die Bedingungen, unter Welchen eine Gleichung nur Wurzeln mit negativen reellen Teilen besitzt, *Math. Ann.*, **46**: 273–284 (1895).

46. Ince, E. L.: On a General Solution of Hill's Equation, *Monthly Notices Roy. Astron. Soc.*, **75**: 436–448 (1915); **76**: 431–442 (1916); **78**: 141–147 (1917).

47. Ince, E. L.: Researches into the Characteristic Numbers of the Mathieu Equation, *Proc. Roy. Soc. Edinburgh*, **46**: 20–29, 316–322 (1925–26); **47**: 294–301 (1926–27).

48. Ince, E. L.: Tables of the Elliptic-cylinder Functions, *Proc. Roy. Soc. Edinburgh*, **52**: 355–423 (1931–32).

49. Jacobsen, L. S.: On a General Method of Solving Second-order Ordinary Differential Equations by Phase-plane Displacements, *J. Appl. Mech.*, **19**: 543–553 (1952).

50. Jarominek, W.: Investigating Linear Systems of Automatic Control by Means of Determinant Stability Margin Indices, *Autom. Remote Control (First Intern. Congr. IFAC)*, vol. 1, 76–84, Butterworth Scientific Publications, London, 1961.

51. Jeffreys, H.: On Certain Solutions of Mathieu's Equation, *Proc. London Math. Soc.*, **23**: 437–448 (1925).

52. Kalman, R. E., and J. E. Bertram: Control System Analysis and Design via the Second Method of Lyapunov, *J. Basic Eng. (Trans. ASME)*, **82D**: 371–400 (1960).

53. Kauderer, H.: "Nichtlineare Mechanik," Springer-Verlag, Berlin, 1958.

54. Kochenburger, R. J.: Frequency Response Method for Analyzing and Synthesizing Contactor Servo Mechanisms, *Trans. AIEE*, **69**(I): 270–284 (1950).

55. Kryloff, N., and N. Bogoliuboff: "Introduction to Nonlinear Mechanics," translated from the Russian by S. Lefschetz, *Ann. Math. Studies*, no. 11, Princeton, N.J., 1947.

56. Ku, Y. H.: "Analysis and Control of Nonlinear Systems," The Ronald Press Company, New York, 1958.

57. La Salle, J., and S. Lefschetz: "Stability by Liapunov's Direct Method with Applications," Academic Press Inc., New York, 1961.

58. Le Corbeiller, P.: Two-stroke Oscillators, *Trans. IRE Circuit Theory*, **CT-7**: 387–398 (1960).

59. Lefschetz, S. (ed.): "Contributions to the Theory of Nonlinear Oscillations," **1** (1950), **2** (1952), **3** (1956), **4** (1958), **5** (1960), *Ann. Math. Studies*, Princeton University Press, Princeton, N.J.

60. Letov, A. M.: "Stability of Nonlinear Control Systems," Gostekhizdat, Moscow, 1962 (in Russian); English translation, Princeton University Press, Princeton, N.J., 1961.

61. Liapunov, M. A.: Problème général de la stabilité du mouvement, *Ann. Fac. Sci. Univ. Toulouse*, **9**: 203–469 (1907); *Ann. Math. Studies*, no. 17, Princeton University Press, Princeton, N.J., 1949.

62. Liénard, A.: Étude des oscillations entretenues, *Rev. Gen. Elec.*, **23**: 901–912, 946–954 (1928).

63. Loud, W. S.: Locking-in in Perturbed Autonomous Systems, *Symp. Nonlinear Oscillations (Intern. Union Theoret. Appl. Mech.)*, Kiev, 1961.

64. Lowan, A. N., and others: "Tables Relating to Mathieu Functions," Columbia University Press, New York, 1951.

65. Ludeke, C. A.: The Generation and Extinction of Subharmonics, *Symp. Nonlinear Circuit Analysis*, **2**: 215–233 (1953), Polytechnic Institute of Brooklyn, New York.

66. Ludeke, C. A.: Nonlinear Phenomena, *Trans. ASME*, **79**: 439–444 (1957).

67. Lurie, A. I.: "Some Nonlinear Problems in the Theory of Automatic Control," Gostekhizdat, Moscow, 1951 (in Russian).

68. McCrumm, J. D.: An Experimental Investigation of Subharmonic Currents, *Trans. AIEE*, **60**: 533–540 (1941).

69. McLachlan, N. W.: "Theory and Application of Mathieu Functions," Oxford University Press, London, 1947.

70. McLachlan, N. W.: "Ordinary Nonlinear Differential Equations," 2d ed., Oxford University Press, London, 1955.

71. Malkin, I. G.: "Some Problems of the Theory of Nonlinear Oscillations," Gostekhizdat, Moscow, 1956 (in Russian); English translation, OTS Atomic Energy Commission, Washington, D.C.

72. Mandelstam, L., and N. Papalexi: Über Resonanzerscheinungen bei Frequenzteilung, *Z. Physik*, **73**: 223–248 (1932).

73. Mathieu, É.: Mémoire sur le mouvement vibratoire d'une membrance de forme elliptique, *J. Math.*, **2-13**: 137–203 (1868).

74. Mathieu, É.: "Cours de physique mathématique," Gauthier-Villars, Paris, 1873.

75. Minorsky, N.: "Introduction to Nonlinear Mechanics," J. W. Edwards, Publisher, Incorporated, Ann Arbor, Mich., 1947.

76. Minorsky, N.: "Nonlinear Oscillations," D. Van Nostrand Company, Inc., Princeton, N.J., 1962.

77. Nag, B. R.: Ultraharmonic and Subharmonic Resonance in an Oscillator, *J. Brit. IRE*, **19**: 411–416 (1959).

78. Nag, B. R.: Forced Oscillation in an Oscillator with Two Degrees of Freedom, *Proc. Inst. Elec. Engrs. (London)*, Part C, **108**: 93–97 (1961).

79. Paynter, H. M.: Methods and Results from M.I.T. Studies in Unsteady Flow, *J. Boston Soc. Civil Engrs.*, **39**: 120–165 (1952).

80. Pedersen, P. O.: Subharmonics in Forced Oscillations in Dissipative Systems, *J. Acoust. Soc. Am.*, **6**: 227–238 (1935); **7**: 64–70 (1935).

81. Poincaré, H.: Sur les courbes définies par une équation différentielle, *J. Math.*, **3-7**: 375–422 (1881); **3-8**: 251–296 (1882).

82. Poincaré, H.: Sur les courbes définies par les équations différentielles, *J. Math.*, **4-1**: 167–244 (1885); **4-2**: 151–217 (1886).

83. Poincaré, H.: "Les Méthodes nouvelles de la mécanique céleste," vol. 1, Gauthier-Villars, Paris, 1892; reprint, Dover Publications, Inc., New York, 1957.

84. Poincaré, H.: "Oeuvres," vol. 1, Gauthier-Villars, Paris, 1928.

85. Pol, B. van der: On Relaxation-oscillations, *Phil. Mag.*, **7-2**: 978–992 (1926).

86. Pol, B. van der: Forced Oscillations in a Circuit with Nonlinear Resistance, *Phil. Mag.*, **7-3**: 65–80 (1927).

87. Pol, B. van der, and M. J. O. Strutt: On the Stability of the Solutions of Mathieu's Equation, *Phil. Mag.*, **7-5**: 18–38 (1928).

88. Pol, B. van der: Nonlinear Theory of Electric Oscillations, *Proc. IRE*, **22**: 1051–1086 (1934).

89. Pontriagin, L. S.: "Ordinary Differential Equations," Fizmatgiz, Moscow, 1961 (in Russian); English translation, Pergamon Press, New York, 1962.

90. Rakhimov, G. R.: "Ferroresonance," Izdat. Akad. Nauk Uz. SSR, Tashkent, 1957 (in Russian).

91. Rayleigh, Lord: "Theory of Sound," vol. 1, Macmillan and Co., Ltd., London, 1894; reprint, Dover Publications, Inc., New York, 1945.

92. Riabov, Yu. A.: On the Evaluation of the Applicability Limits for the Small Parameter Method in the Theory of Nonlinear Oscillations, *Symp. Nonlinear Oscillations (Intern. Union Theoret. Appl. Mech.)*, Kiev, 1961.

93. Riasin, P.: Einstellungs- und Schwebungsprozesse bei der Mitnahme, *J. Tech. Phys. (USSR)*, **2**: 195–214 (1935).

94. Rouelle, E.: Contribution à l'étude expérimentale de la ferro-résonance, *Rev. Gen. Elec.*, **36**: 715–738, 763–780, 795–819, 841–858 (1934).

95. Routh, E. J.: "Dynamics of a System of Rigid Bodies," adv. part, Macmillan and Co., Ltd., London, 1892; reprint, Dover Publications, Inc., New York, 1955.

96. Rüdenberg, R.: Nonharmonic Oscillations as Caused by Magnetic Saturation, *Trans. AIEE*, **68**: 676–684 (1949).

97. Schnyder, C.: Druckstösse in Pumpensteigleitungen, *Schweiz. Bauzeitung*, **94**: 271–273, 283–286 (1929).

98. Schultz, D. G., and J. E. Gibson: The Variable Gradient Method for Generating Liapunov Functions, *Trans. AIEE*, **81**(II): 202–210 (1962).

99. Sorensen, K. E.: Graphical Solution of Hydraulic Problems, *Proc. Am. Soc. Civil Engrs.*, **78**: separate no. 116 (1952).

100. Stoker, J. J.: "Nonlinear Vibrations," Interscience Publishers, Inc., New York, 1950.

101. Strutt, M. J. O.: Lamésche, Mathieusche, und verwandte Funktionen in Physik und Technik, *Ergeb. Math. Grenzg.*, **1-3**: 207–322, Springer-Verlag, Berlin, 1932.

102. "Studies in Nonlinear Vibration Theory," Institute for Mathematics and Mechanics, New York Univ., New York, 1946.

103. *Symp. Nonlinear Circuit Analysis*, **2** (1953), **6** (1956), Polytechnic Institute of Brooklyn, New York.

104. *Symp. Nonlinear Oscillations (Intern. Union Theoret. Appl. Mech.)*, Institute of Mathematics, Acad. Sci. Ukr. SSR, Kiev, 1961.

105. Thomson, W. T.: Resonant Nonlinear Control Circuits, *Trans. AIEE*, **57**: 469–476 (1938).

106. *Trans. IRE Circuit Theory*, **CT-7**: 366–553 (1960).

107. Trefftz, E.: Zu den Grundlagen der Schwingungstheorie, *Math. Ann.*, **95**: 307–312 (1926).

108. Urabe, M.: Periodic Solutions of van der Pol's Equation with Damping Coefficient $\lambda = 0$–10, *Trans. IRE Circuit Theory*, **CT-7**: 382–386 (1960). See also *J. Sci. Hiroshima Univ.*, **A-21**: 193–207 (1958); **A-23**: 325–366 (1960); **A-24**: 201–217 (1960).

109. Wada, T.: Graphical Solution of $dy/dx = f(x,y)$, *Mem. Coll. Sci., Kyoto Univ.*, **2**: 151–197 (1917).

110. Watson, G. N.: Convergence of Series for Mathieu Functions, *Proc. Edinburgh Math. Soc.*, **33**: 25–30 (1914–15).

111. Weber, E.: Nonlinear Physical Phenomena, *Symp. Nonlinear Circuit Analysis*, **2**: 1–27 (1953), Polytechnic Institute of Brooklyn, New York.

112. Whittaker, E. T.: General Solution of Mathieu's Equation, *Proc. Edinburgh Math. Soc.*, **32**: 75–80 (1913–14).

113. Whittaker, E. T., and G. N. Watson: "A Course of Modern Analysis," Cambridge University Press, London, 1935.

114. Young, A. W.: Quasi-periodic Solutions of Mathieu's Equation, *Proc. Edinburgh Math. Soc.*, **32**: 81–90 (1913–14).

Index

Index